THE STRATEGY
OF LIFE

THE STRATEGY OF LIFE

Teleology and Mechanics in Nineteenth-Century German Biology

Timothy Lenoir

The University of Chicago Press

Chicago and London

The University of Chicago Press, Chicago 60637
The University of Chicago Press, Ltd., London

© 1982 by D. Reidel Publishing Company, Dordrecht, Holland
All rights reserved. Originally published 1982
University of Chicago Press edition 1989
Printed in the United States of America

98 97 96 95 94 93 92 91 90 89   5 4 3 2 1

Library of Congress Cataloging in Publication Data

Lenoir, Timothy, 1948–
  The strategy of life : teleology and mechanics in nineteenth century German biology / Timothy Lenoir.
     p.   cm.
  Reprint. Originally published: Dordrecht, Holland ; Boston, U.S.A. : D. Reidel Pub. Co. ; Hingham, MA. : Sold and distributed in the U.S.A. and Canada by Kluwer Boston, c1982.
  Includes bibliographical references and indexes.
  1. Biology—Germany—History. 2. Biology—Germany—Philosophy—History. I. Title.
[QH305.2.G3L46   1989]   89–4646
574'.0943—dc19   CIP
ISBN 0-226-47183-7 (alk. paper)

∞ The paper used in this publication meets the minimum requirements of the American National Standard for Information Sciences—Permanence of Paper for Printed Library Materials, ANSI Z39.48–1984.

For Linda

TABLE OF CONTENTS

Preface	ix
Introduction	1
Chapter 1: Vital Materialism	17
Chapter 2: The Concrete Formulation of the Program: From Vital Materialism to Developmental Morphology	54
Chapter 3: Teleomechanism and the Cell Theory	112
Chapter 4: The Functional Morphologists	156
Chapter 5: Worlds in Collision	195
Chapter 6: Teleomechanism and Darwin's Theory	246
Epilogue	276
Notes	281
Name Index	309
Subject Index	313

The original edition of *The Strategy of Life* contained an unfortunate number of printer's errors which both the author and the editors at the University of Chicago Press had hoped to remove in preparing this paperback edition of the book. Unfortunately the changes required would have defeated the purpose of providing a paperback edition of the book at an affordable price. We have therefore had to compromise by making changes only where the errors in the original edition affected the sense of the argument.

PREFACE

Teleological thinking has been steadfastly resisted by modern biology. And yet, in nearly every area of research biologists are hard pressed to find language that does not impute purposiveness to living forms. The life of the individual organism, if not life itself, seems to make use of a variety of strategems in achieving its purposes. But in an age when physical models dominate our imagination and when physics itself has become accustomed to uncertainty relations and complementarity, biologists have learned to live with a kind of schizophrenic language, employing terms like 'selfish genes' and 'survival machines' to describe the behavior of organisms *as if* they were somehow purposive yet all the while intending that they are highly complicated mechanisms.

The present study treats a period in the history of the life sciences when the imputation of purposiveness to biological organization was not regarded as an embarrassment but rather an accepted fact, and when the principal goal was to reap the benefits of mechanistic explanations by finding a means of incorporating them within the guidelines of a teleological framework. Whereas the history of German biology in the early nineteenth century is usually dismissed as an unfortunate era dominated by arid speculation, the present study aims to reverse that judgment by showing that a consistent, workable program of research was elaborated by a well-connected group of German biologists and that it was based squarely on the unification of teleological and mechanistic models of explanation.

In the course of describing the development of this research tradition, I hope to dispel two further misconceptions that prevail in the literature treating early nineteenth century biology. In keeping with the dogma that in the life sciences one is either a vitalist or a mechanist, it is usually assumed that the proponents of teleology are vitalists. My study attempts to show that in the early nineteenth century the central issue was not vitalism as such, but rather the more interesting problem of causality in biology. To the German biologists who fashioned the research tradition discussed here teleological relationships offered themselves as more sensible patterns for investigating the causal relations of organic form and function.

A second assumption challenged by my study is that persons who defended

PREFACE

...ical thinking in the life sciences were fundamentally motivated by ...us concerns. Several of the figures treated here, such as Immanuel ... and Johannes Müller, certainly were deeply religious men. Some, such as Karl Ernst von Baer, were not religious in an orthodox sense but held a profound, almost pantheistic reverence for nature. But the issue that motivated them to adopt and tenaciously defend teleological thinking in the life sciences was not religion; it was, I hope to show, a concern for good science.

Once the problems of biocausality as perceived by early nineteenth century German biologists are sorted out, the outlines of a powerful program of empirical research begin to take shape. This book traces the main lines in the development of that program. Among the topics explored are its ramifications for work in functional anatomy, embryology, animal systematics, and physiology, particularly the cell theory, early work on heat production in animals, metabolism and muscle action.

It is my hope that this study will also contribute to scholarship on the history of evolutionary theory. One of the objectives of the biological tradition discussed here was to construct a theory to account for the apparent interrelations between organisms. The model eventually devised was a limited form of transformation confined to major organizational types. The principal architect of this view was Karl Ernst von Baer. Von Baer's 'non-Darwinian' theory of evolution should be of some interest to contemporary theorists, who in recent years have proposed a 'punctuated' model of evolution to account for evidence indicating that most species survive for a hundred thousand generations or even more without significant change and for the fact that evolutionary divergence appears to take place rapidly. Although his model was based squarely on the premises of a teleological framework of explanation which led him to propose a different means for explaining evolutionary change, von Baer called special attention to the inadequacies of a gradualist model, and his own 'non-Darwinian' model was based specifically on evidence from the fossil record which indicated that groups of organisms emerge rapidly and undergo little change for many generations before becoming extinct.

If there is a central figure in my study, it is von Baer, and one of the principal aims of this essay is to call attention to von Baer's role as a major biological theorist during the nineteenth century. In most historical treatments of nineteenth century biology von Baer is mentioned for his empirical discoveries, such as the mammalian ovum, and for his unparalleled skill as a microscopist. Not infrequently Johannes Müller is cast as the leading theoretician of the teleological and morphological tradition in early nineteenth century German biology. But at nearly every major turning point in the historical development of the research tradition discussed here I have found the ideas of von Baer in the background. He was a constant source of bold ideas for placing the various branches of the life sciences on consistent, unified foundations.

Few studies have ever profited more from the assistance and advice of teachers, friends and colleagues than this one. Having been emboldened to undertake an investigation of German science in the Romantic era several years ago by Frederick B. Churchill, whose enthusiasm for German culture proved to be as infectious for his students as his concern for the history of the life sciences, I had the subsequent good fortune of making the acquaintance of three persons who have been constant sources of insight and encouragement for this project. They are Phillip R. Sloan, William Coleman, and Reinhard Löw.

For two years while we were colleagues together at Notre Dame, Phil and I teamtaught a course based on primary materials in which we contrasted the teleological and mechanistic approaches to the problems of physiology and evolution. For the purposes of that course I defended the methodological position of the Newtonian mechanist, while Phil adopted the view of Aristotle, Harvey, Owen, Haldane and Polanyi. Those were some of the most delightful and intellectually stimulating debates I have ever been party to; and for me at least they have continued, giving birth to the present book. The students in those classes will, I am sure, not be surprised to learn that Phil's view has prevailed!

My understanding of the issues regarding teleological and mechanistic approaches within German biology has also been shaped by an almost continuous dialogue with Bill Coleman. The Bell System has profited immensely from this book, but I have been a grateful beneficiary of Bill Coleman's expertise and guidance in this area.

Reinhard Löw has not only been a valuable critic of various versions of this book as it has progressed over the last two years, but he has also assisted in locating and calling my attention to several important manuscript sources. I am especially indebted to him for transcribing sections of Kielmeyer's manuscript, "Allgemeine Zoologie", and for locating the Leuckart-von Baer correspondence discussed in Chapter 4. More important by far, however, has been his encouragement of the project and the many pleasant hours we have spent discussing Kant's philosophy of biology. My appreciation of the nature and importance of teleological explanations in biology is due in large part to my association with Reinhard Löw and his important work in this area.

Several persons have offered valuable criticisms of portions of my work as it has progressed. Frederick Gregory read and thrashed the first draft. Many revisions have been due to his helpful criticisms. A series of major revisions in my interpretation of von Baer's embryology and its relation to what I call his theory of limited evolution have been due to the careful consideration of the thesis I am advancing by Dov Ospovat. I am not sure that Dov would agree with my present formulation of the problem; but as it attempts to incorporate his valuable insights, I hope that it would have pleased him. Thanks are also in order for Paul Farber and Gerd Buchdahl for criticisms they have offered of several aspects of my approach to both the historical and philosophical problems treated here.

Portions of this book have appeared in different forms in two articles: 'The Göttingen School and the Development of Transcendental Naturphilosophie in the Romantic Era', *Studies in History of Biology*, 5 (1981), 111–205, published by Johns Hopkins University Press; and 'Teleology without Regrets. The Transformation of Physiology in Germany, 1790–1847', *Studies in History and Philosophy of Science*, 12 (1981), 293–354, published by Pergamon Press. Grateful acknowledgement is made to Johns Hopkins University Press and to Pergamon Press for permission to reprint these previously published materials.

I am grateful to the Würtembergische Landesbibliothek and the Justus von Liebig Universität-Giessen for allowing me to quote from manuscript sources in their collections. Without the assistance of a NATO Postdoctoral Fellowship and a research grant from The National Science Foundation this research could not have been undertaken.

Thanks alone surely could not repay Marilyn Bradian for her incredible secretarial skills and her perseverance in preparing the manuscript in its many different drafts. Hopefully her nightmares in German will soon pass.

Finally, I owe a debt of gratitude beyond calculation to my wife and family. They are the true teleologists behind this study, for they have selflessly supplied the means to its end, which has often seemed hopelessly far from realization. For their trust and support I am forever grateful.

THE STRATEGY
OF LIFE

INTRODUCTION

Confronted with an explosion of discoveries of new plant and animal forms, both living and extinct, and inspired by developments in chemistry and physics to probe the inner workings of the biological organism, anatomists, physiologists and natural historians of the late eighteenth century consciously sought a unified theory of life and its history.. Designated by a variety of terms such as 'zoologie generale', 'zoonomie', 'organology', and 'Biologie', interest in a synthetic theory of life arose simultaneously and independently in England, France, and Germany.[1] Eventually the term 'biology', proposed originally by Jean Baptiste Lamarck and Gotthelf Reinhold Treviranus in 1802, came to identify the new discipline, whose object was to be "the different forms and phenomena of life, the conditions and laws of their existence as well as the causes that determine them".[2]

Having stated their vision of biology as a unified program for investigating the organic world, the 'biologists' were forced just as quickly to retreat from it. The history of pre-Darwinian biology in the nineteenth century has, accordingly, appeared enigmatic in retrospect. The period is universally acknowledged as extremely fertile in the growth of empirical knowledge, which included the beginnings of modern embryology, paleontology, studies of the geographical distribution of forms, organic and physiological chemistry, and the emergence of experimental physiology. According to conventional wisdom, however, this unprecedented growth resulted from the development of techniques and viewpoints for isolating and investigating special problems of life, by placing in abeyance the search for a unified theory of organic form, function and transformation rather than employing a unitary approach as a guide for carving out these research specialties.[3] For according to our present understanding, the grand science of life envisioned by Treviranus and others was but the object of bold speculation until 1859, when Darwin set forth the principle of natural selection as the unifying thread for a comprehensive theory encompassing the diverse relations between plants, animals and their changing environment. Even then the problems of form and function had to await the development of modern genetics, molecular biology, and thermodynamics before they could be integrated with evolutionary theory into a comprehensive science of life, the theoretical foundations for which are only being laid in our own day. In light of what we now know to have been required before the most modest steps could be taken toward the realization of Treviranus' goal, the principal achievement of biologists in the early nineteenth century appears to be this: Turning away from broad speculation and importing the methods of physics and chemistry along with a massive infusion of experimental technique and technology, they succeeded in preparing the ground for a comprehensive theory of life by eliminating the main conceptual stumbling blocks to genuine scientific advance in biology; namely, vitalism and teleological thinking.

While I do not dispute the validity of this description of the main outlines of advance in nineteenth and twentieth century biology, it does not, in my view, adequately characterize the richness of biology in the early nineteenth century, particularly in Germany. There a very coherent body of theory based on a teleological approach was worked out, and it did provide a constantly fertile source for the advance of biological science on a number of different research fronts. The purpose of this study is to describe that research tradition. My principal thesis is that the development of biology in Germany during the first half of the nineteenth century was guided by a core of ideas and a program for research set forth initially during the 1790s. The clearest early formulation of those ideas is to be found in the writings of the philosopher Immanuel Kant. While a conscious concern with ideas set forth by Kant provided the unifying thread of later research, and while explicit recognition of Kant's contribution accompanied by frequent re-examination of his philosophy of biology is a constant theme in the work of the biologists considered herein, I do not claim that German biologists discovered a program of research in Kant's writings which they set out to realize in practice. The evidence indicates rather that in the latter part of the eighteenth century a number of biologists were seeking to establish a foundation for constructing a consistent body of unified theory for the life sciences which could adapt the methods and conceptual framework of Newtonian science to the special requirements of investigating biological organisms. Kant stepped into this ongoing dialogue and set forth a clear synthesis of the principal elements of an emerging consensus among biologists.

In a penetrating analysis of the different structure of causal relations obtaining between inorganic phenomena and the sort of mechanisms encountered in biological organization, Kant argued that the life sciences must ultimately rest on an explanatory framework uniting the principles of both teleology and mechanism. Among the persons who discussed Kant's examination of the problems of scientific explanation the most important for the subsequent development of biology in Germany was a group of colleagues and students associated with Johann Friedrich Blumenbach and Georg Christoph Lichtenberg in Göttingen. It was through Blumenbach and his students that Kant's special brand of teleology entered biology. But others were no less impressed by Kant's biological thought. In the first phase of its development, Treviranus, Carl Friedrich Kielmeyer, and Johann Friedrich Meckel were among those committed to this research tradition, while in its later, more mature stages Karl Ernst von Baer, Johannes Müller, Carl Bergmann and Rudolph Leuckart were its leading practitioners.

Even the partial list of contributors to this research tradition mentioned above raises the question of how such an extensive program encompassing an impressive array of empirical research and including so many important figures has escaped the notice of historians of biology. In part, of course, it has not been neglected. Attention was called to its importance by E.S. Russell[4] and more recently by Hartwig Kuhlenbeck.[5] But it has not become part of the main body of the history of biology for two reasons: First very few recent historians have attempted to understand fully the nature of teleological explanation and its heuristic power in biology; and, in the case of German biology in the early nineteenth century, an assumption has been made that the main impetus for the development of biological thought came from a monolithic, idealist philosophy of nature known as *romantische Naturphilosophie*.

Due to the magnitude of his achievement, Darwin's work has appropriately received the lion's share of attention from historians of the period. This focus on Darwin, however, has resulted in a distorted image of much important pre-Darwinian as well as a considerable body of significant post-Darwinian biology. In particular, the emphasis on Darwin is in large part responsible for the fact that we have overlooked a significant, valid alternative approach to biological phenomena during the early nineteenth century. In emphasizing the revolutionary character of Darwin's approach to biology historians have been led quite naturally to train their attention on those aspects of his thought in which Darwin broke with previous traditions. Since Darwin did succeed in unifying several disparate fields of biological investigation, it is erroneously assumed that his was the *only* approach in the nineteenth century capable of succeeding at that goal. The development of Darwin's views has become our window to the views of others, and the evaluation of other approaches is measured by their relationship to the Darwinian synthesis.

An important aspect of our image of nineteenth century biology that has been shaped by the triumph of the Darwinian method is the role of vitalistic and teleological explanation. Darwin, it is well known, struggled long and hard to establish materialistic principles of analysis and natural mechanisms that would enable him to escape vitalistic or teleological notions such as perfect adaptation of the organism to its environment or the argument from design.[6] But is must be recognized that the notion of teleology held by Darwin's contemporaries in England, and by Darwin himself for that matter, was an extremely impoverished one.[7] Due to the rather special circumstances of the English context, teleological arguments tended to be the principal weapons of persons defending religious belief such as William Paley and Samuel Wilberforce. They were bent upon demonstrating that the universe was constructed according to a rational plan. Theirs was an anthropomorphic brand of teleology which assumed that adaptation of an organ to its function requires foresight and deliberate action on the part of some agent. Darwin inveighed against this metaphysical and anthropomorphic use of teleological thinking, and he was correct in doing so. But that was not the only sort of teleology capable of being brought to bear on problems of form and function.

The German tradition I will examine here never made use of the design argument or the notion of a purposeful divine architect. This position was expunged by Kant as having no place in natural science in the formative stages of the program. His defense of teleology emerged rather from critical analysis of mechanical causality and the limits of mechanical explanations in biology. Darwin seems to have been unaware of this more sophisticated type of teleology, and most historians have followed Darwin's lead. The unfortunate conclusion has prevailed that since Darwin's arguments were directed against those who adapted teleology to the needs of religion, defense of a teleological approach to life was impossible on purely internal scientific grounds. Accordingly, those who persisted in defending teleology did so out of a commitment to religious views that have no place in science.[8]

Recent scholarship has shown that even within England alternative teleological approaches to biological organization were pursued. Biologists such as William Carpenter, Martin Barry, Richard Owen, and even T.H. Huxley before his conversion to Darwin's views defended teleology on strictly scientific grounds without recourse to religious considerations.[9] It has been suggested that this alternative approach was of German origin, possibly tracing its source to von Baer and Johannes Müller.[10] We have looked in the wrong places, however, for an interpretative framework for understanding the biological theories of these men.

There has been universal agreement that the main impetus for the growth of biological research in early nineteenth century Germany was supplied by *romantische Naturphilosophie*. There were good reasons for making this assumption. On the one hand, that was the view bequeathed to us by Helmholtz, DuBois-Reymond and others, who in their popular lectures characterized the intellectual life of their fathers' generation as under the complete and utterly stifling influence of *Naturphilosophie*. The fact that the Naturphilosophen proposed an organic approach to the entire spectrum of natural sciences and emphasized the importance of a unitary, historical conception of nature lent further credence to the view that *Naturphilosophie* provided the model for unifying the life sciences sought by German biologists in the pre-Darwinian period.

Suspicious that claims of the complete vacuousness of Naturphilosophie may have obscured its potential contribution to important later developments, a number of historians began to call for more detailed studies of *Naturphilosophie*. Many came to assume that any attempt to understand the development of German science in the early nineteenth century will have to come to terms with the contribution of *Naturphilosophie*. In a now famous article forming part of a volume on critical problems for future research in the history of science Thomas Kuhn suggested that Naturphilosophie may have been a fertile source for the development of ideas on energy conservation.[11] Others plunged into the subtleties of Romantic Science in an effort to make clear the views of the Naturphilosophen toward the mathematization of nature and experiment.[12] Where it has not proven possible to establish a direct positive contribution toward the impressive development of German science at mid-century, the search has been for residues of *Naturphilosophie* in the approach of scientists who, supposedly under its influence during their early careers, subsequently broke with its central dogmas.

The present study originated in an attempt to explore the contribution of the *Naturphilosophie* of the young Schelling and Hegel to the development of the life sciences in Germany. Just as others had begun to chart its influence on the course of the inorganic sciences, I intended to pursue its impact on biology. I was quickly diverted from that venture, however. On closer inspection the view that German biology in the early nineteenth century was shaped by *Naturphilosophie* turned out to be illusory. It dissolves as soon as one begins to examine the major biological writings of the period. One would expect that the works of Kielmeyer and Treviranus, for example, would be a good place to begin to trace the role of *Naturphilosophie* in shaping biological thought in Germany. While often citing their work as examples of

Naturphilosophie, few historians have actually read Kielmeyer's writings, however, and even fewer have bothered to examine Treviranus' six volume work. The effort is revealing; for Treviranus is cautiously critical while Kielmeyers is violently opposed to *romantische Naturphilosophie*.[13] The same pattern is repeated by others reputedly influenced by *Naturphilosophie* such as Johann Friedrich Meckel, Karl Ernst von Baer, and Johannes Müller. I have discovered, however, that a common core of natural philosophy does run throughout the works of these individuals; it is a philosophy of biology proposed by Immanuel Kant.

The Kantian tradition in German biology has remained largely unexplored for two reasons. Its superficial similarity to positions advanced by the Naturphilosophen has certainly been a contributing factor. Convinced that German biology in the early nineteenth century was dominated by an arid philosophy of nature, historians can hardly be blamed for not wishing to pursue lines of disagreement within it. It has simply not seemed worth the effort. The primary cause, however, resides in a failure to appreciate the nature and variety of teleological thinking in biology; for while both traditions insisted upon a teleological orientation to organic phenomena, there were fundamental differences between Kant's teleology and that of the Naturphilosophen.

THE PERSISTENCE OF TELEOLOGICAL THINKING IN BIOLOGY

Historians are fond of characterizing the cutting edge of advance in biology as the mechanization of life, which is usually interpreted as the elimination of teleological thinking through reduction of vital phenomena to the laws of physics and chemistry. But few indeed are those who have examined the phenomena of life without claiming to see a complex tapestry of purposive organization. At every major turning point in the history of biological thought there have been individuals who, thoroughly knowledgeable of the best physical theories available to them, have rejected the possibility of reducing life to the laws of inanimate matter.

There has been a remarkable constancy in the reasons advanced for this position. Foremost among them has been the overwhelming presence of pattern and order in biological phenomena. In the almost continuous debate with Democritus throughout his biological writings, for example, Aristotle points to the fact of harmoniously interconnected sequences of patterned events as the distinguishing characteristic of beings endowed with life. Blind chance, he argues, could not have been the source of processes which demonstrate such a high degree of determinacy and precision. Against a mechanistic interpretation which views organization as the result of accidental concatenations of processes obeying physical laws, Aristotle argued that the *telos* or goal of organization itself, namely the whole organism functionally adapted to its environment, must be the overarching cause of order in biological phenomena.[14]

The advances of modern biochemistry and molecular biology have not rendered this position less relevant to the problems of biology. Rather than rejecting it, some biologists have in fact been led to refine and reformulate Aristotle's objections concerning the reducibility of life. Adopting a form of teleological thinking in many respects closely akin to Aristotle's, many modern embryologists and cellular biologists argue that while organization must be seen as resting on properties inherent in individual molecules, those properties can only find expression in collective interaction. Paul Weiss, in particular, emphasizes that supra-molecular structures and organelles such as cilia and mitochondria appear only as quantal units with properties and faculties that are characteristic of the particular collective but are not manifested by individual components in the unassembled state.[15] As Michael Polanyi has shown, this sort of finding has more interesting implications than the mere fact that the joining of parts may produce features which are not observed in the separate parts themselves. He discusses the rationale behind holist positions by drawing an analogy between the hardware of a machine, a computer, for example, and the blueprint or program which specifies the boundary conditions for harnessing the physical and chemical processes locked into its components. Polanyi's position is that the growth of the blueprint may not be specifiable in terms of the laws of physics and chemistry; rather it is an irreducible principle which does not interfere with the laws of physics and chemistry but which operates *in addition* to them.[16] An exactly parallel phenomenon obtains in biology: biological principles establish hierarchies of organization in which each level relies for its working on the levels below it while being irreducible to these lower principles. Biological principles in this view, therefore, control the boundary conditions within which the forces of physics and chemistry carry on the business of life. Or, employing language similar to Aristotle's, we might say that the whole is functionally prior to the parts.

The persistence of teleological thinking in biology is, of course, no criterion by which to judge its merits. The most meaningful way to evaluate an approach to nature is to examine its success in defining problems and guiding research toward their solution. On this issue the consensus

INTRODUCTION

teleological approaches has been in support of the opinion
[b]y Jacques Monod: Recourse taken by biologists to teleological
[analysi]s draws its justification not from precise knowledge or definitive
observations, but from present day ignorance alone.[17] According to this interpretation the progressive developments in biology were those that led to the reduction of biological phenomena to the laws of chemistry and physics. Each advance has been accompanied by a corresponding retreat of teleological explanations. They have been, in short, simply a veil for the domain of vital phenomena inaccessible to then-current physicochemical explanations. Progress in biology has been the result of the elimination rather then a concerted commitment to teleological thinking.

There are two flaws in this line of argument. First, it does not agree with the facts; and secondly, it rests on a fundamental confusion. The factual insufficiency of the view becomes apparent as soon as the work of, for example, the eminent biologist E.S. Russell is consulted. Looking at the growth of biology since Aristotle in his classic work, *Form and Function*, published in 1916, Russell argued that the progress of the life sciences had been due precisely to the pursuit of teleological and functional explanations and that mechanistic reductionism was simply a passing fad associated with the predominant materialistic attitude fostered by late nineteenth century industrial society.[18] In defense of his claims Russell assembled in that book a most impressive array of successes, particularly in embryology, embodying the teleological approach to biological organization.

The fundamental confusion in evaluations of teleological explanations is due to an implicit assumption that *only* reductionistic forms of explanation are capable of generating a quantitative treatment of biological phenomena. Furthermore, it is generally assumed that anyone who defends a teleological approach to biology is opposed to quantitative, experimental science. Both assumptions are incorrect. D'Arcy Thompson, for example, was a stout defender of teleological approaches to the problem of organization; and yet the sole purpose of his book, *On Growth and Form*, was to correlate the outward phenomena of organic growth and structure in terms of mathematical statement and physical law.[19] In an important recent work in which she defends the anti-reductionist thesis of Polanyi, Koestler and others, Lila Gatlin has proposed a quantitative model based on game theory for elucidating the biological principles postulated by Polanyi as the source of the boundary conditions for biological evolution.[20] The work of such persons makes it plain that viewing biological organization as incapable of reduction to the laws of chemistry and physics is in fact compatible with a commitment to quantitative rigor as the touchstone of good science.

VARIETIES OF TELEOLOGY

Several non-religiously motivated types of teleological argument have been defended on scientific grounds. Each of them incorporates in some fashion a claim concerning the special status of biological phenomena and their ultimate irreducibility to physics and chemistry. A central concern of each is the apparent priority of the whole over its parts. Teleological forms of explanation in biology offer a middle ground between vitalism and reductionism. Varieties of teleology occupy a spectrum of positions between these two poles.

One version of teleology differs only slightly from vitalism. Vitalistic positions assume in some form or other the existence of an agent which actively selects and arranges matter in the organism. Some vitalist approaches assume furthermore that this agent, which may be a rational soul, can exist separately from matter and that the organism is in a healthy, functional state so long as the vital agent remains in control. Such a position was defended by Georg Ernst Stahl, for instance. A vitalistic position similar to this renounces the role of a rational, purposeful agent but continues to acknowledge special vital forces, such as sensibility and irritability. These forces are characterized as the organic analogues of Newtonian forces. This form of vitalism was popular in the eighteenth century and in fact its 'scientific' status depended very much on the prevailing confusion in the characterization of Newtonian forces; for many capable scientists treated forces as active powers superimposed upon inert matter. In a similar 'Newtonian' vein, Albrecht von Haller and Xavier Bichat excluded purposive agents from their physiology and 'reduced' physiology to certain blind forces present in different tissue types.

This position was only slightly removed from one version of teleological argument. All that was required to transform Haller's vitalism to a teleological approach was the further restriction that a specific vital force depended completely upon the organization of the contsituent elements of the organ exhibiting it. The force was not to be conceived as some independent entity but rather as an emergent property dependent upon the specific order and arrangement of the components. Such a position, which I have called 'vital materialism', was worked out by Johann Friedrich Blumenbach and Johann Christian Reil, two figures important in the formative period of the research tradition to be treated here.

A second form of teleology was one dear to E.S. Russell's heart. Generally designated as functionalism, this form of teleology invokes neither purposive agents nor emergent vital forces. While permitting no other natural forces than the forces of physics and chemistry, it does not, however, reduce biological organization to these forces. Rather, functional requirements establish boundary conditions within which the laws of physics and chemistry are to be applied. The ends of life establish a hierarchical framework of ordering principles within which physicochemical forces operate. This position is similar in many respects to that of D'Arcy Thompson outlined above, but some of its foremost practitioners were nineteenth century physiologists such as Claude Bernard, and as we shall see, Carl Bergmann and Rudolph Leuckart.

While disagreeing over the existence of special biological forces, both forms of teleology discussed above reject the vitalistic notion of purposive activity and attempt to explain such apparent activity in terms of natural physical causation. Furthermore both recognize a dichotomy between inanimate matter and biological organization based on the irreducibility of certain biological properties to the laws of physics and chemistry. A third form of teleology does not recognize this dichotomy, however. According to this position the universe is fundamentally biological. Not only is each part subservient to the organization of the whole, but there are only 'biological' laws. There are no independent physical laws according to this view. The laws of mechanics are simply a subset of laws arrived at by placing limitations and constraints upon the biological laws driving the universe.

This form of teleological thinking is encountered, of course, in the works of Aristotle. In his system of nature the laws of free fall and projectile motion were conditioned by the overarching organization of the cosmos. Each part had its proper position relative to the whole which determined its natural motion. The same viewpoint was adopted by the Naturphilosophen, particularly Hegel. Nature is fundamentally organized for the Naturphilosophen, and the highest form of organization is selfconsciousnss. For them too the fundamental laws of nature are biological, which they depict as developmental in character. The universe is accordingly represented as following a sequenced developmental path toward the realization of consciousness. In fact some naturphilosophic writings employ the metaphor of the ontogenetic stages of the mammalian ovum as a schema for describing this cosmic evolution.[21] As in Aristotle's system, the laws of physics and chemistry are subservient to the higher biological laws of nature. Hegel in fact attempted to show that physical and chemical laws are imbedded in the structure of the developmental laws governing the evolution of the universe.

In the twentieth century a further variant of the holist position has emerged. Although it too denies the reducibility of life to the laws of chemistry and physics, it frequently qualifies this assertion by adding 'current' chemistry and physics. Implied thereby is the notion that present day physical theory is simply a limiting case of physical laws. Just as investigation of processes at the micro-level resulted in the quantum revolution which left the old macroscopic physics of Newton to be reinterpreted as a limiting case, so, the argument goes, investigation of the phenomena of organisms endowded with consciousness will lead not only to a reinterpretation of physical laws but rather to a complete change in the laws of physics. As Eugene P. Wigner has noted, "All major extensions of physics, to encompass added sets of phenomena were possible only on the basis of new concepts and, one may add, increasingly subtle and abstract concepts. It is not likely that the extension to the phenomena of life will be an exception".[22]

Many biologists have turned to general systems theory and game theory for the extension of physical concepts needed to generate models for the 'blueprints' and 'strategies' employed by biologic systems in evading the law of entropy and in establishing orderly, hierarchic patterns of living structures. The proponents of this view do not reject physics but seek to extend classical notions through the addition of new mathematical knowledge. Jacques Monod is incorrect, in my view, when he describes modern holists who turn to systems theory as "more or less consciously or confusedly influenced by Hegel". For in contrast to Hegel these modern theorists do not see the structure of mind as permeating the universe. Rather than legislating the issues of concern to older teleologists out of court as biological pseudo-problems, these modern theorists seek a causal explanation for such notions as internal principles of organization and morphotypes employed frequently by vitalists and teleologists. The fact that they are not yet able to explain these phenomena is interpreted as due to the inadequacy of current physical theory. They seek, accordingly, a new physical theory, and hence the theoretical foundations of the biology they envision will be reductionistic.[23] But not in the old sence of 'reductionistic'. For their 'new physics' will not attempt to construct biological organization out of micro-processes alone but will subordinate them to a hierarchy of controls. Physical theory, in short, will involve a new conception of the relation of wholes to parts.

Neither Hegel nor Aristotle would have acquiesced in such a view, however. For them life has a primacy over physical laws and gives order and determinacy to nature. Only in a superficial sense does a certain similarity exist between the views of the informationist school and Hegel's explanation for the order of nature. In accounting for the agreement between the orderly patterns of nature and the patterns of human logic, for instance, one recent theorist has suggested that the four basic patterns of organic order – standard part, hierarchy, interdependence, and traditive inheritance – are also the preconditions for thought.[24] Hegel would have reversed the order of emphasis, placing thought as the precondition for organic order.

While not intended as an exhaustive classification of the richness of different approaches to teleology, the preceding discussion serves to focus some of the issues German biologists faced in the early nineteenth century. The lines between these different forms of teleological explanations are by no means rigid. Changes in the scope of problems considered and improvements in chemistry and physics could and did lead to the modification of one position and the adoption of another. This is particularly true of the first two types discussed above. As we shall see, German biologists in the first decade of the nineteenth century relied heavily upon the notion of emergent vital forces, such as the 'developmental force' (Bildungstrieb, or Gestaltungskraft), based on the organization of chemical-physical forces. By the late 1840s they had ceased to attribute a special status to vital forces and turned their attention instead to interpreting organization in terms of functional constraints placed on natural forces in a physiological context. This shift of emphasis was brought about by advances in organic and physiological chemistry as well as through improved understanding of energy conversions. This progressive development was guided, however, by a common framework of ideas which I have characterized as teleo-mechanism, the research tradition which Kant helped to initiate in German biology.

THE TELEOMECHANIST PROGRAM

Throughout this discussion I have characterized the Kantian teleo-mechanist approach alternately as a 'research tradition' and as a 'research program'. These terms, of course, have quite precise meanings to philosophers of science, being associated with two related but different theories of scientific development.[25] It is not relevant to my purpose here to discuss the merits of these theories, for I do not intend to employ the full complement of categories of either in the present study.[26] Certain aspects of the methodology of scientific research programs, however, do provide convenient terminology and a useful framework for interpreting the materials of the present study.

Two features of Lakatos' methodology of scientific research programs are particularly relevant to my discussion of the development of biology in Germany. First, a research program is not a scientific theory. It is a set of guidelines for the development of specific theories. These guidelines consist of (a) a "hardcore" of fundamental assumptions which set forth the central principles of an approach to nature, and (b) a set of heuristic guidelines for developing specific models that simulate reality.[27] According to Lakatos the hardcore can never be the object of empirical refutation, and it cannot be abandoned without repudiation of the program.

Research programs are historical in character. The hardcore need not leap onto the stage of history fully developed. It may take time to nurture and articulate the conventions that establish the conceptual framework of the program. Moreover, while the hardcore may never be abandoned, shifts in the heuristic guidelines may contribute to the theoretical progress of the program. This feature is related to the second and most important aspect of the methodology of scientific research programs. The aim of a progressive research program is to generate a succession of refutable variants $P_1, P_2, P_3,$ ... each one of which criticizes, refutes some aspect of, and ultimately extends the explanatory domain of its predecessor.[28] As is evident from the historical character of the hardcore described above, the full content and power of an approach to nature may not be entirely evident in the first attempt to apply it. Several different improved versions of the program may be required before its content is completely unpacked. Accordingly, each successive variant of the program, while refuting its predecessor, results in a clarification of the original adumbrated core of the research program as well as in the extension of its problem domain.

As I have indicated, the principal claim of my study is that the life sciences in Germany between 1800 and roughly 1860 should be viewed as progressing on the basis of a coherent and relatively constant body of doctrine regarding biological organization. The two features of Lakatos' model discussed above provide a useful tool for understanding how it was possible for a unified approach to both guide and absorb such a diversity of empirical research, particularly since some of the specific theories of early workers were contradicted and overturned by later researchers who nonetheless viewed themselves as working within the same tradition.

Several persons contributed to the teleo-mechanist approach to the life sciences, but, in terms of Lakatos' categories, its hardcore was stated by Immanuel Kant in his *Critique of Judgement* (1790). As I shall demonstrate in the first chapter, Kant also sketched some guidelines for applying this approach. He introduced a notion which was to become a central feature of all later teleo-mechanist approaches, namely, the morphotype. The first chapter goes on to discuss the attempt by Johann Friedrich Blumenbach,

Johann Christian Reil, and especially Karl Friedrich Kielmeyer to construct the first fledgling version of the teleo-mechanist program. Following Oswei Temkin's perceptive discussion of similar biological theories on the frontiers between vitalism and mechanism in the early nineteenth century, I have called this first version of the program 'vital materialism'.[29]

While Kant, Blumenbach, Reil and Kielmeyer had sketched the potential applications of the teleo-mechanist program, they did not work it out in sufficiently explicit detail. Specifically, in the language of the research program model, they had not settled on an adequate positive heuristic. In particular, although they had stressed the need for employing the notion of a morphotype or organizational plan as a guiding principle in theory construction, they had not resolved upon a satisfactory means for accomplishing this. The second chapter deals with the resolution of this problem by several persons, including Johann Friedrich Meckel, Karl Ernst von Baer, Heinrich Rathke, and Johannes Müller. Von Baer and Müller worked out the first really succesful and comprehensive version of the program by defining the notion of an organizational plan in terms of very definite embryological criteria and morphological principles, such as that of homology. I call this powerful version of teleo-mechanism 'developmental morphology'. It had an illustrious career.

One of the central features of both vital materialism and developmental morphology was their commitment to the notion of special emergent vital forces, such as Blumenbach's *Bildungstrieb*, or von Baer's *Gestaltungskraft*. This notion came under careful scrutiny between 1838 and 1842 as a result of advances in organic chemistry, and it comes under scrutiny here in chapter three. The primary issue around which discussion over this matter centered was the cell theory. I have followed the development of the cell theory and its perceived relation to the debate concerning vitalism and mechanism in the works of several persons, particularly Theodor Schwann, Carl Vogt, and Rudolph Virchow. I have also traced the acceptance and interpretation of the cell theory by the developmental morphologists, particularly Johannes Müller. Out of these discussions surrounding the cell theory the basis was being laid for the formulation of a more consistent and theoretically more powerful version of teleo-mechanism. Vitalism was dead, but teleological thinking continued to exert a strong hold on biological theory.

Chapter four explores the evolution of this improved version of the program, which I call "functional morphology". Two persons were especially responsible for laying its critical foundations, the Göttingen physiologist-philosopher, Hermann Lotze, and the magnate of organic chemistry, Justus

Liebig. These men both emphasized that the teleo-mechanical hardcore of the program must be preserved but that the assumption of vital forces had to be rejected. Carl Bergmann and Rudolph Leuckart, two younger colleagues of these men, took up the challenge and attempted to construct a new variant of the program. Out of the fruitful collaboration of these men functional morphology was born, a program which acknowledged the need for a teleological framework, but a framework completely divested of developmental forces. While continuing to incorporate the notion of morphotypes, they now sought to explain them in terms of functional criteria consistent with their rejection of vital forces.

Central to a research program or research tradition conceived as a historical entity is the notion that those practitioners who engage in constructing a new improved version of the program perceive themselves as building upon while at the same time modifying the conceptions and methods of their predecessors. As we shall see, close personal and intellectual ties existed between the practitioners of each of the variants of the teleo-mechanist program. Early developmental morphologists were reared in the lecture halls and at the disecting tables of the vital materialists; and in concluding my examination of the functional morphology of Bergmann and Leuckart, I present evidence concerning the close relationship these men saw between their own approach and the developmental morphology of von Baer. They clearly perceived themselves as following in his footsteps.

No sooner had the functional morphologists begun to publish some of their ideas than they found themselves under heavy siege together with their fellow teleo-mechanists, the developmental morphologists. They were surprised, shocked, and as we shall see, outraged. Their opponents, of course, were the reductionists led by Carl Ludwig and Emil DuBois-Reymond, whose own considerable skills were soon to be joined with the formidable talents of Hermann Helmholtz. In chapter five I have followed the stages in the acrimonious debate that broke out between these two schools, particularly evident in their reviews of one another's work and in their public addresses and popular lectures. I have also followed the argument between these two groups at the level of their substantive disagreements over theoretical issues concerning biological organization. The central issue over which they divided concerned their views on the role of teleology and the power of mechanical explanations in biology. I have followed the genesis of the debate through a reconstruction of Helmholtz's path to the discovery of the conservation of energy. As one of the main conclusions of my study, I attempt to show that contrary to the claims of Helmholtz and DuBois-Reymond, the principle of the conservation of energy did not, and indeed could not, deliver a fatal blow to teleological thinking in biology.

With the conservation of energy as their primary weapon the reductionists attacked the assumption of teleological principles of explanation at both the micro-level of cellular organization and the macro-level of biological evolution. They argued that inter-atomic processes are completely sufficient for understanding the organization of the cell and that the principle of natural selection demonstrates how to apply this same line of reasonings to the fine-tuning and adaptation of the organism to its environment. No one better understood the role of teleological thinking in biology than Karl Ernst von Baer, and as one of the principal architects of the teleo-mechanist research program no one was better able to defend its claim to legitimacy against the attacks of the reductionists than he was. Von Baer devoted the last years of his life to re-examining the presumed sufficiency of mechanical explanations in accounting for biological organization. As part of his defense of teleology von Baer submitted the evidence for and against evolution by natural selection to careful examination. Von Baer's exposition of the core elements of teleo-mechanism as a strategy for the life sciences and his elaboration of an alternate theory in terms of them to account for the evidence in support of biological evolution is the subject of the final chapter in this study. As we shall see, the view that biological organization is the blind result of chance and necessity never met more vigorous opposition than in the person of Karl Ernst von Baer.

CHAPTER 1

VITAL MATERIALISM

BLUMENBACH, KANT AND THE TELEOMECHANICAL APPROACH TO LIFE

Perhaps no area of endeavor better exhibits the hold that the notion of goal-directedness as a mode of natural causation has over the human mind than the writing of history. It seems we are never quite able to disregard the advice that, according to Herodotus' account, was given to Croesus by Solon. With incomparable wealth and renown that reached the four corners of the earth, Croesus deemed himself the happiest man on earth; and in order to confirm that conviction he sought the judgement of Solon, who had the reputation of being the wisest of men. Solon could not be certain, however, until Croesus had died. For life, after all, is a chancy thing; and one's apparent prosperity and competence may simply be a matter of luck. "Look to the end", Solon advised, "no matter what it is you are considering. Often enough God gives a man a glimpse of happiness, and then utterly ruins him". Solon was no fool, and neither was Herodotus who put those words into his mouth. Through his character Herodotus was speaking in fact for the majority of historians who, because of the unique vantage of hindsight, are able to judge the 'real' merits of a man's ideas by seeing what happened to them after he died.

Take the case of Johann Friedrich Blumenbach, for instance. Who among the stellar cast of scientists assembled in Göttingen in 1825 for Blumenbach's jubileum would not have numbered him among the happiest men of science? At least half of the major contributors to German biological science during the early nineteenth century were his students or at least claimed to have been inspired by the man. Among his most notable students were Alexander von Humboldt, Carl Friedrich Kielmeyer, Georg Reinhold Treviranus, Heinrich Friedrich Link, Johann Friedrich Meckel, Johannes Illiger, Rudolph Wagner and many others. Among the numerous persons directly and consciously influenced by Blumenbach in their scientific work are no less than Goethe, whose work on the intermaxillary bone grew out of anatomical investigations undertaken with Blumenbach[1]; von Baer, whose anthropological thought had direct roots in the writings of the Göttingen zoologist; and Friedrich Tiedemann, whose work in embryology and teratology was directly inspired by Blumenbach. No less a talent than Schelling acknowledged that Blumenbach was a seminal figure in the revolution in physiology underway at the turn of the century.[2] And, as we shall see, Immanuel Kant regarded Blumenbach as one of the most profound biological theorists of the modern era.

But Blumenbach's luck has not held. If he is mentioned at all in modern histories of biology it is for his work on the races of man, particularly his defense of the idea, somewhat novel for the time, that the different varieties of man are actually members of the same natural species. He is remembered disparagingly as a vitalist and as the compiler of one of the most successful textbooks of natural history of the late eighteenth and early nineteenth centuries, which, as every hard-nosed professional is aware, does not provide one with a passkey to the portals of immortality. Blumenbach's ideas on species and on classification as well as his vitalistic tendencies had the misfortune of not forming part of the royal road leading to Darwin. He is not, accordingly, remembered as an important biological theorist. But he was. Very early on in his scientific career Blumenbach envisioned a unified science of life which would encompass the best features of the thought of Buffon, Linneaus, and Haller. His vision and the methods he attempted to develop in order to realize it proved to be a powerful incentive to an entire generation of German biologists.

The foundational ideas of Blumenbach's plan for the life sciences were laid in his short treatise *Über den Bildungstrieb und das Zeugungsgeschäfte*. This work is usually praised for its strong rejection of the pre-formationist doctrine and its advocacy of an epigenetic theory of development. In point of fact the book went far beyond this relatively modest achievement by offering a completely new approach to constructing a theory of organic form with implications for several areas of research. To appreciate this work fully and its consequences for later developments, it is necessary to consider it in the context of the problem of constructing a natural system, and of the species-race relation in particular, from which it emerged.

Blumenbach's theory of epigenesis emerged from the context of discussions concerning the construction of a proper theoretical framework for natural history that had been going on in Göttingen scientific circles since the days of Albrech von Haller.[3] In his earliest published work, his dissertation, entitled *De generis varietati humani nativa* (1775) Blumenbach had explicitly acknowledged that in the theory of generation he followed the views of Haller in all details. Like Haller he asserted that "the embryo is contained in the maternal egg, and the female provides the true stamina of the future foetus".[4]

The sole function of the sperm, he argued, is to awaken the germ from its eternal slumber "by the subtle odor of its parts which are particularly adapted to causing irritation".[5]

Implicit in this theory was the conclusion that the paternal contribution in generation is miniscule, merely setting in motion the development of structures already present in the egg, while the greater part of the form in animals is derived from the mother. Although hybrids might be possible, due to the minor contribution of the male, the offspring would return to the maternal stock after several generations. Consequently the preformationist model led to the conclusion that "the offspring at last brought to light, ... ought to go on forever like their first parents".[5] For the staunch defender of Haller's theory of development and organic form, then, the problem that leaped clearly to the fore, the problem that generated the substance of Blumenbach's dissertation, was to account for the manifestly varied forms of species, the human species in particular:

What is it which changes the course of generation and now produces a worse and now a better progeny, at all events widely different from its original progenitor.[7]

In order to explain the origin of the races of man Blumenbach adopted in his dissertation a model popular in his day; namely, he argued that the various races were all degenerations of the caucasian race, resulting from climatic variation produced by migration, changes in nutrition corresponding to the differences in the new habitat, and difference in the mode of upbringing due to cultural differences.

By the apperance of the first edition of his *Handbuch der Naturgeschichte* in 1779 Blumenbach had already begun to sense the incompatibility of the preformationist theory with his own developing views on natural history; and within a few months, by early 1780, he had rejected the theory altogether. Two sets of facts led to this development. First it appeared to him that the preformationist theory could not account for the fact that the offspring of mixed parental racial stock always leads to a progeny demonstrating a *blend* of parental traits. According to the preformationist theory, the offspring should resemble the maternal form. Furthermore, the results of Kölreuter's experiments, which had appeared between 1761-1766, now seemed in light of this observation to provide convincing refutation of the theory. For Kölreuter had not only succeeded in producing fertile hybrid offspring by crossing *nicotina rustica* and *nicotina paniculata*; more importantly, he had succeeded in reverting the hybrid offspring to the paternal form (paniculata) after several generations.[8] This flatly contradicted Haller's scheme.

In spite of his decision to support the epigenetic theory, Blumenbach did not want to abandon what he considered to be desirable features of the preformationist account. In particular, he wanted to retain the idea fundamental to Haller's preformationism, that the fact of biological organization could not be accounted for in terms of physico-mechanical causes, but had to be treated as primary. On the other hand, he wanted to avoid treating biological organization as a result of the super-imposition upon inorganic matter of a separate force, a *Lebenskraft*, or a soul, which exists independently of a material substrate. In attempting to steer a course between the Scylla of materialistic reductionism and the Charybdis of vitalism, Blumenbach treated the agent responsible for organic structure as a kind of Newtonian force for the biological realm, which he called the *Bildungstrieb*.

The basic model for the *Bildungstrieb* grew out of Blumenbach's experiments on the polyp.[9] What was particularly striking about that organism was not only that it could regenerate amputated parts without noticeable modification of structure but that the regenerated parts were always *smaller* than their originals.[10] Upon closer inspection this seemed to be characteristic of the reproduction of injured organic parts generally. In cases of serious flesh wounds, for example, the repaired region was never completely renewed but always retained somewhat of a depression. Such observations led to two conclusions:

First that in all living organisms, a special inborn *Trieb* exists which is active throughout the entire lifespan of the organism, by means of which they receive a determinate shape originally, then maintain it, and when it is destroyed repair it where possible.

Second that all organized bodies have a *Trieb* which is to be distinguished from the general properties of the body as a whole as well as from the particular forces characteristic of that body. This *Trieb* appears to be the primary cause of all generation, reproduction, and nutrition. An in order to distinguish it from the other forces of nature, I call it the *Bildungstrieb*.[11]

That the *Bildungstrieb* was conceived as intimately linked to a material basis can be seen from the manner in which Blumenbach claimed to have been led to the idea; namely that while the polyp always regenerates a lost part, that regenerated part is always smaller. Having lost a substantial portion of its primary generative substance, the force of the *Bildungstrieb* had been weakened. Though its force could be diminished, if it had sufficient strength it would always bring forth the whole structure associated with it:

No small evidence in support of the theory of the *Bildungstrieb* consists in the fact that the shape and structure of organic bodies is much more determinate than either their size, length or other such qualities, ... not only in the case of waterplants but also in the case of animals and even man, the size of many parts, even the most important tissues of the stomach and brain and the length of the intestines, can vary enormously, while the variation in their structure and organization is seldom ever encountered.[12]

Two features of Blumenbach's *Bildungstrieb* are extremely important to bear in mind. The first is that it could not be reduced to the chemical constituents of the generative fluid. Blumenbach repeatedly emphasized the immanent teleological character of his conception of vital force. The formative force existed in the organization of the *Zeugungssaft* as a whole: change any of its constituent elements and the organization of the whole was not just altered; it was completely destroyed. On the other hand, it is to be emphasized that this teleological agent was not to be conceived as a soul superimposed on matter. This form of vitalism is what Blumenbach found objectionable in Buffon's concept of the *moule interieur* and in Wolff's conception of the *vis essentialis*.[13] For Blumenbach the *Bildungstrieb* did not exist apart from its material constituents, but it could not be explained in terms of those elements. It was an emergent property having a completely different character from its constituents. This is why in later, more mature formulations, he portrayed the concept as an organic version of a Newtonian force:

The term *Bildungstrieb* just like all other *Lebenskräfte* such as sensibility and irritability explains nothing itself, rather it is intended to designate a particular force whose constant effect is to be recognized from the phenomena of experience, but whose cause, just like the causes of all other universally recognized natural forces remains for us an occult quality. That does not hinder us in any way whatsoever, however, from attempting to investigate the effects of this force through empirical observations and to bring them under general laws.[14]

Fashioned in the language of the 'General Scholium' to the *Principia*, this passage revealed Blumenbach's goal of doing for organic bodies what Newton had accomplished for inert matter. Fort each class of organized beings there was a specific *Bildungstrieb* which gave rise to its determinate structure. And just as Newton had succeeded in finding the universal organizing force of inert matter by constructing a model which succesfully unified Kepler's laws, Galileo's law and a host of other observed regularities under a single plan, so it was the task of the naturalist to reconstruct the *Bildungstrieb* for each class of organism by unifying the regularities found in reproduction, generation, and nutrition under a general law.

A second feature associated with the *Bildungstrieb* important to emphasize is its explicit commitment to the notion of functional adaption. This is evident in Blumenbach's general definition of the term 'Lebenskraft'. 'By Lebenskraft', wrote Blumenbach, "the animal organization maintains its receptivity to stimulating impressions and the ability of setting its organs in motion".[15] It was this feature of the concept, which stressed the adaptability of the organism to its environment, that enabled the *Bildungstrieb* to function as an explanatory concept for natural history. Although Blumenbach did not think that different species could produce fertile offspring, and this primarily from evidence based on the comparative anatomy of the generating organs as well as differences in the periods of fecundation for different species,[16] he did assign to the organism an ability to make slight modifications in its structure in order to adapt to its environment. There were severe limits placed on this adaptive power by the original organization of the *Bildungstrieb*, however. A variation in a single part entailed, according to what he termed the law of *homogeneity*, a correlative variation in other parts of the organism.[17] Such variations could not proceed beyond certain extremes without destroying the economy of the organism itself. In accordance with this adaptive ability of the organism, Blumenbach argued that the forces of the external environment, through gradual shifts in climate and nutrition, could produce variations in the formative force and that after many generations these variations could take root in the generative fluid itself thus becoming a permanent structural feature of the organism.[18] In terms of this mechanism Blumenbach accounted for races and varieties as degenerate forms produced by physical causes. In order to apply this causal theory to the construction of the system of nature, he urged the empirical investigation of the laws regulating the variations in the formative force through studies of teratology and the effects of climate and nutrition on generation.

A major contribution toward calling attention to the basic idea underlying the theory of the *Bildungstrieb* and the possibility it held for a general theory of animal organization was made by Immanuel Kant. Kant was quick to seize upon the *Bildungstrieb* as exemplifying exactly what he intended by a 'regulative principle' in theory construction.[19] Elsewhere I have attempted to document in detail the relationship between these two men and the extent to which Blumenbach incorporated Kant's work into the mature formulation of his ideas.[20] The importance of Kant's work did not consist in proposing hypotheses or a system of organic nature for which Blumenbach attempted to provide empirical support. It cannot be argued that Blumenbach fashioned himself a follower of Kant. Rather the work of the two men was mutually

IMMANUEL KANT AT AGE 67.

supportive of the same program, the program that I am calling teleomechanism. Although not deficient in original ideas about how to improve biology, a point to which we will return, Kant's main contribution to Blumenbach's work was in making explicit the quite extraordinary assumptions behind the model of the *Bildungstrieb*. As we have seen in our discussion of Blumenbach's idea, the theory of the *Bildungstrieb* tottered precariously on the brink of accepting an out and out vitalism on the one hand and a complete reductionism on the other. It was difficult to see how in Blumenbach's view the formative force could be completely rooted in the constitutive materials of the generative substance, to the extent that altering the organization of these constituents would result in the production of different organisms, and still somehow be incapable of reduction pure and simple via chemical and physical laws to the constitutive material itself; how it could be both dependent on and independent of the materials constitutive of the generative substance. Blumenbach always seemed to skirt this issue by invoking a parallel to Newton's refusal to entertain a mechanical explanation for gravity. Kant explained clearly and forcefully why this was not an *ad hoc* strategem; how biological explanations could be both teleological and mechanical without being occult. Kant's own reason for doing this was that he had encountered difficulties in attempting to extend to the organic realm the categorical framework of the *Kritik der reinen Vernunft*, which had seemed to work perfectly for the purposes of establishing the conceptual foundations of physics. Blumenbach's conception of the Bildungstrieb did not resolve that difficulty; but it did permit the construction of a theory that acknowledged the special character of organic phenomena while at the same time limiting explanations in biology to mechanical explanations. Thus in the accompanying letter with the copy of his *Kritik der Urteilskraft*, which Kant sent to Blumenbach in August, 1790, he wrote: "Your works have taught me a great many things. Indeed your recent unification of the two principles, namely the physico-mechanical and the teleological, which everyone had otherwise thought to be incompatible, has a very close relation to the ideas that currently occupy me but which require just the sort of factual basis that you provide".[21]

The essential problem, which necessarily requires for its solution the assumption of the Bildungstrieb or its equivalent, according to Kant, is that mechanical modes of explanation are by themselves inadequate to deal with the organic realm. Although even in the inorganic realm there are reciprocal effects due to the dynamic interaction of matter, which result in the deflection from a norm or ideal – as for instance the departure from a smooth elliptical orbit by the introduction of a third body – nevertheless such phenomena are incapable of being analyzed in some way as a linear combination of causes and effects, A→B→C, etc. This is not the case in the organic realm, however. Here cause and effect are so mutually interdependent that it is impossible to think of one without the other, so that instead of linear series it is much more appropriate to think of a sort of reflexive series, A → B → C → A. This is a teleological mode of explanation, for it involves the notion of a 'final cause.' In contrast to the mechanical model where A can exist and have its effect independently of C, in the teleological model A causes C but is not also capable itself of existing independently of C. A is both cause and effect of C. The final cause is, logically speaking, the first cause as Aristotle might have expressed it. Because of its similarity with human intentionality or purpose, Kant calls this form of causal explanation *Zweckmässigkeit* and the object that exhibit such patterns, namely organic bodies, he calls *Naturzwecke*, or natural purposes:

The first principle required for the notion of an object conceived as a natural purpose is that the parts, with respect to both form and being, are only possible through their relationship to the whole [das Ganze] ... Secondly, it is required that the parts bind themselves mutually into the unity of a whole in such a way that they are mutually cause and effect of one another.[22]

Now that such 'natural purposes' exist is an objective fact of experience according to Kant. Two sorts of evidence, both of which I have already discussed in connection with Blumenbach, confirm this. First, notes Kant, it is impossible by mechanical means such as chemical combination, either empirically or theoretically to produce functional organisms.[23] Secondly, the evidence of generation, even in the cases of misbirths, indicates that something analogous to 'purpose' or final causation operates in the organic realm, for the goal of constructing a functional organism is always visible in the products of organic nature, including its unsuccessful attempts.

It might be objected that Kant (and Blumenbach) were overly hasty in asserting the impossibility of constructing organized bodies via mechanical means. In fact both Kant and Blumenbach were willing to admit this as a possibility. Kant was willing to admit, indeed he was strongly committed to the notion, that all natural products come about through natural-physical causation. Similarly, Blumenbach grounded the Bildungstrieb in the material constitution of the generative substance. But what Kant insisted upon is that even if nature somehow uses mechanical means in constituting organized bodies, and even if the process is capable of technical duplication, we are nevertheless incapable of understanding that constitutive act from a theoretical scientific point of view. The reason lies not in nature but in the limitations of the human faculty of understanding. The problem is that the human faculty of understanding is only capable of constructing scientific theories that employ the 'linear' mode of causation discussed above. The types of objects that nature constructs in the organic realm, however, involve physical processes which require the teleological mode of causation. Since human reason is only capable of theoretically constructing (or reconstructing if one likes) objects that depend upon 'linear' types of causal relation, the organic realm at its most fundamental constitutive level must therefore necessarily transcend the explanatory or theoretical constructive capacity of reason. Accordingly, the life sciences must rest upon a different set of assumptions and a strategy different from that of the physical sciences must be worked out if biology is to enter upon the royal road of science.

To be sure, there is a certain analogy between the products of technology, according to Kant, and the products of nature. But there is an essential difference. Organisms can in a certain sense be viewed as similar to clockworks. Thus Kant was willing to argue that the functional organization of birds, for example – the air pockets in their bones, the shape and position of the wings and tail, etc. – can all be understood in terms of mechanical principles,[24] just as an *a priori* functional explanation of a clock can be given from the physical characteristics of its parts. But while in a clock each part is *arranged* with a view to its relationship to the whole, and thus satisfies the first condition to be fulfilled in a biological explanation as stated above, it is not the case – as it is in the organic realm – that each part is the *generative cause* of the other, as is required by the second condition to be fulfilled by a biological explanation according to Kant. The principles of mechanics are applicable to the analysis of functional relations, but the teleological explanations demanded by biology require an active, productive principle which transcends any form of causal (natural-physical) explanation available to human reason.

In order to understand the basis for Kant's position regarding biological explanations it is necessary to consider the argument set forth in the *Kritik der Urteilskraft*. This argument is extremely important for understanding the different biological traditions of the Romantic era, for vital materialism and, as we shall see, its successor program of developmental morphology can be considered as having accepted the position outlined by Kant, while the system of nature constructed by the Naturphilosophen originated with the attempt to solve the problem concerning the theoretical construction of organized bodies that Kant had claimed must remain forever intractable. The special form of these Romantic theories, their employment of concepts such as polarity, unity, metamorphosis, and ideal types, as well as the structure of the system of nature constructed from them, were determined by their stand with respect to this Kantian problem and its resolution.

In order to understand the basis for Kant's position regarding biological explanations it is necessary to consider the argument set forth in the *Kritik der Urteilskraft*. Judgement is the faculty, according to Kant, that subsumes particulars given in sense experience under general concepts or rules. It can fullfil this task in two different ways: if the rule, law, or concept is already given *a priori*, then judgement is *determinate* [*bestimmende Urteilskraft*]; if the particulars only are given and a general rule is sought among them, then judgement is *reflective* [*reflectierende Urteilskraft*].[25] In the first case the faculty of judgement is constitutive when applied to nature, while in the latter it is merely regulative; i.e., in the first case it is objective, in the second it is subjective.

These distinctions are important to bear in mind. Both are necessary conditions of experience but in different senses. In the *Kritik der reinen Vernunft* and in the second part of the *Prolegomena*, Kant showed that the basis for any possible objectively valid experience of nature rests ultimately upon an *a priori* component supplied by the categories of the understanding. Concepts of the understanding, such as 'substance', 'cause', and 'interaction' for example, provide formal universal rules of relation for performing determinate syntheses of appearances. These different modes of relation are 'laws' that the understanding is bound to follow in connecting given appearances into judgements of experience. Hence they are constitutive of experience; for although they in no way generate the content of experience, they do provide its form, namely through the injection of formal rules that render objective experience of nature possible. In this particular sense, the understanding, therefore, is according to Kant, the 'lawgiver of nature'. Since it establishes the framework for possible objective experience, the system of the categories of the understanding provides a universal physics, 'which precedes all empirical knowledge of nature and makes it possible'.[26]

But there are other, necessary conditions for experience which are not, as such, constitutive of objects. According to Kant, it is necessary, for example, that we seek systematic unity in experience, that we seek to unite as many different experiences as possible under the fewest number of principles. The aim of science is, thus, to produce systematic experience of nature. Accordingly, a principle must be sought in terms of which the systematic unity of the laws of nature can be achieved. This requirement is subjective and regulative. It concerns the rules that must be followed in the employment of reason and the empirical understanding. They are 'subjective' rules which are not constitutive of objects. Thus such maxims as 'Nature always follows the shortest path' do not say anything about what actually happens; i.e., according to which rule our cognitive powers play their game [ihr Spiel wirklich treiben] and come to an actual determination, but rather how they *ought* to go about it.[27] Such principles, then, are merely subjective guidelines and the results of their application cannot be accorded objective reality. This is no less true of one of the highest of all regulative principles, the principle of mechanism itself, the notion that all products of nature can be judged in mechanically deterministic form. Although suggested by the manner in which the categories of the pure understanding function, this principle is capable of serving as a guiding thread for introducing systematicity into the corpus of empirical laws.

This distinction having been made, the question is whether the concept of *Naturzweck* or natural purpose, which as we have seen is necessary for interpreting our experience of organized bodies, is a concept belonging to the *bestimmende* or to the *reflectierende Urteilskraft*. From the preceding discussion we see that the solution to this problem lies in determining whether the notion of *Naturzweck* is capable of generating *a priori* deductive statements constitutive of experience.

In order to prepare the ground for deciding this issue, Kant considers several examples. The laws whereby organic forms grow and develop, he notes, are completely different from the mechanical laws of the inorganic realm. The matter absorbed by the growing organism is transformed into a basic organic matter by a process incapable of duplication by an artificial process not involving organic substances. This organic matter is then shaped into organs in such a way that each generated part is dependent on every other part for its continued preservation: The organism is both cause and effect of itself. "To be exact, therefore, organic matter is in no way analogous to any sort of causality that we know ... and is therefore not capable of being explicated in terms analogous to any sort of physical capacities at our disposal. ... The concept of an object which is itself a natural purpose is therefore not a concept of the determinate faculty of judgement; it can, however, be a regulative concept of the faculty of reflective judgement."[28]

The result of these considerations is that it is not possible to offer a deductive, *a priori* scientific treatment of organic forms. Biology cannot reduce life to physics or explain biological organization in terms of physical principles. Rather organization must be accepted as the primary given starting point of investigation within the organic realm. In order to conduct biological research it is necessary to assume the notion of *zweckmässig* or purposive agents as a regulative concept. These are to be interpreted analogously to the notion of rational purpose in the construction of technical devices, but it is never admissable to attribute to this regulative principle an objective existence as though there were a physical agent selecting, arranging and determining the outcome of organic processes. At the limits of mechanical explanation in biology we must assume the presence of other forces following different types of laws than those of physics. These forces can never be constructed *a priori* from other natural forces, but they can be the object of research. Within the organic realm the various empirical regularities associated with functional organisms can be investigated. Employing the principles of technology as a regulative guide, these regularities can be united after the analogy of artificial products. Restraints must always be exercised in attributing to nature powers that are the analogs of art, of seeing nature as a divine architect, of imposing a soul on matter. We cannot know that there are natural purposive agents; that would be to make constitutive use of a regulative principle. In order to satisfy all these requirements it is necessary, therefore, to unite the teleological and mechanical frameworks as Herr Hofrat Blumenbach had done by assuming a special force, the *Bildungstrieb*, as the basis for empirical scientific investigation of the organic realm:

In all physical explanations of organic formations Herr Hofrat Blumenbach starts from matter already organized. That crude matter should have originally formed itself according to mechanical laws, that life should have sprung from the nature of what is lifeless, that matter should have been able to dispose itself into the form of a self-maintaining purposiveness – this he rightly declares to be contradictory to reason. But at the same time he leaves to natural mechanism, under this to us indispensable principle of an original organization an undeterminable and yet unmistakeable element, in reference to which the faculty of matter is an organized body called a *formative force* in contrast to and yet standing under the higher guidance and direction of that merely mechanical power universally resident in all matter.[29]

According to the position developed by Kant in the *Kritik der Urteilskraft*, therefore, biology as a science must have a completely different character from physics. Biology must always be an empirical science. Its first principles must ultimately be found in experience. It must assume that certain bodies are organized and the particular form of their organization must be taken as given in experience. The origin of these original forms themselves can never be the subject of a theoretical treatment. This contrasts sharply with physics.

Whereas in physics, for example, it is possible, knowing the law of attraction between all particles of matter, to deduce the shape of the earth, it is not possible, knowing the elements of organic bodies and the laws of organic chemical combination to deduce the form and organization of plants and animals actually existing.

From his analysis of the teleomechanical framework that must underpin the life sciences, Kant went on to draw several methodological consequences. A principal feature of Kant's conception of natural science is that a mechanical explanation is always to be pursued as far as possible. In the organic realm, however, purposive [zweckmässige] organization has to be assumed as given. This primitive state of organization was then to serve as the starting point for constructing a mechanical explanation. Of methodological significance, therefore, was the question of exactly how in practice the mechanical framework was to be related to the teleological framework, and secondly, at what level of investigation a primitive state of organization no longer asccessible to analysis by mechanical models had to be assumed. Kant set out to answer these questions in section 80 and 81 of the *Kritik der Urteilskraft*. These sections contain some of his most significant reflections on biology, reflections which contain in embryo the biological theory of what might best be termed 'transcendental Naturphilosophie.'

One strategy would be to assume that species are the most primitive natural groups united by a common generative capacity. Indeed Kant had alrcady announced in an earlier essay that: "I deduce all organization from other organized beings through reproduction".[30] Using this definition of natural species Kant had gone on to provide a mechanical model in which races were distinguished as members of the same species but adapted to different environmental circumstances. The source of this adaptive capacity was presumed to lie in the original organization of the species, in a set of *Keime* and *Anlagen* present in the generative fluid. In certain environmental circumstances particular combinations of these structures and capacities would be developed while others would remain dormant. Prolonged exposure to the same climatic conditions over many generations would cause these suppressed capacities of the original form of organization to remain permanently dormant. In the case of races, the characters affected were external, such as the structure of the epidermis, hair, nails, etc., while internal organization and the capacity to interbreed and leave fertile offspring remained unaffected.[31]

In the *Kritik der Urteilskraft* Kant expanded upon this model. Perhaps it might be possible, he mused, to find other types of organic unities containing the generative source of several related species. Such an idea had no doubt crossed the mind of every perceptive naturalist, he observed in a footnote, but only to be rejected as a fantasy of reason, since it was no more acceptable to permit the generation of one species from another than it was to permit the generation of organized beings by mechanical means from inorganic means. But the hypothesis he was proposing was not at all of this sort for this was a *generatio univoca* in the most general sense, insofar as organized beings would still be assumed to produce other organisms of the same type, but specifically different in some respect.[32] Kant preferred to think of his model of generation as 'generic preformationism'.

The path to these more fundamental organic unities lay in comparative anatomy and physiology:

The agreement of so many species of animals in a particular common schema, which appears to be grounded not only in their skeletal structure but also in the organization of other parts, whereby a multiplicity of species may be generated by an amazing simplicity of a fundamental plan, through the suppressed development of one part and the greater articulation of another, the lengthening of now this part accompanied by the shortening of another, gives at least a glimmer of hope that the principle of mechanism, without which no science of nature is possible, may be in a position to accomplish something here.[33]

The correctness of such hypothetical unities, Kant argued, would have to be established through careful archaeological investigation of the remains of previous revolutions. Beginning from the common forms that had been provided by comparative anatomy and physiology, the archaeologist must

in accordance with all the known or probable mechanisms available to him determine the generation of that large family of creatures (for they must be conceived as such *i.e.*, as a family, if their presumed thoroughly interconnected interrelatedness is to have a material basis).[34]

Analogous to the reconstruction of the real unity at the basis of the phenomena of races of the same species, an entity which he called the *Stammrasse*, Kant was now encouraging the construction of larger common groupings of species, which he called *Stammgattungen*. Just as a common set of structures and adaptive capacities [Anlagen] were thought to ground the purposive organization of the species, so a similar plan of organization and common set of organs would underpin the purposive organization of several species. When exposed to varying external circumstances, including climate and, as we shall see, other organisms, this original form of organization would be capable of manifesting itself in several different but closely related ways, each being a different species of the same natural family.

There are important differences between the model proposed for identifying races belonging to the same species and that for identifying species belonging to the same family. Members of the same species can be identified with certainty verified by experiment. Any two organisms capable of interbreeding and leaving fertile progency belong to one and the same natural species, according to Kant. The reconstruction of natural families cannot proceed by direct experiment, however. Resting on evidence of comparative anatomy, physiology and archaeology, it is much more hypothetical in character. We are introduced here directly to one of those regulative unities that must characterize biology as an empirical science. What is important, however, is that here the approach is empirical and capable of (limited) test.

A question that must immediately occur, particularly to anyone familiar with the modern Darwinian theory of evolution, is whether Kant meant to infer from his model that the form here being described as the generative source of different species is an actual historical, ancestral form. The answer is unequivocally no. Such an assumption can only be consistent with a completely mechanical and reductionistic theory of organic form in Kant's view. To understand what he meant it is important to recall the model of the *Stammrasse* once again. While this *zweckmässige* organization is the source of all members of the same species, it is not itself represented in an actual historical individual. Kant strenuously denied the thesis then common among contemporary naturalists, including Blumenbach, that the various races of man are modifications of an ancestral race, which most took to be the caucasian race.[35] What Kant had in mind is a distinction much closer to that between a genotype and its phenotypical representations.[36] For he describes the *Stammrasse* as a generative stock containing all its potential adaptive variations. This is important to bear in mind when considering the generalization of the model at the level of families. Were it the case that the *Stammgattung* has an actual representative, say in the fossil record, then in passing from this individual to others of the same family new and different characters would have to be *added* to the existing stock; and this addition would have to occur by means of some mechanical agency. Such an account, in short, runs strongly counter to his teleological conception of biology. According to Kant it can never be argued that an organism *acquires* its ability to adapt to its changing environment. That adaptive capacity must already be present in the organism itself, in the original purposive organization that grounds it. How that purposive organization came originally to be constructed lies forever beyond the reach of scientific treatment. What the archaeologist must presume is that the same *Stammgattung*, which is in reality a complex interrelation of organic forces potentially capable of generating numerous adaptive responses to the environment, underlies a group of forms having both current and extinct representatives. The earlier representatives will, in Kant's view, necessarily be less complex. Once he understands this regulative unity in terms of comparative anatomy and physiology, it will appear to the archaeologist that these forms have been pressed together with single organisms, forms that have been broken up and distributed among many organisms in later periods. Due to the increasing demands of the environment, the potential originally present in the Stammgattung is 'unpacked', appearing as differentiated into more complex representatives. The role of archaeology is to provide an empirical test and guideline for the correctness of the hypothetical or regulative unities constructed through comparative anatomy and physiology. In any given epoch the same forces reign, giving rise in the end to the manifold of nature. The task of biology is to uncover the laws in terms of which those forces in the organic realm operate.

Rather than seeing these organic unities reconstructed by comparative anatomy as potential historical ancestors, it is more appropriate to view them as *plans of organization*, as the particular ways in which the forces constituting the organic world can be assembled into functional organs and systems of organs capable of surviving. Under different circumstances these *zweckmässige Ordnungen* are capable of various adaptive manifestations; that is, the forces that underlie these plans are capable of assuming various expressions in achieving their effect, which is the production of a functional organism. Only under the conditions of a dynamic interpretation of form can we understand how, in Kant's view, it is possible for the fossil record to reveal an ever increasing complexity of forms having the same generative source, while at the same time assuming that this complexity is not the result of an addition of characters:

The archaeologist can let the great womb of nature, which emerges from the original chaos as a great animal, give birth to creatures of less purposive form, those in turn to others which are better adapted to their birthplace and to their inter-relations with one another; until this womb has petrified, fossilized and limited its progeny to determinate species incapable of further modification, and this manifold of forms remains just as it emerged at the end of the operation of that fruitful formative force. But in the end, he must attribute the imposition of the original purposive organization to each of these creatures to the Mother herself.[37]

From this passage we see that the system of nature Kant envisioned was a dynamic one which runs through a cycle of birth, a fruitful period of growth and development of the potential organic forms stored in it originally, maturity and finally ultimate decay. From the undifferentiated potential of the entire system but governed by certain organic laws of adaptive combination that are expressed in definite organizational plans, the first primitive organisms emerge. Each of these purposive organizations has associated with it a reserve of energy. Like Blumenbach's polyps, this *Bildungstrieb* can be used up in regenerating duplicates of the same organisms, or it can be partitioned out so as to produce adaptive variations on the same theme. Originally these organisms are simple and, as Kielmeyer will demonstrate for us, the simplicity of structure is compensated by the enormous fecundity of the organisms themselves. These organisms are governed by a *zweckmässige* generative force; hence, they are capable of adapting to their physical environment as well as to the relationships which emerge between them and other organisms, "but only by taking up into the generative substance those materials alone which are compatible with the original, undeveloped Anlagen of the system".[38] The result is the alteration of the formative force and the alteration consists in a modification of complexity in structure. Each such divergence of the Bildungstrieb must be compensated in some fashion, as for instance in the loss of the ability to produce numerous offspring or in the ability to regenerate lost parts. There are limits on the extent to which these forces can vary and still retain their functional integrity, however. When this occurs, the period of growth is over and all species then in existence continue unchanged into the future. A revolution of the globe, or perhaps even a gradual but continuous change, can lead to the destruction of this system and its replacement by an entirely new set of dynamically interrelated organisms.

Both Kant and Blumenbach realized the potential of this general scheme. It was explored explicitly by Blumenbach's student Christoph Girtanner in a work, which was dedicated to his teacher, entitled *Über das Kantische Prinzip für die Naturgeschichte* (1796). There he introduced the notion of the *Stammgattung*, defined as a generative stock of *Keime* and *Anlagen* which determined certain limits of structural adaption, and which under appropriate environmental conditions became manifest as different but related species (*Gattungen*). Whereas in the case of the production of races a variation of the *Bildungstrieb* was assumed to generate structural modifications which left the ability to breed and leave fertile progeny unaffected, in the case of species, modification of the Bildungstrieb led to a structural modification affecting the possibillity of mating in all but the most closely related species of the same *Stammgattung*. The task of natural history, wrote Girtanner, is to teach "how the original form of each and every *Stammgattung* of animals and plants was constructed, and how species [Gattungen] have gradually been derived from their *Stammgattung*".[39]

Blumenbach himself explored the implications of his generalized notion of the generative stock for paleontology. Discussing fossil forms of marine life in his *Beiträge zur Naturgeschichte* (1790) Blumenbach, who before assimilating Kant's ideas into his own work had earlier denied the possibility of transforming species, interpreted the fossil record as indicating that the *Bildungstrieb* for certain groups of organisms had been altered, giving rise to new species.[40] He also suggested utilizing the degree of divergence among ancestral forms as a means for dating strata, and he made extensive studies of fossil forms near Hannover as well as in Groningen in the Netherlands.[41]

JOHANN CHRISTIAN REIL AND THE LEBENSKRAFT

Although Johann Christian Reil had studied in Göttingen during 1779 and 1780, where he certainly came into contact with the young Blumenbach, he was completely independent from and possibly more original than Blumenbach when he published his treatise 'Von der Lebenskraft' in the first volume of his *Archiv für die Physiologie* in 1795. Reil's discussion of the vital force was presented in a Kantian framework of epistemology and scientific method and it drew explicitly upon Kant's treatment of 'force' in his *Metaphysische Anfangsgründe der Naturwissenschaft*. Reil stressed more emphatically, however, the need for formulating the term *Kraft* in its application to organic phenomena in such a way that it preserved the sense of its signification in the inorganic realm. More importantly, he stressed the need for investigating the *material* basis of the *Lebenskraft* and he set forth more explicitly the materialist assumptions that form the basis of the theory of generation and its relation to questions of taxonomy.

According to Reil the foundation of all structure and function in animal bodies lies originally in the nature of the physical relations of matter itself:

Structure and organization is, ... the appearance and effect of matter itself.[42]

In a Kantian vein, Reil argued that '*Kraft*' designates the relationship between appearance [Erscheinung] and the qualities of matter which are the cause of the appearance. Accordingly, *Kraft* cannot be conceived as something separate from matter:

We are inclined to think of *Kraft* as something separate from matter and matter as merely the vehicle of *Kraft*, although its appearance is inseparable from matter and is a result of the qualities of matter itself. Matter is nothing other than *Kraft*, ... alkalai and acid combine to form a neutral salt, because this is the quality of these things in combination which cannot be separated from them. In addition to the alkalai and acid there is no third thing which effects this combination.[43]

In this context he argued that were it possible for human reason to grasp every natural body as it really is, "to grasp the nature of all its elements and the relations between them, their form and mixture [Form und Mischung], there would be no necessity for emplying the concept *Kraft*".[44] Relying upon the same framework within which Kant and Blumenbach defended the use of teleological principles, Reil thus argued that in the present state of the chemical art, and in particular in consideration of the need for a better understanding of organic chemistry, resort must be taken to the concept of a *Lebenskraft* as the cause of the appearances resulting from organization, even though it is understood that the basis of that organization rests ultimately upon the laws of chemical affinity:

To be sure organic matter belongs as such only to the organic realm and is never encountered in inert nature [in der toten Natur]. But the prime origins of organic matter lay with certainty completely prepared in the womb of the inorganic realm. What is required is a means to bring them together into a purposive combination [zweckmässigen Ordnung] namely a seed [Kern] or an ovary [Stock] of an organized being – upon which the inorganic materials can be assembled.[45]

In the present state of science the cause of biological organization could not be explained; resort had to be taken to the pre-existence of some already organized body which transmutes the affinities of the inorganic realm to those more complex affinities characteristic of the organic realm. Nevertheless, the task of physiology is to press on in understanding these relations within the framework of physical science:

The new discoveries in chemistry and physics teach us that we may be able to discover a great deal more than we already know. The effectiveness of matter is not only determined but increased by its chemical combinations. If the simple elements are not completely capable of bringing forth particular phenomena, why is it not possible for mixtures of these simple combinations?[46]

In order to carry this materialistic approach to the phenomena of organization further consistently, Reil argued that the notion of a purposive organization [zweckmässigen Form] must be understood as designating the general law or rule constituting the animal and that it is determined by the chemical affinities between the organic materials, "just as the seed [Kern] of a salt crystal attracts particles according to a particular law in which the basis of its cubic shape is to be found".[47] Like Blumenbach, he argued that at the basis of each species is a particular *Bildungstrieb* or formative force; but he emphasized that by this 'force' is to be understood a system of regularities determined by a particular order, not to be explained mechanically, among the chemical affinities of the organic materials which constitute the generative substance of the organism. Each organ must be viewed as somehow constituted from a particular arrangement of affinities laid in the generative substance.[48] An increasing complexity of animal structure in nature is accompanied by increasingly complex, distinctly ordered sets of chemical affinity.[49]

Although Reil supported an epigenetic theory of development, his version of generic preformation differed from that presented by Blumenbach and Kant. To be sure, it was never explicitly stated by Blumenbach that both parents make material contributions to the germ, but it was an obvious implication of his treatment of the question of race. The male must make a material contribution in Blumenbach's scheme in order to account for the phenomena of hybridization and race, characterized by the blending of paternal traits. This important insight was not incorporated into Reil's scheme, and his treatment of generation proved to be paradigmatic for the subsequent development in the tradition we are discussing here.

In Reil's view the germ is generated by the mother, and "its matter operates according to the same laws of affinity".[50]

The germ slumbers without developing, probably because its organization has too little irritability [Reizbarkeit]. The father enhances the animal force of the dormant germ perhaps through the addition of the fluid of his semen to the matter of the germ.[51]

Reil's version of the theory captures the literal sense of Kant's term 'generic preformationism' more completely than the idea behind Blumenbach's *Bildungstrieb*. According to Reil all the *Keime* and *Anlagen* of the future organisms are embedded in the system of chemical affinities present in the unfertilized egg. The contribution of the sperm is to function as a catalyst for awakening these affinities into action. This idea, we will see, was to have an illustrious future.

CARL FRIEDRICH KIELMEYER AND THE PHYSIK DES TIERREICHS

This same emphasis upon treating the organization of a material substrate as the primary given from which an investigation of the laws of animal organization were to be conducted also formed the core of Carl Friedrich Kielmeyer's plan for a systematic zoology. Having studied previously at the Hohen-Karlsschule in Stuttgart, Kielmeyer moved to the Göttingen where he studied with Blumenbach, Gmelin and Lichtenberg from 1786 to 1788. He returned to the Karlsschule from 1790-93, during which time he lectured on comparative zoology as well as chemistry and natural history. He returned once again to Göttingen for several months during 1794.[52] Kielmeyer was thus a participant and, as we shall see, a lively contributor to the intense discussions on the construction of a general theory of animal form going on in Blumenbach's circle during the late 1780s and early 1790s.

In his lectures at the Karlsschule Kielmeyer assembled into a grand and comprehensive program the various aspects of the approach to constructing a general theory of animal organization that I have sketched from the writings of Blumenbach, Kant and Reil. Although these lectures were never published, their contents were widely known, and copies of the lectures must have circulated. In a letter to Windischmann, Kielmeyer mentions that copies of these manuscripts were circulated. Cuvier's correspondence with Kielmeyer's student Christian Heinrich Pfaff, demonstrate that while Cuvier did not receive copies of Kielmeyer's manuscripts he was following the development of Kielmeyer's thought in these lectures.[53] In a paper on the advances made in physiology since Haller, Ignaz Döllinger confirmed that Kielmeyer's lectures circulated amongst a select group; and as we shall see, von Baer's limited theory of evolution closely resembles the view presented by Kielmeyer.

In addition to stating the conditions for a materialistic interpretation of the teleological-mechanical conception of the phenomena of organization we have sketched from the works of Reil and Blumenbach as well as stating the implications for generalizing the model for constructing a natural system, Kielmeyer's lectures made an essential contribution by describing a path for beginning to implement these ideas given the present state of biological and chemical science. Two essential problems demanded solution: First, although ultimately the proposed scheme required that the basis of each type of organism lay in the system of organic chemical affinities embedded in the first instance in the generative substance, the analysis of organic materials had only just begun; and although the French chemists in particular had made some advances in this area, still no satisfactory application of chemical methods to the general theory of animal organization could be expected in the foreseeable future. Kielmeyer, who made extensive and substantial contributions to the development of *Pflanzenchemie*, the beginnings of organic chemistry, was deeply sensitive to this problem.[54]

The second problem concerned the actual construction of the natural system viewed as a genealogical system based on the laws of generation and reproduction. As Blumenbach had noted, even though the natural system must be based on generation as a theoretical principle, the practical application of the breeding criterion is circumscribed within certain definite limits.[55] Although different races of the same species are theoretically capable of interbreeding, slight differences in periods of fecundity and differences in behavioral characteristics might set up natural barriers to interbreeding even among members of the same species.[56] Moreover, the breeding criterion was obviously useless for higher taxonomic levels. Blumenbach proposed as the solution to this problem the use of multiple characters in classifying organisms: based on comparative anatomical and physiological investigation animals were to be grouped together in accordance with their agreement in *total* number of characters. Kielmeyer built upon this idea.

In his lectures on comparative zoology Kielmeyer set forth a plan for constructing what he called the *Physik des Tierreichs*. Its design was to develop methods for revealing the laws of organic form through comparative anatomical studies of animals, birds, amphibians, fish, insects and worms. The program consisted of a multifaceted investigation of animal organization, first through a comparative study of the chemical basis of organization. This was to be followed by a comparative anatomy and physiology of basic organs as they exist fully developed and in various periods of embryonic development. Here attention was devoted to three groups of organs: First those which concerned the relation of the organism with its *external* environment: namely digestive organs, the lymphatic system, circulatory system, the brain and nervous system; also included in this group was a comparative study of sensory organs and the investigation of systems of motion; namely muscles, bones and their "analogues" in various animal forms. The second group of organs for study were those concerned with the regulation of the *internal* functions. Here Kielmeyer included comparative studies of the kidneys and the various 'regulatory' glands of the animal economy. The third and final group of organs to be considered were those which served for the *communication* of the animals with the other members of its species, namely organs of generation. Kielmeyer also included in this group the comparative anatomy and physiology of organs of speech, *Stimmorgane*.[57]

After establishing the 'elements' of structure in the organic realm Kielmeyer's program proceeded to a general theory of the relations between them, or to an *Allgemeine Physiologie der Tiere*. Here Kielmeyer advocated the use of developmental histories of the genesis of the germ and its material constituents, the subsequent development of the embryo, and finally the development of the mature organism and the changes it undergoes in relation to its environment. Since the principles regulating each type of organic form lay locked up in the *Keime* and *Anlagen* of its generative substance, comparative developmental histories would reveal interrelations between different organic systems; nature itself would provide, so to speak, its own experimental laboratory. By systematizing and unifying the patterns through which form is unfolded more general relations would emerge from which general laws could be constructed.

Thus far Kielmeyer had presented the methods for revealing the laws of the 'deep structure' of the internal forms of organization. In turning to an analysis of the external surface elements of form Kielmeyer attached special significance to behavioral studies as a means of understanding the principles of organization. He advocated the construction of a *Psychologie der Tiere*. Its object was to study the activities in terms of which animals (a) seek out nourishment, a proper climate and a suitable habitat; (b) activities through which they defend their position in the economy of nature against enemies.[58] Animal psychology was also to include the investigation of activities which promote the preservation of the species, among which he included mating behavior and the rearing of offspring.[59]

Like Blumenbach and Reil, Kielmeyer believed that a systematic study of the variations to which animal forms are subject and the patterns of these anomalies would provide positive insight into the principles of organization. Consequently, he advocated the construction of a *vergleichende Pathologie der Tiere* as a third methodological tool to be employed in the new science of zoology. Here "permanent, inborn as well as accidental variations of species would be investigated; and chiefly under two classes of variation, (a) malformations, monstrous births, bastards; Variations with respect to geographical location and other (similar) circumstances; inheritable degenerations and permanent, inborn variations induced by climatic and geographic variation; universality of variation, (b) Variations in capabilities of the organs and their stimulation; temperament, both individual natural temperament and characteristic idiosyncracies".[60]

Kielmeyer summarized the various aspects of his *Physik der Tierreichs* and the order of their application as follows:

(a) The number of organs in the machine of the animal kingdom or

the number of animal forms generally and the laws according to which these are divided into different groups. Causes, consequences or purposes [Zwecke].
(b) The relative position of the organs in the machine of the animal kingdom, or the division of the animal kingdom into groups upon the earth (geography) according to different characters. Laws of the differences according to different groups. Causes and effects.
(c) The interrelated formation of organs in the animal kingdom. Gradation of animals and affinities in their formation generally as well as according to groups. Laws, causes, and effects of this gradation.

In the next category Kielmeyer introduced an area of study which he had not previously mentioned in the outline of his lectures, namely paleontological research:

(d) Changes which the animal kingdom and its groups have suffered on the earth. The *developmental history of the animal kingdom* in relation to the epochs of the earth and those probable for our solar system. Symbolized by the parabola.
(e) Changes which the animal kingdom and its groups undergoes repeatedly throughout all epochs. The life of the machine of the animal kingdom or its *physiology*. Symbolized by the circle.[61]

In a concluding section of this manuscript, which Kielmeyer crossed out, the *Physik der Tierreichs* was characterized generally as a kind of Laplacian dynamics of animal organization according to which the series of animals and the elements of their organization were to be viewed as a series of attempts by nature to break up the integral of life into a series of partial fractions.

From this plan of a general science of animal form sketched in his lectures we see that Kielmeyer, in addition to uniting the various elements characteristic of the approach of Blumenbach and the Göttingen school , had begun to introduce a completely new dimension to the discussion, namely the use of the embryological criterion for detecting affinities between animal forms. To be sure this was to some extent implicit in the earlier notion of a generative stock shared by different groups of organisms and the related interest in inheritable degeneration and malformations, but the idea of utilizing embryogenesis as a means for investigating the unity of the generative stock was Kielmeyer's most significant contribution.

Kielmeyer expanded upon his notion of the biogenetic law in a treatise, which like almost all of Kielmeyer's work was never published. It was written in 1793-94 and entitled 'Ideen zu einer algemeinen Geschichte und Theorie der Entwickelungserscheinungen der Organizationen'. Several aspects of Kielmeyer's conception of the relationship between phylogeny and ontogeny presented in this manuscript provide an important context for later developments in Germany.

Kielmeyer began by pointing to a fundamental difference beween the results to be expected from teratology and embryology. Malformations appear to be dependent on external circumstances, such as environment, and while they are probably rooted in the matter of the germ, they are departures from the rule and are not repeated similarly in all individuals. Embryological development, however, always reveals a patterned series of successive changes which is the same for each individual of the same species and patterned differently for different species.[62] These patterns of embryogenesis are, therefore, more dependent on an internal directive force: They tell us more about the internal organizing principles of animals, which as we have seen, depend not so essentially on the chemical conditions of life as much more on the *order* and arrangement of those conditions. For Kielmeyer the beauty of focusing on embryological patterns was that "they demonstrate the path and contents of the system of animal organization as a whole without requiring the assumption of a special directive force existing outside of the individual organism, through which the life and economy of organic nature is maintained".[63] That is, recourse need not be taken to a *Weltseele*, to any supra-material organizing force. Furthermore, although in his view embryological investigation is the most useful means for constructing a general theory of animal organization, it can also aid in the construction of a natural classification, which most 'descriptive' biologists regard as the highest aim of their science:

... insofar as the relationship between the different forces and different forms of manifestations of the same force in different organisms is exactly that which determines the essence of the differences and relationships between species. With the determination of these forces therefore, and the laws they obey, the path toward constructing the natural system would be given at the same time.[64]

In a letter to Windischmann of 1804, Kielmeyer explained the reasoning behind his postulation of an interdependence of the results of embryological and paleontological research in his earlier lectures at the Karlsschule:

The idea of a close relationship between the developmental history of the earth and the series of organized bodies, in which each can be used interchangeably to illuminate the other, appears to me to be worthy of praise. The reason is this: Because I consider the force by means of which the *series* of organized forms has been brought forth on the earth to be in its essence and the laws of its manifestation *identical* with the force by means of which the series of developmental stages in each *individual* are produced, which are *similar* to those in the series of organized bodies ... These forms, however, demonstrate a certain regular gradation in structure as well as similarity to the stages of individual development; therefore it can be concluded that the developmental history of the earth and that of the series of organized bodies *are related to one another exactly* and therefore their histories must be bound together.[65]

Kielmeyer went on to add an extremely important qualification to this thesis. He wanted to emphasize that in his view this 'series' of forms must not be conceived as *continuous*. There are gaps in the developmental series that can never be replaced, not simply because of defects in the fossil record, but because there are different types of organization.[66] Like Blumenbach, Kielmeyer denied the existence of a chain of beings.[67]

Nevertheless, while Kielmeyer denied the existence of a continuous developmental series, he did argue for the transformation of species and the interconnection of forms within the intervals punctuated by the gaps in the developmental series:

Many species have apparently emerged from other species, just as the butterfly emerges from the caterpillar ... *They were originally developmental stages and only later achieved the rank of independent species*; they are transformed developmental stages. Others, on the other hand, are original children of the earth. Perhaps, however, all of these primitive ancestors have died out.[68]

He went on to note that, like Lamarck – and though he is not cited in this context, Blumenbach – he believed that the production of these genetically related but distinct forms "was due to an altered direction of the formative force introduced by changes in the earth".[69] But this alteration of the *Bildungstrieb* did not proceed continuously. In Kielmeyer's view the "paths through which the different series of organisms has been brought forth have been very different during different periods of the history of the earth".[70] Thus, not only were the genetic relations between groups of organisms to be viewed as circumscribed within definite limits due to the internal organization of different types, but the manner in which these fundamental organized plans were worked out in different periods and the (limited) developmental series of organisms descendent from them were dependent upon and circumscribed by the external conditions prevailing within a given geological age.[71]

We might summarize the general theory of natural history emerging from Kielmeyer's works as follows: There are definite epochs of nature, during which a different flora and fauna, specific to that epoch, flourish. Within each of these epochs the same laws regulating animal organization prevail, just as the same laws continue to regulate inorganic phenomena. Each epoch contains a system of interrelated organisms based on a small number of ground plans. Within each epoch gradual transitions occur within the forces of both the inorganic and organic realm. As gradual shifts in environmental circumstances occur within an epoch the *Bildungstrieb* of the primitive forms is modified, giving rise to divergent phylogenetic lines of organisms within the same type. Although the forms of the next epoch are based on the same principal plans, there is no continuation of the previous forms. A chance in one element of the system entails a modification in all the others, for each individual form is related to the whole of organized nature. Each epoch, therefore, is its own complete, closed system; and it is not possible to trace a single phylogenetic line, even within the same ground plan, from the most recent epoch.

The essence of the position developed in his unpublished lecture notes was distilled elegantly by Kielmeyer in his famous lecture delivered at the Karlsschule on February 11, 1793, entitled 'Über die Verhältnisse der organischen Kräfte untereinander in der Reihe der verschiedenen Organizationen: Die Gesetze und Folgen dieser Verhältnisse". This paper, approximately forty pages in length, is one of the milestones of the Romantic era; anyone wishing to understand the biology of this period would do well to examine it carefully.

The lecture begins by discussing the general methodological framework that must be assumed if success is to be achieved in constructing the system of nature. The framework is that set forth by Kant in the *Kritik der Urteilskraft*: The constitutive causes of organic nature cannot be grasped. Nature must be treated as if it employed a technique analogous to purposive action, one which relates means to ends in a teleological fashion. The definition of an organized body, following Kant, is one in which all its parts are reciprocally cause and effect of one another.[72] In a literary vein, but one which reflects both Kant's powerful imagery of the great womb of nature as well as indicating that the most fundamental secrets of nature can at best be reflected in a story conscious of its analogy to purposive human activity, Kielmeyer himself speaks forth as *die Natur*. To underscore the necessity but at the same time the futility of ever penetrating the secrets of organization through teleological judgements, Nature is asked what her intentions were in constructing this multiplicity of forms. Her answer is: "I had no intentions, even though the intermingling of cause and effect appears analogous to the connections your reason makes between means and ends; but you will find it easier to understand these matters if you assume such a linkage of cause and effect as though it were in reality one of means to ends".[73]

Lyonet and Bonnet had estimated at least seven million different organic forms on the surface of the earth. Each of these is represented by at least 10,000 different individuals. Each individual in turn is constructed from as many as 1000-10,000 organs. In order to make a system out of this fullness of life according to Kielmeyer, it is necessary to understand the forces that are united in and generative of these individuals. Next it is important to understand the relationship of these forces with respect to one another in different species of animals and the laws according to which this relationship changes in the series of organic forms. "Finally the task is to understand how both the continuity and change in species are grounded in the causes and effects of these forces".[74]

In answer to the first question – what are the forces united in individuals? – Kielmeyer identifies five forces: (1) sensibility; (2) irritability; (3) reproductive power; (4) power of secretion; and (5) power of propulsion. In order to measure these forces and compare them to one another, he proposes that the strength of vital force be conceived as a compound function of (a) the frequency of its effect, (b) the diversity of this effect (i.e. the number of diverse forms in which it is manifested), and (c) the magnitude of the opposition it encounters from other forces. In the absence of an exact measure, and until one satisfying the demands of this function can be constructed, Kielmeyer notes that in essence a vital force is one that demonstrates "permanence of effects under otherwise constant conditions",[75] a definition which seeks to indentify vital force as the source of regulative maintenance of the organized body. The similarity in formulation to Newton's principle of inertia – the force of inactivity as it was then understood – is strong.

Kielmeyer's plan in the work was to look at each of the five vital forces considered singly, and then compare each of their strengths within different species of animals. Beginning his examination with sensibility, Kielmeyer notes that the capacity for retaining a diversity of types of sensations specifically different from one another falls off in a graduated series beginning with man. In the mammals, birds, snakes, and fish all the same sense organs as in man are present, but the degree of complexity of these organs differs for the different classes and even within the same class. In the insects the organ for hearing is absent, while sensitivity to odors is much enhanced; and even if the eye appears multiplied a thousandfold in these animals, it is for the most part immobile and only capable of admitting light in a few species. In the worms, finally, all the diverse organs of the other species are replaced by a single sensibility to touch and light. It must not be overlooked, Kielmeyer tells us, that when in the series of organic forms one sense organ is lost, hence diminishing the diversity of the effect of the force of sensibility – component (b) of the function above – greater opportunity for the development of one of the other senses is afforded; and when one sense is less developed, another will be more sensitive, its organ more delicately structured:

> From these observations we derive the following law: The diversity of possible sensations falls off in the series of organic forms in proportion to the increase in the fineness and discrimination of the remaining senses within a limited domain.[76]

A little reflection revealed that this law is not exactly correct, that even within the same class of animals the reduction in capacity of one of the senses is not always compensated by an increase in another. The ground for departure from this first law Kielmeyer sought to find in the law governing the effects of the second force identified above, namely irritability. In contrast to sensibility, irritability manifests variations not only in the diversity of its effect – component (a) of the definition of vital force – but also in the frequency of its manifestation in a given time and in the length of its manifestation under similar circumstances – i.e. components (b) and (c) from the definition.

In the mammals and in the birds, if the trunk is severed from the head, and after separation of the individual members from the trunk, all traces of irritability vanish within a short time. Cold-blooded animals exhibit quite a contrary set of phenomena. Frogs can hop around with their heads removed, and decapitated turtles can move around with their hearts removed for several days.[77] Kielmeyer noted similar observations for spiders and fish.

These phenomena lead to the conclusion that irritability *increases* its strength and independence from the rest of the organic system in the series of organisms beginning with man. Looking toward other characteristics associated with this phenomenon, Kielmeyer notes that most of the animals that tenaciously preserve this power of irritability are animals in which either very few irritable organs are present or ones in which the muscles are separated from one another. Mussels, for instance, which exhibit a high degree of irritability, have at most two or three distinct muscles.[78] Fish, while possessing numerous muscles, have only a small number of different types of muscles, in contrast with man, where there are relatively few muscles but a great variety of muscletypes and complexity. Moreover, those animals capable of preserving irritability in the highest degree are also those that move the slowest. From all of these observations Kielmeyer derives the following law:

Irritability increases in the permanence of its manifestation in the same proportion as the speed, frequency or diversity of its type and as the multiplicity of different types of sensation decrease.[79]

The second law, therefore, provides the needed corrective factor to the first law, for we see that in the series of different organic forms deficiency in sensibility is compensated by an increase in irritability. But it provides only part of the needed correction Kielmeyer tells us. The force of irritability cannot be preserved as long in mussels, or even in plants, as it can in amphibians. Another force must be sought which affects irritability, accounting for its departure from the norm in certain forms.

Kielmeyer finds the needed modification in the force of reproduction. As a first approximation to the law of the reproductive force, he notes that the mammals normally produce one to fifteen offspring, while birds produce many more than fifteen, and some species of amphibians produce at least one hundred thousand. Examining these phenomena more closely, Kielmeyer observes that the animals that bring forth the fewest offspring in each class are those having the largest bodies. Thus rats give birth to from ten to fifteen offspring at once, while whales produce only one calf. Furthermore, it appears that the less prolific animals are also those having more complex structure and the ones whose offspring require the most time to come to term. "Thus it takes nature two years to make an elephant, while only a few weeks suffice for contructing a rat".[80] These observations result in the following law for the reproductive force:

The more the reproductive force is expressed in the number of new individuals, the smaller are the bodies of these new individuals, the less complex are they, the smaller is the period required for their production, and the shorter is the active period of this force itself.[81]

As in his discussion of the previous laws, Kielmeyer went on to point out several exceptions to this one. The exceptions in this case, however, were only apparent. Thus, while some insects are less prolific than certain fish, it is exactly these insects that exhibit the greatest number of metamorphoses or possess the capacity for regeneration in the greatest degree. Similarly the least prolific amphibians, namely the lizards and snakes, are also the ones capable of achieving the largest body size. Also, according to Kielmeyer, the least prolific mammals and birds are exactly the ones that exhibit the greatest degree of difference in their sexual organs; species of insects and worms exhibiting unlimited growth and high capacity for regenerating damaged parts are also the ones in which sexual differentiation is either absent or in which both sexes are very similar. Kielmeyer was, however, willing to acknowledge certain exceptions to the operation of the law of the reproductive force, but he thought they could be clarified by determining the influence of the external medium in which the animal lives and also the effect of temperature on the reproductive force.[82] These considerations led finally to a reformulation of the law of the reproductive force:

> The more so we find all the different modes of reproductive force united in a single organism, the sooner do we find sensibility excluded, and the sooner also does even irritability disappear.[83]

Having made a comparative study of the three most important forces in his original list, Kielmeyer turned to a consideration of their relations with respect to one another. The system implied by his preceding analysis is obviously a dynamic one. Taken together his three laws imply that in the series of organic forms sensibility is gradually superceded by the reproductive force. Irritability too is finally superceded by the reproductive force, the increase in one of these forces being compensated by a decrease in one of the others. These are the *internal* forces giving rise to animal form and function, and while they do not operate independently of external forces such as the medium, temperature, etc., they are the only sources of animal structure. These forces alone, the same forces operating in every individual, give rise to the entire structure of the organic realm. This point, as we have seen, was essential to the Göttingen program, and it was especially emphasized by Kant: a purposive unity of forces must give rise to the organic realm. The same forces must operate at all levels of differentiation bringing forth families, races, varieties, and ultimately individuals. The individual carries in it the organic forces that differentiate it as a member of each of these higher collective unities. This differentiation cannot at all come about as a result of accidental external modification of inorganic nature. Rather the conditions for bringing forth specifically different types of organisms must always lie in those organisms themselves, in the purposive interrelation of the organic forces productive of organic bodies. External factors provide the conditions for expressing now one permissable expression of these forces and then another, but the true source of this manifold of diversity lies in the internal forces of organization.

Fundamental to Kielmeyer's conception, therefore, a point which he emphasized at the beginning of his lecture, is that the same set of forces united in every individual, though expressed in different degrees, are also the forces that give rise to the entire system of organic nature. This led to the major claim of the paper, and to Kielmeyer's greatest contribution to the Göttingen program; namely that the order in the appearance of these forces in the generation of an individual is the same as the order of appearance of these forces in the system of nature. Ontogeny recapitulates phylogeny:

> The simplicity of these laws becomes evident when one considers that the laws according to which the organic forces are distributed among the different forms of life are exactly the same laws according to which these forces are distributed amongst individuals of the same species and even within the same individuals in different developmental stages: even men and birds are plant-like in their earliest stages of development, the reproductive force is highly excited in them during this period; at a later period the irritable element emerges in the moist substance in which they live – (according to experiments which I have made on chickens, geese, and ducks) even the heart is possessed of almost indestructable irritability during this period – and only later does one sense organ after another emerge appearing almost exactly in the order of their appearance from the lowest to the highest in the series of organized beings, and what previously was irritability develops in the end into the power of understanding, or at least into its immediate material organ.[84]

This principle – that the distribution of forces in the series of organized beings is the same as the division between different developmental states of the same individual – offers a means for constructing the system of nature. According to it the lowest classes are the ones in which the reproductive force is most pronounced. These we might call *Reproductivtieren*. Being characterized by a prolific reproductive and regenerative capacity, this class will contain among all other classes the greatest number of species. Included in this class will be the worms and insects. Similarly there will be *Irritabilitätstieren* and *Sensibilitätstieren*, these classes corresponding to the invertebrates, amphibians, mammals and birds. Within these various classes of animals the same pattern will be repeated; animals possessing the greatest reproductive power will stand first (or lowest) and so forth.

An important aspect of Kielmeyer's theory is that in neither the lecture on the series of organic forces nor in any of his other lecture materials did he ever assert that the series of beings is linear, so that the ontogeny of any single organism recapitulates the phylogeny of the entire animal kingdom. Although he never explicitly developed the system in detail, the evidence of his writings seems to suggest that he regarded each class of animals as having various interconnected sets of organs as the material expression of the system of forces grounding them. In the *Sensibilitätstiere*, for example, the same organs were the predominant organizing principle of the class, although it is clear that being the highest class, all the organs of the other classes must also be available to them. These animals would then specialize in the development of each of these organs. Due to the dynamic interrelation of all the organic forces, the particular preponderance of one (or more) sense organs would entail a corresponding functional arrangement as compensation among the other groups. The system resulting from this scheme would not be linear but rather radiate in structure. At the core of the stem for each group must be imagined not an actual animal, but the specific purposive combination of organic forces (the five named above) containing *in potentia* all the organs and combinations of organs which will be developed by the different species of the group. Different species will correspond to the developmental grades of this primary functional unity. The series of forms developed will not be such that each developmental grade of a particular organ or closely related system of organs follows upon one another in a tight temporal series. Much more consonant with Kielmeyer's view that all animals forms limit one another is the notion that several different species of the same family develop simultaneously, each one representing a developmental grade specialized on a different organ system. Viewed in this manner, Kielmeyer's system is quite compatible with that sketched by Kant in the *Kritik der Urteilskraft*, but in it one can see rudimentary traces of ideas that would be developed more clearly and systematically later on by Karl Ernst von Baer.

The characteristic feature of the application of the Kantian conception of teleology by the vital materialists was their common assumption of a *Lebenskraft* or *Bildungstrieb* as the organizational principle of each organized body. The earliest systematic presentation of arguments for the necessity of assuming this conception were set forth most clearly by Kilemeyer in a course of lectures entitled 'Allgemeine Zoologie; oder Physik der organischen Körper' delivered in 1806. Like Blumenbach, Kielmeyer argued that theoretical zoology has the same task in the realm of organized bodies as physics in the inorganic realm: namely, "to investigate the most universal phenomena of matter and the special classes of phenomena which are not further reducible to others; theoretical zoology is limited to the study of organic bodies, and is included as part of theoretical physics in so far as organic bodies are themselves isolated special classes of phenomena incapable of further reduction".[85]

In order to establish boundaries for the physics of organized bodies Kielmeyer undertook a comparative study. Both inorganic and organic bodies are constructed from similar and dissimilar parts, he writes, and both are the results of mixtures [Mischungen] caused by the 'affinities of matter'.[86] Moreover, organic and inorganic bodies are constructed from forces of cohesion and adhesion; both can be considered in a certain sense mechanical aggregates. "Even the smallest infusorian is chemically mixed as well as mechanically aggregated".[87]

In spite of these similarities between the two types of physical body, there are important differences. Although no elements appear in organic bodies that cannot also be found in inorganic nature, it is noteworthy that only a relatively small portion of the fifty (then) known elements appear in organic bodies. Moreover, elements common to organic bodies, such as carbon and calcium always seem to bear the traces of an *organic* origin when they are met in the inorganic realm. Thus carbon is found in strata bearing evidence of former vegetation while calcium is found in strata rich in coral and shellfish.[88] Kielmeyer drew the tentative conclusion that organic bodies were thus apparently factories [Werkstätten] for the production of certain elements central to the inorganic kingdom. In addition the elements typically found in organic bodies are usually capable of combustion and are of lower specific gravity than inorganic elements, and hence "more suited for motion and development".[89]

The essential difference between organic and inorganic materials appeared to lie in a difference in the laws of chemical affinity characteristic of each:

All mixtures in the inorganic realm are pure chemical works, capable of being explained merely by the laws of chemical affinity as products of the affinity of matter. The mixtures in the organic realm on the other hand are either contrary to the laws of affinity which are observed to hold outside of organized bodies, or at least they are not formed according to them. The only exception to this general observation occurs in cases where the material of the organic body is expelled as a dead substance as in urine and even in lifeless bones. Here in the excreted parts of the organic body the normal affinities begin to reappear.[90]

From these considerations Kielmeyer drew the conclusion that the affinities from which substances in an organized body are composed must stand under the guidance of a "dominating force" which prevents the normal chemical affinities of the physical elements from taking effect so long as the organism is alive or in a healthy condition. In Kielmeyer's view this 'dominating force' [dominierende Kraft] could not be the property of some particular material substance peculiar to organized forms, for in that case the only difference between organic and inorganic bodies would be in the integration of materials; "but the assumption that the force dominating organic affinities is the property of some special matter and this in turn subject to its own series of affinities in organic bodies expresses itself through structural effects which are not explicable in terms of chemical affinity".[91] Thus, for example, while chemical analysis may be in a position to identify the elements composing nerve fibre, the properties of those substances and their affinities for one another would never enable one to predict that together they would constitute a particular structure having a specialized function in the animal organism. Like Kant, Kielmeyer stressed that this property of matter is inexplicable in terms of causal-mechanical explanation; it is only intelligible in terms of the concept of an organizing whole.

Other empirical evidence seemed to strengthen the need for assuming a *Lebenskraft* as the organizing principle of organic bodies. In many instances it appeared that variation of the chemical constituents in organized bodies produced corresponding variation in the organized forms themselves. Thus, as Humboldt had shown, plants deprived of light normally contain more oxygen and fluids [Feuchigkeit], and correspondingly they exhibit variations in leaf and stem structure. Similar correlations between chemical constitution and form appear in the phenomena connected with race.[92] But this correspondence was not universal, for in fact it appeared that identical combinations of elements in different organic bodies are associated with completely different structure. The exact correlation between form and chemical combination did not always hold in the inorganic realm – in crystal structures for example – but the analogy between form and mixture was even less exact in the organic realm:

The cause of this difference in the forms of organized bodies having a high degree of similarity in chemical mixture appears to be grounded in a force which controls and delimits the chemical affinities, a force which at the same time effects the formation of the organized body. In inorganic bodies on the other hand the force which produces the structure is closely associated with the forces of chemical affinity and is for all practical purposes identical with the force of chemical affinity. Therefore in inorganic bodies variation in chemical mixture must produce different structures; in organic bodies on the otherhand, the force which effects the chemical mixtures and organized structures must be different from laws of mere affinity.[93]

In discussing this conclusion Kielmeyer was emphatic in cautioning against interpreting the *Lebenskraft* as a purposive natural agent: first, because even in the case of certain inorganic bodies difference of form is associated with unity in chemical mixture; and secondly, "because organic bodies are judged to be purposive [zweckmässige] organizations *before* a conception of their chemical mixture has been obtained".[94] Expressed in the language of Kant's

Kritik der Urteilskraft this last statement went to the methodological core of the concept of the *Lebenskraft*. Incapable of forming the notion of a purposive organization from the principles of a causal-mechanical explanation, reason must take organization as its starting point in understanding the mechanical interaction of the parts. It is because nerve substance is first experienced as an organized body with particular functions that an investigation of the electro-chemical mechanisms at the basis of its functional organization can take place, not vice versa.

In the writings of Kielmeyer we see the first evidence of a tendency that was to surface in the 1840s as the chief conceptual problem for the further development of vital materialism. For Kielmeyer 'Lebenskraft' has a double significance. On the one hand it is a methodological principle having a philosophical significance only. This was in fact the sense of the term that Kant had recommended. It was to serve as a framework principle setting the limits within which organisms were to be investigated. In this sense *Lebenskraft* was not to be taken as an actual force or unitary matter existing in organisms. This was a *constitutive* use of a principle that was merely *regulative*. Kant had emphasized that reason could not know that natural purposes, or in other words teleological agents, have an actual, physical existence in the organic realm. His point was that in order to investigate biological organization within a causal-mechanical framework reason must assume such agents based on an *analogy* with the unitary organization of rational intentionality, but this organizing agent was to remain an hypothesis incapable of being accounted for in terms of the causal–mechanical framework of the categories of the understanding. The passages we have cited from Kielmeyer, particularly tha last passage cited above, indicate that he was deeply sensitive to the methodological nature of this principle. But on the other hand, we see that Kielmeyer as a biologist found it difficult to remain consistent with this regulative use of the principle. As a biologist he felt it impossible not to assume as a fact demonstrated by all the evidence of the best chemical analysis that an actual unifying organizing force different from that ordering chemical affinity is present in the organism investigated by the zoologist. In this, however, he had overstepped the valid limits of the concept of teleology as Kant had formulated it; he had begun to make a *constitutive* use of the *Lebenskraft*.[95]

CHAPTER 2

THE CONCRETE FORMULATION OF THE PROGRAM: FROM VITAL MATERIALISM TO DEVELOPMENTAL MORPHOLOGY

The second major phase in the development of the program sketched by Blumenbach, Reil and Kielmeyer spans roughly the period between 1806 and 1834. It is bracketed at one end by Kielmeyer's lectures on the 'Allgemeine Theorie der Zoologie' in Tübingen in 1806-1807 and the first introduction of Cuvier's works in Germany by Johann Friedrich Meckel; and it is completed by the great embryological work of Karl Ernst von Baer, Heinrich Rathke and Johannes Müller between 1828-1834. The major impetus of change in the early period came from Cuvier. He stimulated a group of young teleomechanists – some of whom worked directly with him in Paris – with powerful new methods for further expanding their approach to nature. By the end of the period von Baer, Müller and Rathke had succeeded in integrating those methods into an improved version of the teleomechanist program which I call 'developmental morphology'.

Second generation teleomechanists were concerned with extending the investigation of special internal biological laws governing morphogenesis, and in particular with improving the criteria for specifying the morphotype. Through the distinction between 'analogous' and 'homologous' structures von Baer, Rathke, and Müller defined the valid limits of application of the embryological criterion and its importance was underscored for determining the interrelations between different groups of organisms. At the same time that these important methodological improvements were being introduced, evidence accumulated which strengthened the 'epoch theory', the view that there were several ages of the earth, each constituting a complete, self-contained system governed by the same set of organic laws. This evidence was contributed mainly by Cuvier and his students, but German zoologists were keenly interested in integrating them into their own unified theory of life. By 1834 when von Baer departed for St. Petersburg these various pieces of theory and empirical evidence had been fashioned into the structure of a new version of the teleomechanist program.

Though not himself a member of the research tradition discussed here, Georges Cuvier was responsible for deepening and extending the teleomechanist program of the vital materialists. In the 1790s all roads led to Göttingen, where the young men in Blumenbach's inner circle were envisioning a comprehensive approach to organic nature. Between 1800 and 1815, however, particularly after the German states had largely become satellites of France and many German universities were closed, those roads led to Paris, which for German zoologists meant they led to Georges Cuvier. During these years., when Alexander von Humbolt made Paris his home base for exploring the world, a small German colony sprang up around Cuvier. These young Germans had several features common in their backgrounds. They were either students of Blumenbach and Reil, or, through contact with their students, were about to become enthusiastic converts to the teleomechanist program. A second feature they all shared in common was their opposition to *romantische Naturphilosophie*, feelings which were certainly intensified by contact with Cuvier, who was even less tolerant of it than he was of the views of his rivals, Lamarck and Geoffroy Saint-Hilaire.

Friedrich Tiedemann's early career illustrates well the general patterns followed by the men who were the principal architects of developmental morphology. Tiedemann was a student in Würzburg in 1804, where he became the close friend of Ignaz Döllinger, then serving as physician to the poor in that city. Tiedemann attended Schelling's lectures on Naturphilosophie at the university in 1804, which he, like everyone else who heard the young magician, found spellbinding. He was also deeply impressed by Gall's speculative theory of the skull at this time. But Tiedemann became disenchanted with Naturphilosophie. In 1807 at the urging of Döllinger and Samuel Thomas Sömmering he went to Paris to study with Cuvier. In Paris he learned the principles of te new science of comparative anatomy. Tiedemann, however, did not remain solely within the framework of Cuvier's natural history when he began to map out the terrain of his own research. Rather he attempted to synthesize Cuvier's methods with the dynamic approach of the Göttingen school.

Tiedemann's *Anatomie und Bildungsgeschichte des Gehirns* (Nürenberg, 1816) illustrates the common orientation of the young German zoologists. In this work, which was dedicated to Blumenbach, Tiedemann argued that a single morphotype underlies the organization of the skull in each of the classes of the vertebrate type. The various vertebrate skulls were presented as a series of distinct degrees of structural complexity of this ground plan. The less complex forms were treated as arrested developmental stages of the morphotype of the vertebrate skull. The source of these developmental stages was a *Bildungstrieb* arising out of the arrangement of organic materials in the germ substance.

Tiedemann's early work is typical of the young Germans who came to Paris as well as that of other second generation teleomechanists. In sympathy with Treviranus' goal of constructing a unified science of life, they sought a general theory of organic form by uniting Cuvier's laws of structure with the embryological criterion being developed by the Göttingen school. Characteristically, Tiedemann, like the other Germans attempting to work out of this new perspective, adopted the position in his early writings that ontogeny recapitulates phylogeny.[1]

Tiedemann's work also exhibits other features characteristic of the young Germans who sought to extend and improve the program of the Göttingen zoologists. A persistent concern of Tiedemann was to relate questions of form to processes of material exchange within the organism. Like Kielmeyer, Tiedemann viewed physiological chemistry as the ultimate tool in constructing the 'Physik des Thierreiches' that was the dream of the Göttingen school. Together with Leopold Gmelin, he later made great strides toward making that dream a reality.[2]

The rich perspective to be gained from the results of physiological chemistry had to await the development of a variety of new techniques, however. Its impact was only felt much later. In the first three decades of the nineteenth century the primary focus was on the power of embryology as a tool for investigating the structure of the animal kingdom. The principal contributions to the introduction, development, and exploitation of the embryological methods within the teleomechanist tradition by Meckel, von Baer, Rathke and Müller is the subject of the present chapter.

JOHANN FRIEDRICH MECKEL, ORGAN FUNCTION AND THE EMBRYOLOGICAL METHOD

Johann Friedrich Meckel played a significant role in shaping the developments of the second phase, both through his intellectual contributions and through the orientation he gave to the *Archiv für Physiologie*, which he began to edit to 1815. Meckel, whose father was one of the leading figures in the great medical school at Halle and who was a devoted follower of Haller, began his scientific training in comparative anatomy and physiology under Reil in 1798. From 1801-1802 he studied with Blumenbach in Göttingen. There he began to work out his dissertation which discussed cardiac malfunctions through an examination of foetal development and malformations of the foetal heart. After taking his medical degree in Halle in 1803 Meckel studied for several months in Paris with Cuvier and Alexander von Humboldt, followed by a period of study with Johann Peter Frank in

Vienna. Meckel had, therefore, the advantages not only of growing up in one of the most distinguished medical families in Europe but also of having directly studied under most of the major contributors to the new developments of theoretical anatomy and physiology. He was from a very early age at the forefront of the developments of the new 'vital materialism', whose postulates as formulated especially by Reil and Blumenbach, he accepted completely as the foundation for his own approach to discovering the laws of organic form.

In his first major work, *Abhandlungen aus der menschlichen und vergleichenden Anatomie* (Halle, 1806) which was dedicated to Cuvier and written in Paris, Meckel argued that further advancements in both theoretical and practical medicine could only be expected from the construction of a general theory of animal organization, and that recent developments in this area could already provide evidence for this claim. The theory of organic form, he argued must rest ultimately on four areas of investigation: chemical analysis; comparative anatomy and physiology of both vertebrates and invertebrates; pathological anatomy, or the investigation of the conditions under which variations of form occur; and comparative embrylogy. In order to illustrate the practical advantages to be gained from the broader concerns of comparative physiology and embryology, Meckel chose to illuminate the function of the thymus, thyroid gland and the adrenal glands. His examination of these problems blazed the trail for the formation of developmental morphology.

Methods of chemical analysis, he noted, had not advanced sufficiently to illuminate the functioning of the thyroid and adrenal glands. Moreover the use of vivisection was not possible in studying these organs, for surgical technique was incapable of exposing them without causing extensive damage to other tissues and organs.[3] The only means of discovering the function of those organs was a comparative study of their presence or absence, their position and relative size with respect to other organs, and their development in different classes, genera and species of animals.[4] By coupling the constant interrelationships of these organs with other organs in the animal economy together with a knowledge of the manner of life and habitat of the organisms in which they appear, inferences could be drawn concerning the function of the organs in question.

Following the functional method, and utilizing materials supplied by Cuvier, Meckel argued that the thymus must be active primarily during the period of foetal development in man and shortly after birth, for it is very large in the embryo and is very small in adults.[5] The adrenal glands must, he argued, be connected to the organs of generation. He drew this inference from a comparative study of malformations during foetal developments. His subjects were primarily female. While the surrounding organs in the lower half of the body might be normally developed, if the adrenal glands were malformed or deficient, so were the sexual parts.[6] He noted exceptions to this pattern, but thought the connection to be supported by the evidence of numerous anatomies. Anatomies of guinea pigs done by himself and corroborated by Daubenton appeared to strengthen this correlation, for in the embryonic development of that organism the adrenal glands and sexual organs appear simultaneously and retain their relative proportions to one another throughout development. In other rodents Meckel found that if the sexual parts were relatively large, so were the adrenal glands.[7]

Although Meckel suspected that both the thymus and thyroid are somehow connected to sexual development, he thought they had other functions which the comparative method could specify even more precisely. Both of these glands have a larger size relative to other organs in foetal than in post–natal development. Pallas had reported that those glands, particularly the thyroid, are also enlarged in hybernating animals during the period of hybernation. Moreover, Meckel's own dissections of rodents had demonstrated the thymus and thyroid glands to be similarly enlarged. This led to the hypothesis (previously proposed by Autenrieth) that the thyroid and thymus are connected somehow with the process of respiration; for in animals living in oxygen-deficient environments, and for man during the foetal period, they become prominent, indicating an increased function in Meckel's view. Evidence from pathological anatomy seemed to confirm this theory, for according to Meckel in the greatest number of cases where an enlarged thymus was detected, "those persons also suffered from some sort of respiratory disorder, exhibiting either a deterioration of the respiratory organs themselves or some cardiac illness".[8]

Through careful dissection and comparative anatomies under the methodological guidance of the principles of the new vital materialism Meckel hoped to advance the cause of medical science while simultaneously advancing the cause of the budding science of zoology, and it was such goals that he hoped to foster as he took over Reil's *Archiv* in 1815 under the joint editorial direction of Kielmeyer, Blumenbach, Döllinger, Autenrieth, and Kurt Sprengel. The new journal was to be devoted primarily to the investigation of organic form, Meckel wrote in the introduction, and speculation unsupported by emperical evidence was not to be tolerated.[9] Nevertheless, in spite of his strong stand against speculative thinking in biology, Meckel himself overstepped the bounds for the legitimate use of the embryological criterion as it had been proposed by Kielmeyer; and this former student of Cuvier and translator of his *Leçons d'anatomie comparée* violated the principle of the master in drawing unwarranted inferences based on a plentiful and imaginative use of analogical reasoning.

Meckel dissented from the view regarding the chain of being we have found in the works of the vital materialists. Instead he argued, it is well known, that during their embryonic development the more complex animals traverse the organizational forms of all lower animal forms, and accordingly the lower forms can be considered as arrested stages in the embryonic development of the higher animals. The cause for this, he asserted, must lay in "certain inborn capacities for modification built into each organ".[10] Accordingly, modifications in the developmental pattern of the brain and nervous system produced one branch in the *Stufenfolge* of nature, modifications of the sense organs another, and similarly for respiratory organs and the organs of motion. In reaching this conclusion Meckel had committed an error that Cuvier had warned against in his *Leçons*; namely Meckel had based his inference on the graduation of *single* organs and organ systems without taking into consideration the interrelation of different systems into a functioning whole organism. Cuvier's dictum was easy to assent to in theory, but in practice it was difficult to follow. There was genius in Meckel's idea, however, and others such as von Baer were quick to see it.

Meckel defended his construction of a single developmental series of animals in terms of analogies between structures as revealed by comparative anatomy. The principle for detecting those analogies was symmetry. Although Kielmeyer had not defended a single developmental series of animals, he too had used exactly this notion of analogical structures revealed through symmetry as a means of detecting stem relations.[11] According to both Meckel and Kielmeyer there is a symmetry, for example, between the upper and lower half of the organism so that 'analogous' structures are produced: "Thus there is an analogy between the respiratory system and the renal system. The lungs represent the kidneys and the bronchial tubes correspond to the ureter ... The sexual organs (eg. ovaries) are represented in the upper body by the thymus and thyroid".[12] The number of bones in the forearm corresponds exactly to the number of bones in the lower leg, and similarly for symmetries in the right and left sides, as well as the dorsal and ventral aspects of the body.

It is true that this concern for symmetry was one of the principal characteristics of the writings of the Naturphilosophen evident in the work of

Schelling, Oken, and Carus; but in the case of both Kielmeyer and Meckel, who were outspoken and vigorous critics of Schelling's *Naturphilosophie*, the source of this concern is to be sought in the internal workings of their vital materialism. At the heart of this system was the belief that the development of the germ occurs through the unfolding of an original set of chemical affinities and is then followed by a further construction of more potent organic chemical relations. These affinities were considered to operate in terms of laws analogous to the principles of electricity. Consider the following description of embryonic development from Meckel's *Abhandlungen*.

> The earliest development of the embryo must occur at the edges of the primitive spinal column analogously to the phenomenon of electricity and can perhaps be conceived best as follows: The two plates which extend from each side of the spinal coloumn, similarly to the leaves of the Diana tree on the surface of a copper plate, must on account of the lateral duality in terms of which they are constructed, have opposite polarity on the surface and on each of the ends; and on account of their polarity the two opposite plates have a tendency to approach one another and unite ... and consequently each part forms a whole just as the poles of a very soft magnet or as easily as a straight line is formed into a circle. This is the most general and primitive manner in which the principal organs [Hauptorgane] of the embryo are formed ...[13]

Through the polar oppositions present in the germ, the embryo receives its symmetric structure. In simple organisms only a single polar axis may be present, while in more complex organisms, due to the greater complexity of the germinal substance, polar axes in several directions may be present; and hence the complex symmetries of higher organisms. As we shall see, a similar conception of developmental mechanics is to be found in von Baer's writings.

From the point of view of later conceptual developments within the teleomechanist tradition, Meckel's contribution cannot be measured in terms of the positive empirical results he offered, for much of what he did was rejected by later workers in the area. This was particularly the case for the form of the biogenetic law presented in his work, which was roundly rejected after the researches of von Baer.[14] But his notion that a set of 'principal organs' formed by the enfolding of the embryonic plates provides the generalized structures for groups of related organisms became one of the leading ideas of developmental morphology. Another aspect of Meckel's approach to embryology that was retained and more fully developed by later teleomechanists was his strong emphasis on similarity in the relative position of organs in related types of animals as a principle for tracing phylogenetic and taxonomic relations. A further element in his approach was concentration on the simultaneous appearance of organs in the embryo as a clue to their functional interdependence, as well as the idea that changes in the relative size of related organs in different developmental stages indicates order within a functional arrangement. Both of these principles came to be of cental importance within the tradition of developmental morphology. The reason for their significance is clear when it is recalled that one of the guiding ideas of the teleomechanist program was the notion that a set of Keime and Anlagen underlays the organization of interrelated groups of animals. The object of empirical investigation in terms of this theory was to discover the principle of order in terms of which the organic elements are organized. Finally, another concern in Meckel's work fundamental to the original formulation of the program was that of discovering the mechanism utilized by the organized body in its functional activity. In order to accomplish this end Meckel advocated a better understanding of chemical affinities and electro-chemical forces as the material basis for the theory of biological organization.

CUVIER'S THEORY OF TYPES

During the next decade interpretations of the comparative method and the embryological criterion more rigorously consistent with the original core of ideas that guided Blumenbach, Reil and Kielmeyer were introduced. This resulted chiefly from fuller practical understanding of the principles of comparative anatomy and physiology developed by Cuvier. In Germany this was accomplished by the efforts of men such as Blumenbach, whose lectures on comparative anatomy and physiology were now famous, but also by others such as Rudolphi and Döllinger. Perhaps the key development in terms of which this conceptual rigor was formulated in a concrete fashion was through the introduction by Cuvier, and independently by Rudolphi,[16] of the notion that four basic plans of animal structure underlay the phenomena of the animal kingdom. Here were the *Urstämme*, the fundamental laws of organization , upon which Kant had mused and upon which Kielmeyer had lectured.

In spite of some major differences in Cuvier's conception of the history of the earth and its inhabitants, there was an agreement on fundamental methodological precepts at the heart of his work which made it readily acceptable to the vital materialists. In fact, in the 'Première Discours' of the *Leçons*, Cuvier himself came very close to adopting the position of teleomechanism.[17] There he speaks of a necessary interrelation between organic and inorganic matter, the irreducibility of the former to the latter, and the need for accepting organization as a primary given:

... Living bodies are hidden from our inspection until formed, in the exercise of life and in the midst of that vortex of which we seek to discover the origin. However small the parts of an embryo or of a seed, when first visible to us they are already in full possession of life and contain the germ of all the phenomena which, through means of life, they are afterwards to develop.

... The vital motions of living bodies have, therefore, their real origin in the parent stock.

Since we are then, unable to trace life back to its origin, we have no other means of ascertaining the real nature of the living powers but by examining the structure and composition of living bodies, for though it be true that these are in some measure effects of the living powers which formed and support them, still it is equally clear that the living powers can have no other source or foundation but in the body in which they inhere. If the chemical and mechanical elements of the body were originally combined by the living power of its parent, there must be the same living power [force vital] in the body itself, since it exercises a similar action in favor of its descendents.[18]

The object of natural history, Cuvier went on to note, must be to connect the phenomena of life with the laws of matter; but since this goal is incapable of being achieved in the foreseeable future, until the time when natural history has found its Newton, other methods must be employed to reveal the laws of organization. These were supplied by the principles of functional anatomy applied to the study of living and fossil organisms. But the guiding principle in applying these methods to the construction of general laws was the same teleological principle that provided the framework for the research program of the vital materialists, the notion that life presupposes organization. This principle is stated more clearly in the *Règne animal* (1817):

There is, however, a principle peculiar to natural history, ... it is that of the *conditions of existence*, commonly styled *final causes*. As nothing can exist without the reunion of those conditions which render its existence possible, the component parts of each being must be so arranged as to render possible the whole being, not only with regard to itself but to its surrounding relations. The analysis of these conditions frequently conducts us to general laws as certain as those that are derived from deduction or experiment ...

The most effective way of obtaining general laws is that of comparison. This consists in successively observing the same bodies in the different positions in which nature has placed them, or in a mutual comparison of different bodies; until we have ascertained invariable relations between their structure and the phenomena they exhibit. These various bodies are kinds of experiments ready prepared by nature, who adds to or deducts from each of them different parts, just as we might wish to do so in our laboratories, showing us herself at the same time their various results.[19]

Just as Kielmeyer in his lectures at the Karlsschule had attempted to define a methodological framework within which the laws of animal organization could be revealed through the various natural products to which they give rise, dissected and compared by nature herself as a sum of partial fractions, 'das Integral des Lebens', so Cuvier, in a less florid and more positivistic tone, emphasized that nature herself provides the laboratory in which the various manifestations of the laws of organization appear as successful experiments in the constitution of whole, functioning organisms.

Two points of agreement between Cuvier's approach and that of the vital materialists are important to emphasize. The first is the concern for finding the general laws of organization and the related belief that the various orders of taxonomic division are real divisions representing different levels of generality in the organizational powers operative in nature. According to this view the most general law is the basic plan of structure followed by nature, and it is at the same time the most essential element of the organism, the principle of order forming as it were the deep structure of the animal interior. As we recede from this core we arrive at levels less essential to the nature of the animal itself; more variation in the structure and appearance of the principal plan are encountered, "until we arrive at the surface of the body, where the parts that are least essential are placed and whose injuries are the least momentous".[20] Thus, as in the generalized Kantian notion of the generative stock, all the various levels of organizational power are united in the constitution of the individual organism. The second point to emphasize, therefore, is that the functionalism essential to this teleological-mechanical framework led both Cuvier and the vital materialists to stress the study of the individual, functionally whole organism as it is found in nature. Accordingly, as Cuvier expressed it, "every animal may be considered as a particular machine, having certain fixed relations to all other machines that together form the universe".[21] This insistence that each organism must be viewed as a functional whole conditioned *simultaneously* by specifically biological laws of internal organization and the external relation of the individual to the conditions of its existence – namely to its environment and to other organisms – led naturally to the view that "there is not a single function which does not require the assistance and cooperation of all the other functions and which is not affected by the degree of energy with which the other functions are exercised".[22]

Upon this mutual dependence of the functions, upon this reciprocal aid which they receive and which they bestowe, depends a system of laws which regulate the relations of living organs and which are themselves founded on the same necessary relations as the laws of metaphysics or of mathematics; for it is self-evident that a suitable harmony among organs which are to act upon one another is a necessary condition for the existence of the being to which they belong, and that if one of its functions were modified in a manner incompatible with the modifications of the other functions, that being could not exist.[23]

The application of this philosophy of organization to the animal kingdom in terms of the implied principles of the *subordination of functions* and the *correlation of parts* had as a necessary consequence the rejection of the possibility of a continuous series of increasingly complex forms. While it is possible to form continuous series of animals from the invertebrate through the vertebrate class by examining single organs, Cuvier argued that the result is that "*this* organ is at its highest state of perfection in one species of animals, *that* organ is most perfect in a different species; so that if the species are arranged after the gradations of each organ, there must be assumed as many scales as there are organs".[24] Combinations of organs that seem possible in the abstract turn our to be impossible in fact; for the functioning of individual organs necessarily implies the co-existence of certain organs and the exclusion of others. Accordingly, it is impossible not to recognize certain abrupt discontinuities in the organization of the animal kingdom. It is well known that this principle led ultimately to the result announced first in a memoir of 1812[25] and developed fully in the *Règne animal*, that four basic plans of animal organization must be recognized for which no transitionary forms are possible; and that even within these 'embranchement' certain gaps appear.

While Cuvier's work thus stated in an elegant, explicit, and systematic fashion many of the fundamental positions that had also been sketched in the writings of the vital materialists and had, moreover, supported these with the weight of empirical evidence, there were important differences between Cuvier's system and that being developed in Germany by the followers of Blumenbach, Kielmeyer and Reil. For these men the hierarchy of different levels of organization present in each animal was not only the manifestation of the laws of functional organization; if was the trace of a historical lineage of *materially* connected forms, the transformation of the animal type within the limits set by the physical conditions prevailing within an epoch. Both the historical and materialist dimensions are completely absent from Cuvier's approach. While he admits the extinction of certain forms, he excludes the emergence of new ones except by separate creation, and he denies the transformation of species.[26] For the vital materialists the type consists in a set of material dispositions which permit the organism certain degrees of freedom in adapting to changed environmental circumstances, and the effects of these changes are manifested in a *Stammbaum* of historically and genealogically related organisms within a given epoch. It was the importance of this historical dimension for their view of animal organization that led the vital materialists, in sharp contrast to

Cuvier, to include comparative embryology as an essential part of their program.[27]

Our previous discussion of Kant's analysis of the problems of explanation in biology reveals why Cuvier's comparative method alone was insufficient for the teleomechanists. Comparative anatomy led, in Kant's terms, to an artificial synthesis. It could not avoid constructing the whole from an external aggregation of its parts. Somehow Cuvier's approach had to be supplemented with an additional method which corrected this situation by having its starting point in the internal unity of the whole itself, empirically unfolding its components in their natural systemic order. The embryological criterion held the key to that method.

IGNAZ DÖLLINGER AND THE BEGINNINGS OF THE EMBRYOLOGICAL SCHOOL

We have seen the emphasis placed on constructing a generative theory of organic form by the Göttingen school. In the works of Blumenbach and Kielmeyer the significance of embryological studies for the construction of such a generative theory was clearly indicated, and the first extensive use of it was made by Meckel. It was Ignaz Döllinger, however, who first set out in an explicit and clear manner the connection between the embryological method and the principles of vital materialism, and it was his students Christian Pander and Karl Ernst von Baer who worked out the practical methodological details of this aspect of the program.

There can be no doubt that in his early career Döllinger was very strongly affected by the ideas of both Schelling and the Göttingen school. The influence of Schelling, particularly Schelling's emphasis on the notion of polarity as an essential conceptual element of physiology, is particularly evident in Döllinger's early work. We must examine those ideas, for while they were later (in the period between 1815-1824) modified in a manner more consonant with the work of the Göttingen school their traces remained in his work, and they also directly influenced von Baer.

In the discussion of Kant's theory of teleology above, it was noted that a striking deficiency remained in the application of teleological judgment as a reflective and regulative concept. According to this regulative or heuristic viewpoint it was impossible, for instance, to distinguish between the unity of a collection of stones and the more integrated unity exhibited by organized beings. This problem formed the starting point for much of Schelling's reflection on the theory of nature. His solution, the conceptual roots of which lay in Spinoza's *Ethics*, Fichte's *Wissenschaftslehre*, and Plato's *Parmenides* was to imagine nature as an originally undifferentiated unity containing within it various dimensions of potency related to one another by polar oppositions. These polarities provided the mechanism for the material genesis of form in Schelling's view. Only in this way, he argued, could one understandf the unity and mutual interdependence of parts forming an organic whole.[28]

Döllinger adopted the viewpoint of Schelling's Naturphilosophie in a short treatise entitled *Über die Metamorphose der Erd und Steinarten aus der Kieselreihe*, published in Erlangen in 1803. The treatise was inspired by Heinrik Steffens' book, *Beiträge zur innern Naturgeschichte der Erde* (Freiberg, 1801) and by Schelling's *Weltseele*. Its intention was to construct a natural classification of earth and mineral forms. Steffens had convinced him that such a classification could not rest on external characters as Werner had proposed. Rather the natural system must be constructed in terms of a material, generative principle, the result of which would be a *Stammbaum* of related forms.[29] Nature is a universal organism, Döllinger wrote, at the basis of which is a primordial activity [handelnde Tätigkeit] – which he characterized as the *Bildungstrieb* – containing an original duplicity, namely a 'positing' activity and a 'negating' activity. These two opposed activities achieve a graduated series of partial equilibrium points, the natural expression of which is a series of limited, finite forms. The interrelated series of these forms is ultimately derived from an *Indifferenzpunkt* or equilibrium. In this Naturphilosophic scheme, all related forms were regarded as a material metamorphosis of their original unity. This basic scheme was to be repeated throughout a hierarchy of levels in nature.

Döllinger applied this schema to the metamorphosis of the siliceous and calcareous earths. Because he viewed nature as a universal organism, he argued that it was acceptable to regard omit these earths as organic products. This was deemed all the more valid since the earliest traces of vegetation seemed to be supported by these substances, and in Döllinger's view, vegetable life had actually sprung from them.[30] According to the prevailing view of the electrochemistry of organic compounds, nitrogen represented the positive while carbon represented the negative 'pole' of all organic substances. Moreover, the differences between the various siliceous and calcareous earths, according to then-current views, consisted principally in the relative quantities of carbon and nitrogen they contained. According to Döllinger the siliceous minerals formed a 'negative' series characterized by increased quantities of carbon, while the calcareous minerals formed a positive series. Both of these found their *Indifferenzpunkt* in the species of clay.

Döllinger's investigation of the relations among these series of minerals led him to the following conclusions:

(1) The differences between the minerals belonging to the entire siliceous series rests upon a metamorphosis by means of which the series is constructed.

(2) As a result of this metamorphosis the structure of those minerals passes through as series which begins with a negative pole and ends in a positive pole.

(3) According to the proximity of a mineral to one of these poles and its distance from the *Indifferenzpunkt*, the more differentiated will it appear with respect to its external characteristics, which are merely the material expression of its internal dynamic relations. ... Herein is to be sought the original ground of the relationship of all the minerals.[31]

Except for the characterization of developmental grades in terms of positive and negative series united in an *Indifferenzpunkt*, the ideas expressed in this passage by Döllinger bear strong resemblance to ideas later developed by von Baer in his work on embryology. The third point in particular, we will see, has an exact analogue in von Baer's schema for representing developmental grades of the same ground plan. While in Döllinger's treatise this schema is closely associated with the notion of polarity as it was developed by Schelling, it is nonetheless important to emphasize that a similar schema was envisioned by Kielmeyer in his '*Rede*' of 1793. This particular idea, in short, could be made compatible with both the vital materialist scheme as well as with *romantische Naturphilosophie*.

If Döllinger's treatise of 1803 exhibited strong ties to Schelling's Naturphilosophie, by 1805 he had already begun to turn away from the speculative approach toward a more empirically oriented physiology. In his *Grundriss der Naturlehre des menschlichen Organizmus* of 1805, Döllinger expressly rejected considerations of speculative philosophy as relevant to the enterprise of physiology.[32] The approach he took throughout his work leaned heavily upon Blumenbach's writings and on the works of Haller. By 1819 he had shifted his allegiance completely to the line of research begun by Haller and the vital materialists. He emphasized this approach in an essay of 1819 entitled *Was ist Absonderung una wie geschieht sie?* In this essay Döllinger explained that he regarded his microscopal researches on glandular secretion as opening a new era of progress in the tradition of physiology begun by Haller, and he was extremely critical of the hindrance Haller's brand of physiology had suffered as a result of fruitless philosophical speculation.[33] Specifically, Döllinger was concerned to develop observational techniques capable of eliminating the unavoidable element of speculation connected with earlier work on tissue and glandular function, the area that Haller had declared the most obscure and difficult part of physiology. The only way to advance knowledge in this field beyond the level to which Haller had brought it, Döllinger wrote, was not through speculation but through careful microscopial investigation of glands, tissues, and vessels *in vivo* with the use of dyes and injections. Döllinger's criticism was directed as much against speculative mechanists as it was against romantische Naturphilosophen. It was not sufficient, he argued, to observe a narrowing of arteries and veins, for instance, and infer on the basis of a mechanical analogy a capillary connection between them. Rather it was essential to *observe* that connection first before constructing a mechanical model of life. Another point of interest for developments then about to launch into full swing was Döllinger's conviction, expressed often throughout the essay of 1819, that the embryology of the chick was an invaluable source for physiological inquiry.[34]

Döllinger's work manifests a strong concern for the same set of problems and the same way of viewing them that we have seen in the work of Meckel and which I have characterized as developmental morphology. It is not just Döllinger's turning away from *romantische Naturphilosophie* that underscores this impression. Rather it is supported by the content of the new physiology he envisioned. Döllinger's work, indeed the work of most vital materialists, always did have and continued to have much in common with romantische Naturphilosophie. Indeed both traditions had the same goal of constructing a dynamical morphology, an 'organic physics' as it was termed;[35] but there developed sharp differences over how this was to be achieved. Döllinger's work is a very good indication of those differences.

It had become an axiom of German biology that an adequate understanding of organic form could only be achieved by penetrating to the inner, dynamic core of life. Life, it was assumed by all parties, is in essence *activity*, motion, reciprocal interaction giving rise to form. To construct an adequate understanding of life, upon which all parties assumed a true medical theory must ultimately be dependent, it was necessary to develop methods appropriate to the subject itself. Those methods had to be dynamic. They had to reveal the changes and interdependencies in the activity of life. How was this to be done? Not by the static, comparative methods of the

Enlightenment. It was not deemed sufficient to examine the anatomical structure of dead organisms and attempt to infer the manner of functioning of a living organism even from a thorough comparative study, for it was also axiomatic during this period that upon death much of the important fine detail of organic structure quickly deteriorates. Furthermore, it was believed that chemical reactions involving organic and non-organic compounds taking place outside a living organism are fundamentally different from those generated within the organized body by the life process itself.[36] It was assumed that in order to understand the functioning of any given organ it was essential to take into account its relation to other organs, for it was believed that organ function is in fact determined by such interdependent connections. On the other hand, it was an accepted fact that surgical intervention and experiment with animals *in vivo* was not an infallible source of insight into organ function. The entrance of air into the organ under investigation, explained Döllinger, often disrupts its normal functioning.[37] Moreover the animal is often so badly damaged by surgery and it is in such pain that no insight whatever can be gained into the normal life process.

Schelling and the romantische Naturphilosophen following his lead thought those problems could be overcome by speculative thought, and for some time Döllinger was in fundamental agreement with them on this point. The philosophical basis for their position cannot be developed in detail here, but according to this Romantic conception, when properly trained in the method of philosophical reflection, the understanding is capable, primarily through a higher faculty of judgement, of penetrating and comprehending the structure of the life process itself.[38] The ground for this was thought to lie ultimately in the fact that we are living beings and that we therefore have direct access to an intuitive understanding of life as an activity. In the realm of natural science during the Romantic period this concept was the direct ancestor to the notion of *Verstehen* later developed as a basis for the social sciences. Because man is cn the highest rung of the chain of being, the fullest expression of the unity at the basis of nature, he is capable through speculation of penetrating and grasping from within the life processes of other animals. By drawing analogies between the structure of other animals and his own body through comparative anatomy and physiology, Naturphilosophie was assumed capable of giving material content to his speculative intuition.

As I have argued in the previous chapter, the vital materialists of the Göttingen school, were similarly concerned about penetrating "ins innere der Natur". From the work of Kielmeyer, however, it is clear that the Göttingen biologists did not regard speculative Naturphilosophie as actually capable of accomplishing this task. Döllinger expressed a similar view in his *Grundriss* of 1805:

> For since pure philosophy is immediate knowledge of the Absolute, it is incapable of determining anything about finite things except that they are not the Absolute, ... and it is incapable of distinguishing one finite being from another, for in order to do that it must be directed toward a knowledge of finite things, which is directly opposed to its whole standpoint. Furthermore, the categories under which a particulate finite thing is to be subsumed can only be provided by experience. ...[39]

In place of the speculative approach Döllinger saw new possibilities for gaining insight into the process of life and for constructing a dynamic morphology. These possibilities seemed to be offered by recent advances in surgical and experimental techniques in physiology by Magendie and Charles Bell. But above all he saw embryology as the clue to the dynamic morphology he had sought without success in the approach of the speculative romantics. He discussed the foundations of this new approach and what he took to be its central research problems in a lecture in 1824 entitled "Von den Fortschritten, Welche die Physiologie seit Haller gemacht hat".

Delivered as a *Festrede* in the Bavarian Academy of Sciences in Munich, long one of the centers of romantische Naturphilosophie, this lecture must have created quite a stir, for it is conspicuous in identifying all the principal advances in physiology with the practitioners of the vital materialist approach, which Döllinger saw, as do I, as stemming ultimately from Haller. Conspicuous are the names of the Göttingen school, Haller, Wrisberg, Blumenbach, Reil, Sömmering, Kielmeyer, Treviranus, Link, Humboldt, and Meckel. Absent are such names as Oken, Windischmann, Goerres, Nees von Esenbeck, and Carus.

Following Bichat, Döllinger defined physiology as consisting of two principal fields of study, morphology, or the general theory of organic form, and histology. "The task of morphology is to examine the origin of life in the embryo and ultimately all the various parts of the human body, not only in the formation of their individual structure and texture, but also how these parts are related to one another into larger groups".[40] Embryology provides the key to this problem, and the embryology of the chick in particular:

> Nature sets boundaries to the exploration of the origin of the human body, which even the most stubborn will is unable to cross. The first origins of the embryo are shrouded in a secretive darkness, and it is impossible to see where the opportunity should arise to gain access to them. If the researcher cannot overcome this difficulty, he seeks to get around it. He directs his attention to animals where the opportunity for observing this event are easier. But even here he encounters difficulties. The mammals most closely approximating the

generative process in man can only be used sparingly, and a determinate result can only be expected after many unfortunate circumstances have been overcome. He must therefore take recourse to that class of animals in which the development of the young takes place in an egg outside the mother's body, and where observation can be repeated with ease as certainty of experience requires. Nearest to the mammals stand the birds, within which, in spite of the difference in external shape and manner of generation, we always encouter the same structures in the same interconnections as in the human body, while only the shape and size are different. With seeming certainty, therefore, if the relationships in the embryological development of the birds are known to us, we can make inferences concerning the formation of the human embryo as well.[41]

In addition to embryology Döllinger saw developments in two other principal areas as essential to the advancement of morphology; vivisection and comparative anatomy. Only from direct observation of living processes would physiology advance, and the opening of live animals was the only way to gain observations concerning the interrelation of organic process.[42] Ultimately, however, these individual researches were to be integrated into a general organic physics. Th groundwork for this general science of organic form was to be found in the *Biologie* of Treviranus, but the principles of the new organic physics has been laid by Kielmeyer:

> Kielmeyer held captivating lectures for many years on the structure of the animal body in which the principles of comparative anatomy were followed throughout, without which all zoological knowledge remains a useless set of disparate facts. Except for his students and those precious few persons who by chance had access to his manuscripts, almost no one learned anything about the new creation then being born. Yet never had ripe seeds fallen upon a more fertile soil: Among Kielmeyer's listeners was G. Cuvier the greatest ornament of our age.[43]

In his lecture of 1824 Döllinger expressed his strongest endorsement of the teleomechanist program developed by the Göttingen school. To be sure, the concerns that had motivated him in 1803 were still widely present. He was still concerned to devise a general theory of organic form, and he still conceived it as a dynamic morphology, one in which form could only be intelligible in terms of processes of generation, maintenance and function. Whereas he had earlier attacked these problems in terms of a speculative and metaphysical use of polarities, however, he now saw morphology as grounded in embryology, comparative anatomy, and direct observation of organ function. Döllinger succeeded in conveying this conception of dynamic morphology to his most illustrious student, Karl Ernst von Baer, to whose work we now turn.

KARL ERNST VON BAER AND DEVELOPMENTAL MORPHOLOGY

The incorporation of the principles of Cuvierian comparative anatomy into the historically oriented framework of vital materialism was accomplished by Karl Ernst von Baer. According to his own admission, von Baer learned the principles of the new comparative approach from Ignaz von Döllinger. But even before studying with Döllinger in Würzburg during 1815-16, von Baer's scientific perspective had begun to be shaped within the framework of vital materialism, chiely through the influence of two professors at Dorpat (today Tartu), Georg Friedrich Parrot and Karl Friedrich Burdach. Parrot had been a schoolmate of Kielmeyer and Cuvier at the Karlsschule, and he retained close personal contact with them afterwards.[44] He was also a personal friend of Döllinger. Von Baer studied physics, particularly electricity and magnetism under Parrot. From Burdach he heard lectures on physiology, which emphasized the importance of *Entwickelungsgeschichte* for constructing a general theory of animal organization.[45] Like Döllinger, Burdach was sympathetic to many of the themes in Schelling's *Naturphilosophie*. But also like Döllinger, he emphasized the importance of a solid empirical foundation as precondition for the 'speculative' construction of a general theory of organization.[46] It was the interest fostered by his contact with Parrot and Burdach in constructing a general theory of animal organization grounded upon an empirical knowledge of individual organisms that brought von Baer to Würzburg.

In working with Döllinger von Baer had the good fortune of acquiring the guidance of a mind with depth and originality on a par with Kielmeyer himself. Moreover, Döllinger was unquestionably one of the outstanding teachers of the period. Far from simply absorbing the philosophical currents of the day into his work, Döllinger, who according to von Baer had studied Kant's writings 'mit Eifer', was as we have seen a careful critic of Schelling's *Naturphilosophie*. According to von Baer, Döllinger disapproved of Schelling's speculative tendency, even though he regarded Schelling to be engaged in attacking the most difficult problems of natural philosophy. Döllinger "expected the construction of physiology to emerge from precise observations, which would then be the material for philosophical reflection".[47] It was not Schelling's Naturphilosophie that von Baer read under the watchful direction of Döllinger and which became his daily companion in Döllinger's lab during the winter of 1816, rather it was Cuvier's *Leçons sur l'anatomie comparée*.[48]

From Döllinger von Baer not only learned to prize observation based on comparative anatomies carefully performed by himself; he also learned a variety of techniques which were to be the key to exploring new frontiers in developmental biology. Döllinger had been a student of Georg Prochaska, the Viennese anatomist and master of the art of preparing specimens for observation under the microscope and simple magnifying glass. In the early years of the nineteenth century staining techniques and serial sectioning had not yet been perfected. Specimens were still prepared under a magnifying glass with fine needles. The art of preparation was passed down from one generation to the next by word of mouth and long apprenticeship at the laboratory table alongside a master. Döllinger had learned this art from Prochaska, and he had become a highly skilled practitioner of it by the time he passed it on to von Baer. But von Baer was also aided in the use of these techniques by a slightly myopic condition which rendered his vision unusually sharp for fine-grained observations.[49]

Von Baer's early work at his first academic post in Königsberg provides ample evidence that he fully absorbed the embryological style being formulated by Döllinger and that he was aware of the difference between Döllinger's evolving conception of morphology and that developed by the *romantische Naturphilosophen*. Although he had not yet begun his prodigious career as an author of scientific papers and monographs, evidence of von Baer's viewpoint at that time is provided by his former teacher and new colleague at Königsberg, Karl Friedrich Burdach, who had been selected as the director of the newly organized Anatomical Institute. For the opening of the new Anatomical Institute at Königsberg in 1817, the design for which he and von Baer had worked out together, Burdach held a lecture entitled 'Über die Aufgabe der Morphologie'. In his lecture Burdach set forth the notion of morphology as it was conceived by the romantische Naturphilosophen. He told his listeners, however, that this conception differed significantly from the approach of his able assistant, von Baer. In the description of the design of the Institute and the principles upon which it was based, which was appended to the published version of Burdach's lecture, the mark of the young lion is clearly visible.

Morphology, Burdach told his audience, is that higher science concerned with the generation of organic structure, and it is essential to understanding disease; for disease is the result of some abnormality in the operation of the formative force of the organic body. As a general theory of the formative powers of organic structure, morphology is therefore a major branch of natural philosophy as a whole. It aims at finding the laws not only of the generation of individual parts, but the laws binding those parts into a whole organism and the relation of that organism to the rest of nature. The philosophically minded physician is permitted access to these internal structuring froms of organic nature, "for we are not blind products of the structuring forces, rather being is an object for us, and in us the world comes to self consciousness, to self intuition [Selbstanschauung]".[50] It is this special relation of man to the structuring forces of nature that enables him to grasp the essence and *Grundform* of the species and see it reflected in a manifold of different individuals. It is this special relation of man to nature that ultimately makes science possible:

All knowledge is insight into the causal interconnection of phenomena ... it penetrates the inner bond that links things together, the unity that forms the ground of the external manifold, it penetrates into the source of meaning of the phenomena.

The unity at the basis of the phenomena lies in the essence of our reason, therefore, prior to and independent of all experience. Reason is precisely the point where nature grasps itself as a whole, where being comes to the highest and truest consciousness of itself, where individual beings are raised to universal being and its innermost essence becomes objectified.

Thus pure science raises that which is immediately given in self consciousness to sensuous intuition; it grasps in accordance with its internal interconnections, that which presents itself to us with necessity, that which emerges from the core and truth of our being, from the external laws of thought.

If Necessity is manifest to us from within, then Reality will speak to us from without. ... The inner world of thought reveals the most universal aspects of Being; while particularity in Being extends before us in the external world of the senses. We grasp Being accordinglyas a whole and recognize the objectification of thought, the inner source of phenomena, the ground of Reality.[51]

According to Burdach's view, therefore, the generation of form and the necessary relationships between different parts of organic systems becomes manifest through an act of intellectual intuition which is at the same time sensuous. The necessary connections between the phenomena lie already preformed in embryo as it were in the structure of universals present in reason. Burdach did not think it possible to spin out the entire structure of the organic world from those ideas without careful and methodically constructed empirical experience, however. Only by immersing oneself properly in the phenomena of organic life would the logical structure latent in the 'idea' of animal organization become fully articulated and manifest, or 'self conscious' as he writes. By presenting the mind with the various external phenomena of organic life, i.e. generation, comparative study of organs and organ systems in different animals, etc., would the internal, necessary bond linking them into a unified whole become manifest. As Hegel described it, only then would the *Begriff* pass over into reality. When this occurred Burdach imagined that a pure, sensuous intellectual intuition would occur:

> When in this way morphology is raised to be pure science, form no longer appears as a dead product [ein Gewordenes], rather it appears as a coming into being, which emerges out of the eternally generative and formative ... Organism, the knowledge of which is our goal.[52]

In this intellectual intuition, which is the goal of morphology as Burdach conceived it, the systematic interconnections between all the various organic elements unfold almost as if one were to have before him a timelapse film of embryological development. Clearly, Burdach intended to awaken in his listeners recollections of Goethe's description of his experience of the *Urpflanze* during his stay in Italy.

Burdach considered his appointed task as instructor of young physicians to consist in preparing them to have this intellectual intuition, for it formed the spirit and essence of medicine itself, rendering the physician truly capable of understanding and treating disease. The anatomical institute and the four-year course of lectures designed by Burdach and von Baer were intended to realize this educational goal.

The anatomical institute was a three-story building with a basement, in which corpses were stored and prepared. The main lecture hall was directly above the cellar and a dissection table on an elevator allowed the corpses to be brought directly into the middle of the lecture hall. There were student laboratories as well as a laboratory for the prosector on the first floor. On the next floor there was an anatomical museum containing more than 1600 different prearations. Represented were the various organs of animal and human anatomy in both normal and diseased states. These were arranged in ascendings states of maturity. The museum was divided into four rooms. Included in them were: a skeletal and skull collection of various animals and man in different stages; a collection of circulatory and respiratory organs; digestive organs; muscles; sensory organs; and a collection of 114 human embryos arranged in sequence beginning with a three-week-old fetus.

Burdach made special note of the difference between his own use of those materials in his lectures on zoology and the use made by von Baer. Von Baer, he said, approached the subject of morphology from a *logical* point of view, while his own approach was *topical*. Whereas Burdach selected a single organ and went through the various forms or shapes that organ takes in different animals, thereby generating an evolutionary or generative transformation of shapes based on a common ideal ground plan, von Baer presented the systematic functional organization of each different type of animal starting from the infusorians and ending with the mammals, "presenting each animal type as a whole". In his lectures von Baer placed emphasis not on the shape of organs, but on their relative position, arrangement and systematic connections in the series of animal types.[53]

The difference between those two approaches to morphology is of some importance. They reflect in fact the two different conceptions of morphology developed in the Romantic period by the *romantische Naturphilosophen* on the one hand and the vital materialists on the other. The speculative tradition, by which Burdach was strongly influenced, stressed the *anschaulich* or intuitive character of the science they envisioned, and for them this implied that developmental sequences of related organs were given by *geometrical* transformation of certain basic *Grundformen*. This is the approach followed by Goethe, for instance, in his representation of the interconnection of plants in the form of the *Urpflanze*.[54] This same approach is followed by Oken, Goethe, Carus, and, of course, Geoffroy Saint-Hilaire in their discussion of the metamorphosis of vertebrate and invertebrate skeletal systems.[55] They always emphasized the geometrical transformation of some particular basic structure, and they always focused on linear series of such transformations in individual organs rather than systematic interconnections between different organs in their approach to constructing the natural system. Burdach provides us with a clue to the relation of this point to speculative *Naturphilosophie* in his lecture:

Shape is the relation of spatial limitation. ... Whatever fills space is the product of some previous activity: The source of shape is motion acting in different determinate directions and terminating in certain definite points: for motion is the externally described, progressive filling of space.[56]

Shape, in short, is the external impression in space and time of the organic formative forces, which for Burdach are the external ideas of the *Weltgeist*.[57]

Von Baer, by contrast, never treated the *shape* of a fundamental organ and its geometrical transformation as relevant to his approach. More important to his perspective was the animal as a functional whole; and for dealing with this issue he followed Cuvier in emphasizing the position and arrangement of organs, for these were the expression of *functional* laws of organization. These functional laws were the source of animal form for him. Moreover, to investigate the forms related to one another in terms of these principles of position and arrangement of organs, von Baer did not have to take recourse to the *Weltgeist* or to an intellectual *Anschauung*, "in which the living forms and their connections swim before his soul".[58] Rather he had merely to undertake careful empirical observations of embryos.

That von Baer's conception of morphology was in agreement with that being developed by Kielmeyer, Tiedemann, Meckel and above all Döllinger is evident from his first major work, *Vorlesungen über Anthropologie*, published in 1824. These lectures are of interest not only because von Baer followed exactly the outline of the subject given in Döllinger's *Grundriss*, but also because he stressed the materialist element of his approach to morphology.

Von Baer began by introducing his students to the elements of human anatomy. First he discussed the symmetry of the external parts of the body, but he noted that attention to symmetry is of little use in discussing the internal structure of the body. "The lungs are double, but they are not identical to one another. The liver, stomach and spleen are parts that only appear singularly, but they do not lie in the central plane of the body, and they do not consist of two identical parts".[59] The proper way to approach the internal structure of the body is to divide the trunk into certain regions, the same regions familiar to modern anatomists. "The regions must have a particular relation to one another", von Baer writes, "in order to produce a strong body".

Von Baer proceeded next to a discussion of the individual parts of the body, and here his approach followed that of Reil, Kielmeyer, and Treviranus in focusing on *Struktur*, *Textur* and *Mischung*. Like these vital materialists, he too stressed the importance of *Textur* and *Mischung* as the most important clue to organic structure. The body consists of a multitude of heterogeneous parts, he notes, among which two types are to be distinguished; some appear as repetitions of structures which are essentially indentical in their inner construction [Textur] and are therefore assumed to be also indentical in function. They are always affected by the same types of disease, which is one sign of their close connection. "The totality of such identical parts is what we call an *organic system* ... These systems form a spatial unity; that is all parts of the same system are physically interconnected. This is the case, for example, with the circulatory system, the nervous system, cell tissue and mucous tissue [Schleimstoff]".[60] There are structures, on the other hand, that appear only singly or in pairs. They are composed from some purposeful combination of the simple homogeneous elements and are united in terms of some principle function which they serve. "Thus respiration is performed by the lungs, the bronchi and the larynx. The entire unit is called an *apparatus*, and each part of it is an *organ*, insofar as it has its own function. Thus the bronchi is an organ of the respiratory apparatus".[61] We see, therefore, that even in his definition of the elements of structure, von Baer follows two paths at once. Some parts are materially connected, by which he means they are constituted identically, while on the other hand some parts are constituted in terms of functional relations. As we shall see, the simultaneous pursuit of both these approaches came to be characteristic of von Baer's work as a whole.

In the opening lecture of his course, von Baer also emphasized the importance of understanding the chemical composition of organic structures. Unfortunately, he observed, there are many difficulties in utilizing chemistry effectively in this enterprise. "The part, which the chemist removes from the plant or animal for the purpose of his investigation, is already torn from the vital context, and from the moment of this separation, processes begin to set in which no longer belong strictly to the process of life. While these changes may only be slight in connection with plants, they are significant with respect to animals. The blood that we remove from the artery of an animal alters in consistency and color in a few minutes and begins to separate. What is given over to the retort is scarcely longer to be called the blood of the animal, but rather a *product* of the blood".[62] While the products of chemical analysis thus stand under the suspicion of being artificially induced through removal from the vital context, von Baer was convinced nonetheless that "we cannot give up our need for chemistry, because it instructs us concerning the greater or lesser similarity of the different parts".[63]

Von Baer's defense of chemical analysis in the face of the objections he had raised earlier was based on the assumption of a constancy in the relation of cause and effect. Even if, he argue, organic substances are altered upon removal from an organic system, they must nonetheless always be modified in the same way if the same conditions prevail: " or in employing the same chemical processes, nature must always produce similar mixtures".[64] Thus, in its present state, chemical analysis may not have been able to provide insight into physiological chemistry, into the chemical workshop of the living organism itself, but von Baer thought it might provide reliable indicators of whether the elements of organic structure he was interested in studying were related and the degree of closeness of the relation.

Von Baer was careful to distinguish his position concerning the role of chemical analysis from any form of absolute physical reductionism, however:

There is no doubt that in the organized body the greatest multiplicity of chemical processes is constantly in progress ... But even though strictly speaking the chemical constitution of the body and its individual parts only differs slightly from one moment to the next, there are nonetheless, strict limits between which the mixtures of the individual parts can oscillate without becoming completely useless for the function they are to serve; and the organized body immediately replaces every deficiency which arises in this fashion if it is provided with the proper materials from without. Thus one is justified in maintaining that these chemical changes are governed by the process of life. The organized body is an even greater chemist than the most famous alchemists in their laboratories, for it can prepare matter, which was not formerly present.[65]

Like Kielmeyer, von Baer subscribed to the view that chemical processes within an organized body operate according to principles other than those that must prevail in the chemical laboratory. The organized body is capable of regulating the linear progression of chemical changes. It is able to maintain the chemical machine within the definite limits of a range of reactions. These limits are prescribed by functional necessity. Moreover, it is important to note that von Baer does not mention a supramaterial vital force or *Lebenskraft* as the source of these altered chemical processes or as a constant attendant supervising and controlling the course of transformation. For him the vital forces lie in the *relations* of organic parts with respect to one another, what Reil had described as *zweckmässige Ordnung*. For von Baer this order is determined by functional laws of organization. Like the vital materialists, he too emphasized the special nature of chemical processes in a physiological setting. Thus, he observed, recent chemical analysis had demonstrated that the quantity of calcium required for the production of the egg shell by the hen is much greater than that taken in through nourishment. The organized body must somehow produce this material from other materials in which it was not formerly present.[66]

Such considerations on the special nature of the products of physiological chemistry implied that the materials from which the different organic systems are composed within the living body must also be specifically different. In von Baer's view if functional requirements provided the guidelines for chemical processes within the organized body, it was also true conversely that the special functional requirements of a particular organ or organic system demanded in consequence a determinate material constitution. There are, according to von Baer, seven basic animal substances or elements from which the animal body is structured: *Eiweissstoff, Gallert, Faserstoff, Schleimthierisches Wasser, Fett, Osmazom* and *Milchsäure*. Von Baer cites the chemical composition of these substances from the works of Gay-Lussac, Thenard and Berzelius. Thus *Eiweissstoff*, or protein, consists of more than 1/2 carbon, 1/4 oxygen, 1/6 nitrogen and 1/3 hydrogen. *Gallert*, or gluten, what Haller and Blumenbach had called *tierischer Leim*, consists of less than 1/2 carbon, and more than 1/4 oxygen, more than 1/6 nitrogen, and almost 1/12 hydrogen. *Faserstoff*, or fibrine, consists of more than 1/2 carbon, 1/5 nitrogen, almost 1/5 oxygen and 1/4 hydrogen.

Von Baer places special importance on the chemical analysis of these substances, for it establishes their close interconnection. In his view they are all derivatives of a fundamental animal substance.

> It is nothworthy that these three animal substances, which are the most prominent substances of the human body, are so similar in their chemical constitution. They appear to be modifications of a *Grundmischung* which we must appropriately call the 'animal substance'.[67]

This, we have seen, was a characteristic feature of vital materialist works, particularly evident in Blumenbach's work on the human skeleton. The same position was also adopted by Döllinger in his work on secretion as well as his *Grundriss* of 1805.[68] Von Baer introduced this notion of a material connection between the different substances of the animal body early in his scientific career, and it always remained a principal feature of his work. We see in this the first beginnings of the idea he would later work out of the histological differentiation of a homogeneous animal substance into the germ layers of the young embryo. Thus in his *Untersuchungen über die Entwickelungsgeschichte der Fische* of 1835, von Baer writes:

> According to Rathke's *Abhandlungen zur Bildungs-und-Entwickelungsgeschichte*, Part II, p. 3, fish eggs are formed in the cellular membrane of the ovary. This cellular membrane has meanwhile a very notable thickness, and is much less firm than membranes are normally; and it consists of exactly the same substance which forms the ovaries of the batrachians. It is in general the most beautiful paradigm that we have of the universal animal *Grundstoff* (Döllinger's *Thierstoff*) without specific tissues, and for that reason the reservoir (Stroma, Kiemlager) for the eggs, which alone can give this material organic significance.[69]

Von Baer's concern with the fundamental materials constituting a body was not just for the sake of giving a complete description, rather it followed directly from his conception of the task of morphology. He regarded the material constitution of an organ as the determinant ground of its functional capacity. In his view, "Similarity or difference among the constituent elements implies similarity or difference in the expression of life. Thus fibrine is an essential component of all muscle tissue. Parts which do not have fibrine do not have the characteristics of muscles, even if certain external similarities between them can be exhibited".[70] The heart and soul of von Baer's conception of morphology was the interdependent relationship between form and function. While he regarded functional organization as prescribing the framework of material processed within the organized body, he nonetheless regarded material constitution as the essential determinant of function. This was one of the central propositions of teleomechanism: form cannot be understood without reference to function; function cannot be understood independently of the material constituents of form and structure. In the introduction to his *Anthropologie*, von Baer explained to his readers: "I have pointed frequently throughout this volume to the purposive construction of the body and spoken as a teleologist. In the second volume of this work I will attempt to deduce this purposiveness [Zweckmässigkeit] from a higher principle of necessity".[71] This second volume of von Baer's *Anthropologie* never appeared. The problem of the relationship between form and function remained the outstanding unsolved problem of the teleomechanist tradition.

The work of his early years in Königsberg demonstrates, that while he had absorbed the principles of Cuvier's functional anatomy, von Baer was far from being a 'Cuvierian' pure and simple in his approach to organization; rather it is evident that early on he assimilated Cuvier's principles into the framework of vital materialism.[72] The evidence for this is contained in two unpublished lectures delivered in Königsberg in 1822 and 1825 entitled 'Vorlesung über die Zeugung' (1822), and Über die Entwickelung des Lebens auf die Erde' (1825), as well as von Baer's last lecture in Königsberg delivered in 1834. In his biography of von Baer, Raikov has provided extensive citations from the manuscripts of these lectures which offer invaluable insight into the developing foundations of von Baer's early thought.[73]

In the lecture on generation von Baer adopts the theory of generic preformationism we have sketched from the works of Kant, Reil, and Kielmeyer. "It first appears that at the moment of fertilization", he wrote, "the new being must arise, as if through an electrical shock or some sort of magical artifice. But no matter how exacting the choice of microscope, no matter how one strains the eye, one sees nothing immediately after fertilization that was not previously visible. Only sometime later is the new plant or animal recognizable, and then it is already caught up in growth. The thought must, therefore, impress itself upon us that the beginnings do not at all coincide with fertilization, but that the fruit lays already formed in the parents, and has now entered into relations in which it can develop more quickly".[74] Von Baer went on to explain that the germinal disc is present in the fertilized egg before the embryo, but the germ [Keim] is present in the egg before fertilization. Fertilization is only the '*anstoss*' for the further development of materials formed in the uterus. At the end of the lecture von Baer drew the consequences of this conception for viewing the relationship between phylogeny and ontogeny:

The idea, or essential vitality of the aminal is eternal and is only transmitted through a series of individuals in which it is always realized independently. Thus nature is ruled by the principle that the different forms of organized bodies always maintain themselves while the individuals always die. Just as we observe a continuous transformation in the development of the individuals, however, and nowhere an absolute beginning, so must the different forms which we call species have been gradually developed out of one another without having been originally constructed in all their present diversity.[75]

In August of the same year von Baer developed this idea further through an examination of the evidence provided by recent paleontological research. Here, in the lecture 'Über die Entwickelung des Lebens auf die Erde', he set forth the thesis that the development of the earth and its forms of life are deeply interdependent. The fossil record reveals three classes of remains: there are species for which it is difficult to determine even class and order in terms of present forms; others which seem to belong to current genera but have no representatives among present species; and finally those which seem to be identical to current forms. The extinct forms are found in layers of the earth which, according to geological evidence, must have been formed earlier, while the forms demonstrating characteristics more closely resembling current forms appear in the most recent strata.[76] In order to explain these phenomena, von Baer proposed a model based on the interdependence of animal organization, means of subsistance and climate:

Some mosses of the European Alps are similar to mosses in the Andes. Even if slight differences between them can be detected, this does not contradict our theory, for there are slight climatic differences as well. ... Similar climatic conditions produce similar animals and plants.[77]

In order to see the scope von Baer was willing to accord these transformationist ideas it is useful to consider the lecture 'Das allgemeinste Gesetz der Natur in aller Entwickelung' which he read in Königsberg in 1834 and which was later included as the second of his *Reden*. Here von Baer brought together the various elements of the view of natural history sketched in his more youthful writings. All organized bodies are in a constant state of change, he wrote. Embryology reveals that these alterations do not occur with the same swiftness in all animals, and the lifespan of different types of organisms is subject to great variation. The question then emerges, how were these different forms originated? As an answer to this question von Baer rejects the idea that new forms can be generated through the fruitful mating of different species; nature erects too many barriers for hybridization to be an effective mechanism. Instead he prefers to see the process generated by the

THE CONCRETE FORMULATION OF THE PROGRAM 83

relationship of the organism to its evironment in the manner proposed first by Blumenbach and Kielmeyer:

On the other hand every type of variation in the growth of the individual [Selbstbildung] is transmitted through reproduction, and we see here the most evident confirmation of the principle ... that generation is only a continuation of Selbstbildung or growth [Wachstum]. If, therefore, external influences change the manner of sustenance, it will have an effect on the process of reproduction, and the longer this same influence is continuously active on numerous generations, the more pronounced will its effect be in later generations, even after the influence has subsided.[78]

Earlier zoologists, such as Blumenbach, had pointed to reports of malformations in the structure of animals transported to new environments (such as certain malformations of European pigs in South America) as evidence for this thesis, and then had proceeded to infer that the fossil record could be explained by a similar process. Von Baer believed he could support this model with more positive evidence:

More instructive is the case of the guinea pig, which is a frequent pet. It is certain that this animal did not exist in Europe before the discovery of America. The zoologists of the 16th century tell us that it was brought over from America; now where this part of the globe is completely explored in all directions, an animal of the same coloration is nowhere to be found. In our housepets it is always spotted, of either two or three colors, namely black, brown and white. In America on the otherhand there is an animal of the same shape and size, but it is always a single color, grey-brown. The *Cavia Aperea* of Linneaus might be considered the *Urstamm* of this animal, but the Aperea prefers damp places, and our guinea pig prefers neither dampness nor cold. Is this not the result of degeneration [Verzärtelung] which has become constantly passed on through reproduction, just as our guinea pig reproduces three times in a year while the Aperea only bears young once a year in the wild. Even the bones of the skull have received a somewhat different shape, and the tame guinea pig will no longer mate with the wild stock; this is therefore, according to zoological principles, actually a new species. This much has been produced in three centuries.[79]

Von Baer went on to note that all the species of apes, of which he knew 150, were probably of common origin. He urged a similar thesis for the wild sheep, goats and cattle.

While he defended a transformationist view of species in these early writings, von Baer did not think that all species were ultimately of common origin. Man for example, could not have emerged from the apes. "How could the higher intellectual capacity of man emerge in the Orangoutan?"[80] The transformation of species must, according to von Baer, be contained within definite limits: "We must conclude", he wrote, "that as far as observation has enabled us to determine, a transformation of certain original forms of animals·

in the course of the generations has with great probability taken place, but only to a *limited degree* [von Baer's italics]; that the complete destruction of many types of animal has occurred and at the same time they only gradually emerged".[81]

It is in attempting to understand what von Baer meant by this 'limited degree' of transformation that we come to appreciate one of the central elements of the biological system advocated by the teleomechanists. Such expressions might lead one to think that von Baer advocated a (limited) form of evolution, perhaps requiring only the mechanism of selection to become Darwinian evolution. To do this, however, one must interpret von Baer to mean that through adaptive response to their environment animal forms *acquire* new, mostly more complex characters; and after being subjected to these environmental pressures for many generations, new species ultimately evolve. This, however, was not von Baer's view; for it is inconsistent with the fundamentally teleological conception of zoology he shared with the vital materialists. Kant had made the essential point: organization can never be understood without presupposing an original state of organization. Similarly, higher forms cannot be generated from lower forms of organization. Organization can only be understood as proceeding within equal degrees of organization, or from a higher to a lower state of organization. Consequently if groups of species are to be viewed as descendants of a single ancestral form, the *Keime* and *Anlagen* for those later forms had to be present already *in potentia* in the original generative stock. They represent the built-in capacity for adaptation – and the limits of that capability – of an original state of organization. To see how this teleological conception could be harmonized with a materialist conception of organization and at the same time form the basis for a transformationist theory of species circumscribed with certain limits we must turn to the principle work of von Baer's years in Königsberg, the *Entwickelungsgeschichte*.

Von Baer, it is well known, regarded the major accomplishment of his embryological studies to have established "how the animal type of the vertebrates is gradually set forth in the structure of the embryo",[82] and that "the Type of organization determines the manner of development".[83] For von Baer, the type is the most essential aspect of the animal; it regulates and determines all subsequent development. Already in the earliest stage of its development the chick embryo manifests the character of the vertebrate type; with the emergence of the allantois it manifests itself as a vertebrate animal which can never live in water; in the third period of development with the construction of the respiratory organs, the beak, and the embryonic wing buds, it joins the class of birds; at this stage only a bird in general, it first manifests itself as a galinaceous land bird with the absence of webbing in its feet, when the gizzard separates off from the muscular stomach, and when the squamous scale appears over the nasal opening; finally the species-characters emerge with the formation of the comb and the characteristic structure of the beak.[84] Thus the animal develops in a pattern leading from the most universal, essential to the more specialized and individual characters. More important for von Baer's perspective, however, is not just the *phenomenal* manifestation of general individuation but that the more general forms *condition and control* the development of the more specialized: "Each and every organ is a modified part of a more general organ, and in this respect we might say that each organ is already contained in all its specificity in the *fundamental organ*. ... The respiratory apparatus is a further development of an originally small part of the mucous tube [Schleimhautröhre]. It was therefore already contained herein, and to be sure, in all its specificity".[85] In von Baer's view comparative embryology not only demonstrates the necessity of regarding the specific characters as already somewhat contained in the more universal, it also demonstrates that the more universal characters must actually *direct* the subsequent pattern of individuation:

The younger the embryos are that we compare the more variation we find in the structure of essential characters which later affect the whole manner of life of the organism. ... It is difficult to grasp how all these variations can lead to the same result and how, alongside of complete, well-formed chickens, there are not numerous cripples. Since, however, the number of cripples among more mature embryos is very small, so must it be concluded that the differences somehow cancel each other out and that every variation, as far as it is possible, is conducted back to the norm. For this, however, it is clear that each temporary condition by itself cannot determine the future state of things; on the contrary the more general and higher relations must dominate the process.[86]

Von Baer's viewpoint, therefore, is radically teleological in its orientation. In fact it is fundamentally the notion of teleology proposed by Kant and the Göttingen school that underlies von Baer's approach.[87] It is not the gradual accumulation of new parts that determines subsequent development. The essence of the animal, what it will ultimately become, determines the sequence of events in its developmental history. The final cause, that for the sake of which all the parts are laid down, determines the order of succession; in short the future is at work shaping the present.

Both his teleological framework and his choice of expressions like 'Idea' and 'variations on a theme' to express the regulative character of the principles of organization have led most historians to characterize von Baer as subscribing to an 'idealist' biology, which has most frequently been interpreted to mean that he was a Naturphilosoph.[88] This approach has failed to emphasize that for von Baer the universal, regulative principles of development did not exist independently of organized matter. Nor were these 'Ideen' superimposed upon blind matter from without. Rather they were intimately united to the manner of organization and were used to express the order and arrangement among the materials that were the basis for life and its characteristic manifestations. An appreciation of this point takes us a step further in understanding the nature of von Baer's type and its relationship to phylogeny.

Von Baer, like Cuvier, thought there were four basic plans guiding the structure of the animal kingdom. Cuvier's conception of the 'embranchement' differs in certain essential respects from that of von Baer, however. Cuvier's formulation rested squarely on the constant association of interdependent groups of characters and interdependent functions as revealed by comparative anatomy and physiology. Von Baer's conception of the type rested on constantly associated *developmental patterns*:

In reality, instead of 'Type' and 'Schema' I might have used a common term expressing both. I have only kept them separate in order to make it obvious that every organic form as regards its type, *becomes* by the mode of its formation that which it eventually *is*. The schema of development is nothing but the becoming type, and the type is the result of the scheme of formation. For that reason the Type can only be wholly understood by learning the mode of development.[89]

Only through a study of development can the order underlying the integration of systems of organs into whole, functional organisms be grasped.

In the dedicatory preface to Pander, von Baer emphasized that while he believed his researches demonstrated the regulative agency of the type, the first prize in zoology was reserved for that fortunate person who succeeded in explaining the structuring forces of organisms in terms of the universal forces of nature.[90] This tendency to see the directing agency of the type as grounded in the material conditions of the germ was also emphasized in the concluding passage of his earlier treatise 'Über die Entwickelungsgeschichte der niedern Tiere', (1826), in which he noted that the form of the type is dependent on the generative conditions [zeugenden Momenten] of the embryo.[91] The same theme reappears in the *Entwickelungsgeschichte* of 1828. In a typical expression of his concern for grasping the material basis of life, he observed that as the egg develops its weight is reduced by 144 grains while an unfertilized egg loses only 29 grains during the same period. He attributed

THE CONCRETE FORMULATION OF THE PROGRAM   87

this to chemical processes transforming the liquid into solid parts accompanied by evaporation.[92] In a similar vein he postulated that the successive alignment of heterogeneous materials along the central axis of the egg might, like a voltaic cell, produce a dynamic process in the direction of the axis, the nature of which "more exact physical experiments might be able to determine".[93] This dynamical-chemical process had the consequence that:

> in the germ, which rests above this axis, newly added material is collected in rounded forms which are wider and thicker on the left than on the right side. ... This might somehow be brought into relationship with electromagnetism [Electromagnetismus].[94]

Von Baer seems to have had a mechanism in mind whereby an electric current set up by the heterogenous materials along the axis of the egg would produce a magnetic field encircling the axis. This in turn would account for the pattern of lateral increase in the germ situated above the axis, and hence located in the magnetic field.

These presumed electrochemical processes of the initial phase of development were the preconditions for an essential mechanism for the second period of development. Von Baer characterized the second period in terms of three principal events: (1) the complete separation of the layers from which the animal and 'plastic' organs would originate, i.e. of the separation of the ectoderm and entoderm; (2) the rotation of the embryo on its left side; (3) the transposition of ingestion from the under surface of the germ to its left side:

> It is evident that these three apparently heterogeneous metamorphoses coincide temporally, and we might already suspect that they are of common origin.[95]

The 'common origin' von Baer seems to have had in mind was a polarized condition of the germ generated through the electromagnetic processes in the egg during the first period and which the germ had received by 'induction' through its presence in the plane of the magnetic field encircling the axis:

> The basis for the transposition of the ingestive function to the left side and the egestive function to the right side of the embryo was originally, i.e. in the first developmental period, turned toward the ingestive pole of the egg. It appears in otherwords that during the earliest stage of its development, while the under surface of the embryo was facing the yolk, the polarity of the egg was gradually communicated to the blastoderm and embryo. ... The rotation of the embryo and the transposition of the ingestive and egestive poles to the left and right side of the embryo are therefore manifestations of one and the same metamorphosis.[96]

If we conjoin this discussion with the definition of 'Type' given by von Baer at the beginning of the 'Fifth Scholium', we obtain a clear understanding of how the type could be conceived as the primary regulative agent in development. There he wrote:

> By 'Type' I mean the relational position of the organic elements and organs in the embryo. This 'relational position' is the expression of certain primary conditions in the direction of the particular relations of life; e.g. the direction of the ingestive and egestive poles.[97]

Thus each of the four types was characterized by a set of primary developmental relations bound intimately to the manner of organization of the material of the germ and the relation to its milieu in the egg. How these 'relational positions' in the germ were first generated lay beyond the limits of the teleological-mechanical framework of explanation.

In emphasizing the relational position of organs in the embryo as a characteristic expression of type, the previous definition introduces another, equally important element of von Baer's approach. Von Baer, it is well known, denied the recapitualationist hypothesis advanced by Meckel; in his view embryos never pass through developmental stages corresponding to adult stages of ancestral forms. Organisms of the same type share groups of developmental patterns and accordingly more mature stages of forms further removed from the type may resemble the embryo of a less differentiated form; but at no stage in its development is a chick embryo, for example, a mature reptile, even though in an early stage of its development it resembles the embryo of a reptile. The closer embryos of the same type are to the initial stages of development, the more they resemble one another, not because they are recapitulating ancestral adult forms, but because they share elements of the same developmental schema. As development progresses, greater is the divergence resulting from histological and morphological differentiation. Thus only in early stages are embryos similar, but these forms could never survive as whole organisms. But in spite of his rejection of Meckel's recapitulation scheme, von Baer did nonetheless insist upon the value of embryology for discovering the relationships between animals in the natural system.[98]

In von Baer's view each animal is a system of organs integrated by the principles of its type. Each type consists of a certain arrangement of a particular number of 'fundamental organs', out of which individual organs are formed through histological and morphological differentiation. The vertebrate type, for instance – von Baer never completed the theory of fundamental organs for the other developmental types – consists of five fundamental organs. They are formed from the germ layers. Building upon

Pander's idea of the germ layers, von Baer distinguished two primary divisions of the germ; an upper layer, which he called the animal layer (Pander's serous layer), and a lower layer, which he called the vegetative or mucous layer.[99] These layers were not separable, according to von Baer, but they could be distinguished by their smooth upper and granulated lower surfaces.[100] In the further course of development each of these primary germ layers differentiates into two layers. The 'animal layer' differentiates into the cutis [Hautschicht] and the muscle layer [Fleischschicht], from which the sensory organs, the skeletal and muscular systems and the nerves belonging to them later differentiate. The vegetative layer differentiates into the "vessel layer" [Gefässchicht] and 'mucous layer' [Schleimhautschicht], from which the mesentery organs including the alimentary canal and the remaining vessels of the body will form. Von Baer's *Hautschicht* corresponds to the ectoderm described by modern embryologists, while his *Schleimhautschicht* corresponds to the modern entoderm. Following a revision introduced by Remak in 1850, the *Fleischschicht* and *Gefässschicht*, which originate in distinct primary germ layers in von Baer's theory, are united by modern embryologists in the mesodermal layer.

In the next stage of development the flat, plate-like germ layers are formed into tubes by the bending and fusing of their outermost edges. The outer tube formed by the cutis is the surrounding covering of the body. The mucous layer forms into the alimentary tube. The vessel layer surrounds the alimentary tube and together with it form the tissues and vessels of the alimentary canal. Within the muscle layer the chorda and neural tube forms, from which the spinal cord, central nervous system and the tubular rudiment of the brain are formed.[101] The resulting shape of the germ layers infolded into 'primitive organs' is a figure eight.

The type forms the basis for an interconnected system of animals in which the grade of development of the fundamental organs produces the system of interrelated characters:

> ... which we call classes. Accordingly now one, now another set of life conditions [Lebensverhältnisse] will be further developed, or more correctly; the development of life in this or that direction produces variations of the major types, which are themselves essentially different in their vital manifestations. ... The classes divide themselves further into lesser variations, which we call families. These not only bear the modification of the major type but also include a particular modification of the class, which forms the characteristic of the family. Modifications of lesser degree in these families give rise to species and similarly for races and varieties.[102]

In the 'Beyträge zur Kenntniss der niedern Thiere', (1834) von Baer characterized the classificatory scheme to which this model gives rise in terms of a branching, seriated system of spheres, each consisting of a densely compressed center and surrounded by a thinly populated atmosphere. The basis for this scheme was the notion that:

> Every type may be manifested in higher and lower degrees of organization, for Type and degree of development [Entwickelungsstufe] together determine the individual forms. This produces, therefore, grades of development for each Type, which in certain instances form series; but not a continuous developmental series and never one which completes all possible developmental grades.[103]

The reason von Baer assigned for the absence of a full complement of developmental grades in each form of organization is extremely important to bear in mind:

> ... just as the smaller number of children in one family is not without a cause, but rather is rooted in the *Zeugungskraft* of the parents, so it is no accident that certain forms of organization are realized in fewer variations. The cause must lie in the essence of the forms themselves.[104]

Thus, von Baer notes, the fact that many birds have bills, while only a few mammals and no frogs have them indicates a certain capacity for modification common to both mammals and birds that is not present in frogs.

> We see, therefore, that through the different animal forms certain principle norms of the animal kingdom are modified, but not in an equal fashion, but rather so that most of the modifications stand nearer to one another than to others. The principle norms can be called the Provinces of the animal kingdom, their essential modifications classes; these are further modified to form families, and so on. But these subordinate modifications of a particular grade are mostly quite similar, and only a few ever appear to be radically different, so that a theme of these organic variations can be compared to a sphere, which consists of a densely compacted center surrounded by thinly populated atmosphere. This same relationship holds in the case of *genus* and *species*. The species consists of a number of individuals, which are never alike in all respects, but which have nevertheless an obvious similarity, ... From this emerges two noteworthy rules: First, that the more compressed the center of the sphere the more limited is the extent of the atmosphere, both in the larger, higher spheres as in the smaller, lower spheres ... Second in each large sphere, spheres subordinate to the center are richer in subordinate forms than those subordinate to the periphery.[105]

The model envisioned by von Baer in this passage is a richer, three-dimensional version of the system we have seen described in the works of the vital materialists. Each of the 'compacted centers' described by von Baer might be compared to the Kantian notion of a *Stamm* consisting of *Keime* and Anlagen. As a result of the potential for variation built into the "center" of

THE CONCRETE FORMULATION OF THE PROGRAM   91

each sphere, an "atmosphere" of different adaptive variations is made possible. The greater the core of interrelated characters, the larger is the sphere of variations on the same theme; hence a larger 'atmosphere' inclusive of more animal forms. An example of such a 'core' with its 'atmosphere' of related forms was provided later by von Baer in his discussion of Darwin's work. There he explained that the antelopes, sheep, and goats have probably all radiated from a common ground form.

The 'atmospheres' are not exactly spheres; rather they are more like a three-dimensional system of rays of differing length all leading from a central point. The direction of each ray is determined (arbitrarily initially) by the character, or system of characters of the 'core' that is developed to a particular degree; and the length of each ray is determined by the degree of histological differentiation, *Entwickelungsgrad*. Since the same organ or system of characters can be subject to several degrees of development, each ray is capable of having several 'points' along it. Finally each of these points, constituting a particular developmental degree of the original core, is able to serve itself as the center for a new sphere. In other words each of these atmospheric points is itself a *Stamm* of *Keime* and *Anlagen*, capable of generating modifications on the theme as well. As von Baer tells us in the concluding sentence of this passage the size of the subordinate sphere, the absolute number of forms it is capable of containing, is determined by proximity of its center to the center of the superior sphere; in otherwords a long ray, corresponding to a high *Entwickelungsgrad*, exhausts the potential for varying the original theme. Hence atmospheres surrounding points several degrees removed from the core will be more 'compressed' and hence will contain fewer forms. Figure 1 is a simplified schema of von Baer's model.

Two features of this model are important to emphasize: First not all peripheral points become centers for subsequent development. Not all potential for variation is actually developed. This results from a combination of internal and external factors. Not all developmental grades of a particular set of potentials are capable of producing viable animals; and, correlatively, environmental factors may not be conducive to the manifestation of those capacities for variation. Second, and perhaps most important, we see the model implies that embryos do not repeat adult stages of ancestral forms, but only their embryos. Thus within the limits of the type there is a developmental series T, C_1, O_1, F_3, G_4, S_{10}. The embryos of S_{10} can never resemble the adult form of S_{12}, however, for their developmental schema diverge at F_3. The same is true for all other species of the same type; they have more or fewer grades

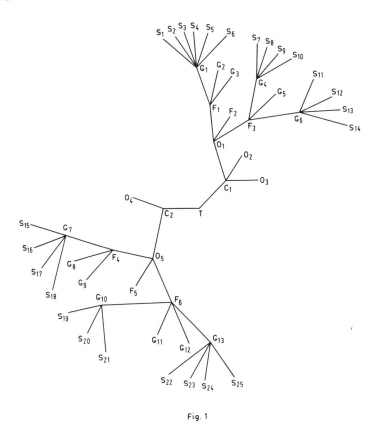

Fig. 1

of development in common but never adult forms.

In order to determine where in such a scheme the various organisms should be placed, von Baer argued that the constant presence of ensembles of characters, as advocated by Cuvier's comparative method, though a necessary condition for inclusion in a particular taxon, was not sufficient to determine the size of the groups and their relative position with respect to one another. To solve this problem, he argued, recourse must be taken to embryology. In particular, he reasoned that insofar as 'the positional arrangement' of the organic elements in the germ in an expression of the type and since the degree of development in a particular direction of the 'fundamental organs' determines the individual organism, the taxonomic grouping of organisms could be accomplished by tracing the developmental path of the *same positions* in the germ layers of embryos

THE CONCRETE FORMULATION OF THE PROGRAM

K.E. VON BAER AT AGE 33.
(Portrait from the special collections of the Museum of Comparative Zoology, Harvard University.)

JOHANNES MÜLLER AT AGE 34.
(From "Johannes Müller, das Leben des rheinischen Naturforschers", edited by Wilhelm Haberling, Leipzig 1924, Akademische Verlagsgesellschaft M.B.H.)

belonging to the same type. That is to say, animals were to be grouped in terms of homologies; and homologies were defined in terms of the developmental sequences of indentical areas of the germ.[106] For example,

> The tracheae of insects certainly are organs for conducting air, but they are not the same organs that we call 'bronchi' in the vertebrates; because the bronchi are developed out of the mucous membrane in the germ (serous layer) while the tracheae in the insects are formed by histological differentiation or invagination of the epdermis [äussere Haut].[107]

Thus structures having the same function are not to be considered as related unless they originate from the same area in the germ; and since each type has its own particular set of organic elements, homologies are only to be traced within the limits of the type.[108] With the introduction of this principle von Baer resolved one of the major problems confronting vital materialism after its initial period of formation. Organisms need no longer be grouped in terms of the inexact recognition of 'analogies', which in the works of Kielmeyer and Meckel were more often unjustified, fanciful creations. At the same time an attempt was made to delimit the proper methodological role of the embryological criterion and the limits of its application to phylogenetic organization. Through his efforts von Baer had succeeded in constructing a much improved version of the teleomechanist program: developmental morphology.

TYPE AND HOMOLOGICAL VARIATIONS ON THE THEME: THE APPLICATION OF VON BAER'S PRINCIPLES by HEINRICH RATHKE and JOHANNES MÜLLER

The methods sketched in von Baer's embryology quickly became the guiding threads for an entire school of researchers, which many commentators have called the 'embryological school'. Foremost among them were Heinrich Rathke, Johannes Müller and Rudolph Wagner. These men had first come into contact with one another and had become aware of their common interests in physiology through the offices of Karl Friedrich Burdach. In 1821 Burdach proposed a multi-volume collaborative work entitled *Die Physiologie als Erfahrungswissenschaft*, which would consist of articles on various aspects of physiology. Von Baer, Rathke, Müller, and Wagner were among the contributors to these volumes. Even before this, however, von Baer and Rathke had become friends. Rathke and von Baer's close friend Pander had studied together in Göttingen during 1817. Rathke later came to Dorpat, where he was appointed as Burdach's successor. When von Baer resigned his position at Königsberg in 1834, Rathke succeeded him there.

Johannes Müller met von Baer and Rathke personally for the first time in 1828 at the *Versammlung Deutscher Naturforscher und Ärtzte*, organized by Alexander von Humboldt in Berlin. Though snubbed by the older anatomists who did not even mention his recent monograph on the mammalian ovum, on the last day of the meeting von Baer had been approached by a group of the younger anatomists, including Johannes Müller, Ernst Heinrich Weber, and Evangelista Purkinje, who asked him to demonstrate the existence of the ovum for them. As he later reported, Johannes Müller was deeply impressed by the demonstration. Müller had himself presented a paper on the embryology of the kidneys in amphibians and birds. The stimulating discussion with von Baer and Rathke afterward led, upon Müller's return to Bonn, to extensive research on the Wolffian Body and its relation to the urogenital system.[109] Müller dedicated that work to Heinrich Rathke.

The principles to be followed in the use of the embryological criterion, often only vaguely present and nowhere collected into an explicit set of guidelines in von Baer's *Entwickelungsgeschichte*, were clearly delineated by Rathke and Müller in two classic monographs. Müller's treatise, concerned with the embryology of the urogenital system, was published in 1830. Rathke's investigation of the homologies of the gill arches was published two years later. Among the most important contributions to embryology in the early nineteenth century, these works provided ample proof of the power of the embryological method.

RATHKE AND THE GILL ARCHES

One of the foundational assumptions we have noted in von Baer's work is the notion that the organizational plan at the basis of systematically connected groups of organisms, such as the vertebrates, consists of interconnected sets of Anlagen for particular structures. Each set of these Anlagen is recognizable through the common developmental sequences of the developmental type linking them together. They obtain their most detailed structural expression in one class of animals and through a difference in degree of development, *Entwickelungsgrad*, they serve as the basis for structural and functional variations of the same plan in related groups of animals. When the question is asked, "What is the basis for the unity among interrelated groups of animals?" The answer von Baer offers is: "The Anlagen of the Type". When next it is inquired into the sources for differentiation of the type, namely the *Entwickelungsgrad*, recourse is taken to two types of consideration; (a) internal, and (b) external. The developmental degree of a particular set of structures cannot be a sort of blind mechanical unfolding of the Anlagen in space and time. A particular developmental grade of one set of Anlagen is always functionally related to compossible developmental degrees for other sets of Anlagen. The internal functional compatibility of different organ systems is what led von Baer, similarly to Cuvier, to stress the importance of the integrated, whole functioning organism as a factor determining the developmental grade. But we have also seen that a central aspect of this scheme is that many more compatible developmental degrees are possible than are actually encountered in nature: there are many more more potential animals forms than are realized. Hence an external factor must be taken into account; for as we have seen in von Baer's view, the environment plays a definite role in determining which compossible systems, as viewed from the internal dynamics of organization, are in point of fact real viable animals.

In his treatise, *Untersuchungen über den Kiemenapparat und das Zungenbein* (Dorpat and Riga, 1832), Heinrich Rathke made these principles explicit in building upon von Baer's approach. Like von Baer, Rathke emphasized that the importance of embryology for the theory of animal organization is that "an organ or system of organs in a fully developed animal which seems confusing in its composition or through its unusual form is often intelligible to us in all its relations and interconnections when we properly follow its development, when we watch its transformation from a simple to a complex structure".[110] Through this approach the formative principles at the basis of animal organization could be discovered.

Rathke had several related objects in mind in the composition of this treatise. Although he claimed that it was only a simple piece of empirical work unattached to any theoretical perspective, the text cannot be read without leaving the impression of a polemical piece directed against the transcendental anatomists, particularly Oken, Geoffroy St. Hilaire, Spix and Carus. In attacking their speculative use of analogical reasoning, which led to the heated debate between Cuvier and Geoffroy in the Academie dès Sciences in 1831, Rathke singled out one among several structures for detailed analysis concerning which numerous fantastic claims had been advanced by the transcendental school. The organ he chose was the gill system. Among the claims that had been made concerning this organ were that the gill arches of the boney fishes are modified ribs, that the bones of the operculum in lower vertebrate forms are repeated as the ossicles of the ear in the mammals, and that the gill arches are repeated in the higher vertebrates in the structure of the larynx and pharynx. In refuting these claims, Rathke wanted to demonstrate that the hyoid bone [Zungenbein] of the higher vertebrates must be considered a primitive form of the gill system which achieves its most complete articulation in the boney fishes. This concrete example was intended to demonstrate the power of the Cuvierian method when supplemented by embryology, and to establish the correct guidelines for identifying the types of a specific organ system and for tracing its variation through related groups of animals. In fulfilling this purpose, Rathke's treatise provides an excellent example of how concepts such as 'Typus', 'Bildungstypus', and 'Entwickelungsgrad' were operationalized in a particular case.

A comparison of the ways in which the term 'Typus' is used in different context by Rathke leads to the conclusion that a 'Type' of an organ or organ system consists of: (a) the positional arrangement [Lagerungsverhältnis] of (b) a determinate set of structural elements, which Rathke calls the 'Anlagen oder Elemente',[111] having (c) roughly the same shape. This is also the sense of the term as used by von Baer. Whenever the same organ type is encountered in different animals, the elements of structure must be connected in exactly the same way; ideally the same number of elements should be present; and they should have the same shape.

As soon as we define the 'Type' we realize that it is at best a scheme that can never be realized in all cases. At best it can be realized in one animal form, such as the boney fishes in the case of the type of the gill system. In this class the gill system achieves its fullest – namely its most complicated – expression; here it has the greatest number of elements, linked together in the most uniform way as we move from one species to another. What the type does then is to lay down a basic pattern upon which all related forms are variations. Two questions, therefore, must be answered in order to operationalize the type concept: In order to recognize the 'variations on the theme' all the elements in the original theme must first be known. How are all the elements forming the gill system, even in its most articulate form, to be identified *as a system*? Secondly, a method is required for identifying a variation as such. Assuming that in related forms the same or fewer elements are present,[112] it is required to find a means of showing that these elements slightly rearranged are (a) the same elements (b) obey the same pattern of relation. How are variations and their limits to be identified?

In order to understand how the elements of the type are identified it is useful to consider two sorts of example; one in which a suspected element is excluded from membership, and one in which a questionable element is included.

At first glance it might appear that the gill arches are repititions of the developmental type of the ribs; for very early in the formation of the embryo several prominences emerge on the external surface of the body parallel to the embryonic gill arches having the same general shape.[113] The mandible, quadrate bone, hyoid, gill arches, and pharyngial arches differentiate from the 'rib type', according to this argument, but they are all modifications of the same plan or type. Rathke denies a homology between the ribs and gill arches on the grounds that:

not to be overlooked is the fact that the ribs arise from the serous layer of the blastoderm, the gill arches and the pharyngial arches, however, arise from the mucous layer: the ribs therefore belong to the nervous skeleton, the bones in question to the visceral skeleton. That is to say, although they have a similar appearance and developmental pattern, they have different origins.[114]

Thus one principle for deciding where to draw the boundaries of a system is a difference of origin of the structural elements in question. But a further point separating the pharyngial arches and gill arches from the ribs is that the former "do not always maintain the same position with respect to the vertebrate column and the skull; that in some classes of fishes, for example, they are without exception pressed together beneath the skull while in others, ... they have their position beneath the vertebrate column".[115] Thus a second criterion is that elements forming a system retain the same relative position to elements in the same system.

Rathke's discussion of the pharyngial bones provides an example illustrating conditions in addition to those mentioned above for including an element in the system characterized by the type. According to Rathke, most anatomists did not include the pharyngial arch as part of the gill system; primarily because it had the appearance of a single bone unattached to the gill arches; and secondly, because in many fish it is armed with teeth. Therefore, it was considered as only accidentally connected to the gill arches by virtue of proximity. Rathke, however, showed that in most fish the pharyngial bone is really capable of being dissolved into two or three small bones closely joined to one another, and that these bones must be considered parts of the gill arches.

Since, therefore, not the least ground exists for separating these skeletal parts which lie directly above the third member of the gill arches and agreeing in respect to their shape, their absence of teeth and their entire constitution with the remaining parts of the gill arches, so must the upper pharyngial arches armed with teeth be considered as special members of the gill arches; and the reason is that between the simplest and the most specialized form armed with teeth, a number of intermediate stages exist, which represent a beautiful series of transitions from one extreme to the other.[116]

Here Rathke was emphasizing that continuity of shape and texture of closely situated structures was a criterion for grouping them as parts of the same system. Secondly, he argued that if a continuous series of intermediate forms could be established between two apparently divergent structures, then they are to be considered elements of the same system.

Rathke went on to strengthen these two principles by adding a third criterion drawn from embryology; namely, serial connection of materially related structures:

That the upper pharyngial bones are elements of the gill arches is also confirmed by developmental history. I have been convinced from the embryos of Blennius viviparus, that each of the bones under discussion together with all the remaining parts of the gill arches arise from one and the same part, namely from a simple jelly-like cartilaginous thread, which runs through the entire gill arch. As calcification sets in, this thread gradually breaks up into several small pieces situated one on top of another, the topmost of which is imbedded in the mucous layer and from which several teeth arise, representing finally what has been called the upper pharyngial bones.[117]

The task of identifying the type is not complete without establishing a method of tracing its variations in related groups of organisms. The problem with the approach advocated by the transcendental anatomists was they had no fixed procedure for identifying which structures in higher vertebrates are truly homologous with structures in the lower vertebrates. They had tended to rely rather on the creative capacity of a "transcendentale Anschauung" for establishing these connections. In contrast to this approach Rathke demanded the construction of a scientific procedure employing methods of direct observation. As an example of how to delimit boundaries of homology (Rathke did not use this term, but it carries the sense of what he sought) Rathke proposed to show that, contrary to the opinion of Geoffroy, the mammalian larynx is not a modification of the gill system found in the fishes. This connection is not valid, argued Rathke, because the developmental series of intermediate stages leading from the gill arches of the boney fishes to the mammalian larynx is not continuous. In the larvae of salamanders, frogs, and toads the beginning of the visceral respiratory apparatus and the embryonic larynx are situated closely behind the gill system. As these larval forms develop, the gill clefts fuse, and the gill tissues and their cartilaginous support structures disappear. Thus from this observation one might be tempted to suspect that the gill arches are transformed into the larynx, for as the gill structures disappear, the laryngial structures emerge in the same area.

Rathke points out, however, that in the tailless Batrachians the cartilages corresponding in both shape and position to the pharyngial bones of the fishes and which are situated in the middle between the larynx and the cartilaginous gill supports (i.e. the corresponding parts of the gill arches in the fishes), remain completely in tact and never contribute to the composition of the larynx. Thus in these forms the required series of intermediate stages passing from the embryonic gill arches to the structural support of the larynx is discontinuous in at least one crucial element:

If, therefore, in the Batrachians exactly those elements of the gill arches which are situated immediately in front of the undeveloped larynx and are in almost direct contact with it remain in tact, never becoming an element of the larynx, and if the larynx then develops completely independently of these cartilages as is actually the case, then it follows that no skeletal part of the gill system of the fishes which has a corresponding part in the gill system of the Batrachians, neither the corresponding pharyngial bones nor the gill arches has the significance of a cartilaginous element in the composition of the Batrachian larynx.[118]

In order to show that this discontinuity indicated by the Batrachians is in fact complete. Rathke next examined the mammalian larynx, which is a more complex structure. Just as in the previous example the gill clefts of sheep embryos are found situated directly above the beginnings of the larynx. But while the gill arches are still present, there arise on both sides of the opening of the trachea two protuberances which later develop into the vocal cords and the arytenoid cartilages along with their openings. "The gill arches then fuse together on top of these, and the larynx develops further without the slightest contribution of the arches to its later enlargement or composition".[119] Furthermore, Rathke noted, if the primitive gill bars are opened up, there are always visible two canal-shaped open spaces between them and the protuberance which at first surround the future epiglottis; "and, furthermore it appears that the arches do not unite *beneath* the opening of the windpipe, but rather *in front of it*. This connection, however, makes it impossible that any part of the arches can later be transformed into a part of the larynx".[120] Thus, according to Rathke, the crucial issue from a methodological point of view is that at no point can a continuous transformation be observed from the gill bars to the larynx, and the impossibility of such a transformation is eliminated by the special separation of the primitive gill system and the parts of the larynx in question. We see, therefore, that in tracing the limits of a type as the basis of homological structures in related animals the key factor was to establish through comparative embryology a directly observable series of intermediate steps demonstrating a material transformation of the elements of the system in question.

Once the prodecures for defining the type and its variations are understood, the meaning of the terms, '*Bildungstypus*' and '*Entwickelungsgrad*', so essential to the framework developed by the Embryological School, become transparent. According to Rathke, the hyoid, gill arches, and pharyngial arches of the fishes all share the same '*Bildungstypus*'. They have the same general shape. The hyoid bone consists of 9 elements, 7 simple and 2 compound; the following three gill arches consist again of 9 elements; the fourth, however, has only five elements; while the pharyngial arch has only three principal elements.[121] These elements form part of the same system, and they must be seen as repetitions of the same schema of formation, the same *Bildungstypus*. The difference between them is that the hyoid is the most complex, or most fully developed; the other bones of the same developmental type are less complex, usually deficient in one or more element, and this degree of complication follows a progressive pattern of decrease. That is to say, the fourth gill arch is just a particular developmental grade, a different degree of complication of the same developmental pattern underlying the whole system.[122] Generalizing upon this same idea and carrying it over to other groups of animals, Rathke describes the gill system of the salamanders and frogs as repeating the same *Bildungstypus* of the fishes, only at a different developmental grade. The hyoid in these animals, for instance, is a "degeneration [Rückbildung] and fusion of the elements of the schema found in the fishes",[123] and he later concludes that the further a class of animals is removed from the fishes, the less complicated is the structure of the gill system.[124] As in von Baer's presentation of the same model, we see that the developmental grade of a system is the determinant of its position in a taxonomy. From such consideration Rathke concluded:

I believe it is sufficient to conclude from all this that
- (1) the system of all the bones and cartilages, which lie between the mandible and the girdle of the pectoral fin in the fishes is the foundation [Vorbildung] for the hyoid bone in the higher vertebrate animals.
- (2) that in none of the classes standing above the fishes is this ever so completely developed, and in the course of development in some animals the later forms even regress from their earlier height.
- (3) that the simple hyoid bone of the higher animals, ... is a repetition – and an imperfect repetition at that – of the foremost arch of the aforementioned system in the fishes.[125]

MÜLLER AND THE WOLFFIAN BODY

Although he had been deeply influenced by the speculative doctrines of Naturphilosopie while a student at the University of Bonn, Johannes Müller, like von Baer, Rathke and others, had rejected that approach to nature by the time he received the *venia legendi* at his old university in 1824. Just as Döllinger had set von Baer in search of an empirical method for unlocking the secrets of biological organization, so too had Johannes Müller been inspired to set physiology on sound experimental foundations by Carl Asmund Rudolphi.[126] In his *Antrittsrede* as professor of anatomy at the University of Bonn in 1824, Müller had declared war on romantische Naturphilosophie as well as on the mechanistic reductionism of the French physiologist François Magendie.

Müller flatly rejected the speculative approach to biology. Biology could not be an *a priori* theoretical science in his view. In agreement with Goethe he asserted that it must be radically empirical, and above all an observational and experimental discipline. Such requirements were not satisfactorily met, however, by the approach to physiology being developed in France by Magendie and his school, an approach which Müller depicted as a caricature [Zerrbild] of actual physiological processes.[127] Müller's objections to this 'falsche Physiologie' emerge most clearly in his consideration of the role of chemical experimentation and vivisection in physiology.

"Physiology cannot be biochemistry", Müller wrote, "but the physiologist's procedures are chemico-physiological".[128] By this statement Müller was underscoring the view, which he held in common with the vital materialists that the mechanical framework alone was insufficient for conducting physiological research and that chemical experimentation in physiology must always stand under the higher guidance of a teleological framework in which organization in assumed as given. Müller illustrated his point by his work on sensory physiology. Each sense organ, he argued, has associated with it a specific sense energy. This sense energy is an active response of the organ to external conditions which serve as a stimulus for initiating its activity. The source of the activity and the special character of the response, such as the sensation 'blue' or the experience of a particular tone, for instance, is the effect of the *organization* of the organ itself. It has nothing, or at most very little, to do with the structure or qualities of the physical stimulus. Thus, for example, a weak electrical current, pressure, or even heat may stimulate the eye to generate the same sensation, 'red', but in Müller's view those physico-chemical stimuli have no direct role in the quality of the sensation itself. It might also be induced by an internally generated nerve impulse or even by a pathological condition. The quality of the sensation, the particular mode of manifestation of the sense energy, is not to be understood as the mysterious action of a vital agent or soul superimposed on matter. Müller conceived the specific activity of the organ to reside in the physical and chemical processes of material exchange. But the key point he wished to emphasize is that the specific sense energy arises from the particular manner in which those physico-chemical processes are arranged in the organ itself. The external stimulus provides the condition for triggering the specific and systematically interconnected set of processes in the organ that are associated with the sensation 'red' for instance. Although its specific action does not depend upon any forces other than those resulting from physico-chemical interactions, it is the mode of organization of those forces that is responsible for the specific sense energy.

This consideration placed certain definite limitations on the ability of even the most carefully controlled chemical experiments to reveal anything conclusive about the internal causal framework of organ function. Müller explained why:

> In chemical experiments the reagent, recognizable in terms of its properties, is contained in the product which it forms with an unknown substance or its parts. The role of the reagent in this case is not simply excitatory. The product consists no less of the reagent as of the material under investigation. What renders experimentation of this sort inadmissable in physiology, however, is that the response of living nature to the effect of the reagent does not contain the properties of the known reagent as an essential part. For all materials, all stimuli acting upon the organism, do not excite in it what they themselves are, but rather something completely different, the vital energy of the organism. Accordingly experiment cannot by itself disclose anything concerning the source of the vital phenomenon. It can only multiply and expand our knowledge of the relation of stimulus as cause to the effect in the organism, which is of a completely different nature; that is, it can familiarize us with a greater number of vital phenomena, the nature of which remains, however, undisclosed.[129]

As Müller explained in his formal lectures on physiology in 1827, experimental physiology of the sort advocated by Magendie in his *Précis elementaire de physiologie* was doomed to failure.[130] The whole enterprise was vitiated by the fact that organ function depends on the internal organization of physical and chemical processes together with the related fact that the response of the organ to an externally applied physical or chemical agent may have no direct causal relation to the internal processes generating that response. Magendie's physiology was 'falsche Physiologie', its "thoroughly empirical, mechanistic, skeptical orientation grating against everything that it turns up, sacrificing sound logic to its fraudulent desire for anything new ...".[131]

The same considerations that led Müller to reject the results of chemical experimentation in physiology as a source of insight into organ function also led him to reject the results of vivisection. Surgical intervention disrupted the natural organization of the system, which was the source of its specific function. To be sure the organism would respond as a result of intervention, but there was little likelihood that what transpired during such experiments was even remotely related to normal physiological function.[132]

Obivously what physiology demanded was a special method which did not make hasty leaps of faith in the power of reason and which was grounded in observation without relying upon a disruptive method of experimentation. In Müller's view there was one phenomenon especially qualified to serve as the basis for constructing an experimentally based physiology: that was embryogenesis. And comparative anatomy supplied the tools for unlocking the secrets it contained for understanding organ function.

Comparative anatomy has the spiritual task [freie geistige Aufgabe] of uncovering the metamorphosis of the organs and the organism in its finite development as Caspar Friedrich Wolff and Goethe recognized and demonstrated.[133]

Müller's *Antrittsrede* was designed to inspire the confidence of young medical students about to embark upon a new era in scientific investigation. The content of Müller's views on method was presented in less flamboyant language some years later in the introduction to a work titled *Bildungsgeschichte der Genitalien*. There he set forth several criteria to be met by a sound experiment.[134] One class of observations met his requirements in the fullest measure:

For when the experiences have attained sufficient extent and the greatest precision I demand that they not be merely thrown together, but rather, just as nature proceeds by the development and preservation of the *organic being*, that from the whole one strives toward the parts, assuming that in the process of analysis one has recognized the particulars and succeeded in arriving at the idea of the whole.[135]

Müller's position was essentially the one von Baer had defended. The special nature of biological phenomena demanded that the functional whole exercise priority over the structural relations of its parts. Accordingly a theory of biological organization could only be constructed on the basis of an experimental method which recognized the special unity of the whole and did not attempt to dissolve it into an accidental collection of physico-chemical processes. Embryogenesis permitted one to become, in Müller's view, "the Galileo of physiology".[136]

Müller exhibited the power of the genetic method by re-examining the problem treated earlier by Meckel which was of considerable importance to scientists during the Romantic era: the origin and development of the human sexual organs. Due to the presence of vestigial organs and the formation of organs by secondary association of parts which have arisen independently and have formerly served different functions, the developemnt of the urogenital system is an extremely difficult piece of embryology. We have already seen the tortures it presented Meckel. Müller unravelled its complexity in what was soon to become a classic monograph. This work joined von Baer's *Entwickelungsgeschichte* and treatise on the mammalian ovum together with Heinrich Rathke's treatise on the homologies of the gill arches as founding documents of developmental morphology.

The whole problem turned around a detailed understanding of the so-called Wolffian body, what modern textbooks in embryology designate as the pronephron and its associated Wolffian or pronephric ducts. First described by Caspar Friedrich Wolff in his *Theoria generationis* and later the subject of intensive investigation by Oken, Meckel, Rathke, and von Baer to name only a few, the Wolffian body had been the fertile source of much speculation. Quite prominent in young avian and mammalian foetuses, it was well known that this structure degenerates as the urogenital system emerges. Hence the received opinion in 1830 was that the Wolffian body was the common source of the kidneys and all of the internal sexual organs. In rejecting this view, Müller demonstrated that while certain elements of the urinary and genital systems are appropriated from the Wolffian body as it degenerates, the suprarenal gland, kidneys, testes, ovaries, and uterovaginal structures have completely separate origins. He demonstrated, furthermore, that in higher vertebrates such as birds, mammals, and man, the function of the Wolffian body is to serve as a foetal kidney, having an analogous relation to the later developing kidneys of the mature foetus that the gill system of larval batrachians has to the lungs of the mature organisms.[137] Comparative embryological studies of the Wolffian body in several vertebrate species, including frogs, salamanders, lizards, snakes, turtles, various birds, sheep, goats, cows, dogs and man, provided the basis for his conclusions.[138]

An important assumption which Müller held in common with Rathke and von Baer was that all organisms of the same developmental type – the vertebrate type in this instance – possess the same common fund of structural elements. Different organisms of the same type differ only in the degree of complication of the fundamental set of organs. 'Higher' or more complex forms of an organ traverse all the lower developmental grades during ontogeny, although this process may be telescoped into a rapid ontogenetic sequence. Thus any difference in the origins and structural relation of organs in 'lower' forms must also hold for higher forms. Another closely related assumption was that any claim of a causal relation between an earlier and later structure had to be established by a sequence of observable structural transformations capable of being shown to be materially interconnected with one another.

Müller relied upon this sort of reasoning to immediately dispose of the notion that the Wolffian body is the source of the kidneys. He pointed out that in salamanders which remain in a larval condition for a relatively long period, the kidneys appear very later in development toward the end of the larval stage. But they appear in a quite different location separated by a considerable distance from the Wolffian body. Similar considerations of frog embryos refuted the claim that the Wolffian body is the source of the testes and ovaries: they too are quite distant from the pronephros in their initial appearance.[139] Müller also confirmed that in chick embryos,[140] as well as in mammals and man, no direct material connection exists between the Wolffian body and the developing kidneys. This is particularly obvious in man, where the mesonephros and embryonic kidney are separated by the quite large suprarenal gland.

Having established that the Wolffian body is not the source of the kidneys, Müller directed his attention to determining its function. Structure provided the clue. The Wolffian body, he showed, consisted of many small tubes, closed at one end and opening at the bottom into a common duct. The duct leads to the cloaca. A second important structural feature he observed was the presence of tiny glomeruli, tightly compacted arterial vessels which when injected looked like a small sphere. These glomeruli were located in very close proximity to the ends of the tubules of the Wolffian body. The relationship between these arterioles and associated tubules was exactly analogous to the structure of the kidney. The glomeruli of the Wolffian body had both the appearance and structural relation to the tubules that the renal glomerulus, or *Corpora Malpighiana*, have to the similarly structured tubules of the kidney.[141]

Müller confirmed the functional correspondence of the nephron and kidney by examining the contents of the nephronic tubules and duct. The avian embryos he was investigating all contained a white substance similar in appearance to a substance found in the allantois and to the yellowish substance of the urine which appears late in the foetal bird in the ureter and renal tubules.[142] Müller did not subject this substance to chemical analysis but he felt confident in drawing the conclusion that it was urine and that it was the filtered extract of the blood passing from the glomerulus into the nephric tubules where it was secreted and emptied into the nephric duct to be transported to the cloaca and excreted into the allantois. The Wolffian body was, in short, the functional kidney of the early foetus.

Continuing his remarkable comparative studies Müller discovered that the Wolffian body plays a quite different function in later foetal life after the appearance of the kidneys. As it begins to degenerate its duct is appropriated by the developing male genital system. Others before Müller, such as Rathke and Oken, had seen this remarkable phenomenon. But they had mistakenly inferred from this observation that the Wolffian body is the source of the internal genitalia and that after the nephron degenerates the Wolffian duct remains behind as the uterine (Fallopian) tube and the *ductus deferens* (semen ducts). This conclusion was only partly correct.

Careful microscopic investigations on birds/eggs throughout the spring of 1829, raided from nests in the woods near Bonn by Müller's assistants, demolished the claim that the male gonads arise from the nephron. His parallel investigations on amphibians had already demonstrated that these structures arise separately, but confusion can arise in higher vertebrates because the gonads first appear on the ventral surface of the mesonephron. The relation is only one of juxtaposition, however, Müller argued; for at no point do any of the mesonephric tubules enter into the substance of the gonads.[143] In Müller's view, the lack of a directly causal material connection was the crucial desideratum in denying the possible metamorphosis of the gonads from the Wolffian body.

Continuing his investigations on mammalian embryos, chiefly sheep embryos of which there was a plentiful supply in the winter and early spring of 1829, Müller showed that a connection does ultimately develop between the testes and the Wolffian ducts. Through a sequence of progressively older specimens Müller established that differentiation of the gonad into ovaries and testes begins at about the same time that the Wolffian body has reached its fullest development and begins to degenerate.[144] After sexual differentiation has commenced a process begins to form on the testes which gradually extends downward and joins with the Wolffian duct. As development of the male organism progresses all that remains of the former foetal kidney is its collecting duct which has now been appropriated for a quite different function, serving as the *ductus deferens* of the internal reproductive system.[145]

The internal male genital tract contains one further element, the epididymus. Because of the outward similarity of appearance between this convuluted structure and the tubules of the mesonephron, several outstanding investigators, most notably Rathke, had regarded the epididymus as a remnant of the Wolffian body. Müller denied this; but as we shall see, he erred in his analysis of its origin.

While the Wolffian body is present in both male and female vertebrates in early foetal life, and while remnant parts are transposed into elements of the male reproductive system, Müller demonstrated that it degenerates almost completely in females.[146] It is neither the source nor a constituent element of the female reproductive system. The uterine tubes, uterus and vagina, according to Müller, arise from completely different structures, the paramesonephric ducts, which he was the first to describe.[147] He did not give them a special name, but they are frequently named after him in modern textbooks. Müller first identified these new ducts in the avian foetuses he was studying and later confirmed their presence in mammals and in man.

Müller observed, and his detailed drawings clearly illustrate, that both the Wolffian and Müllerian ducts are present in the early foetus before the urinary system has developed and prior to sexual differentiation. In females, however, the nephron degenerates much earlier than in males, disappearing almost entirely by the time the chick leaves the egg. Both structures emerge from the cloaca, the Wolffian duct entering the nephron while the Müllerian duct runs parallel along the surface of the Wolffian body. In a careful examination of mammalian foetuses Müller established that the lower portion of the new ducts he had discovered form the vagina and uterus. The upper portion ultimately widens and forms the uterine tube.[148] The cephalic end of the Müllerian duct dilates and eventually develops into the trumpet-like structure of the ostium. Müller explained that the ostium gradually envelops the upper end of the mesonephron, which separates it from the ovary during early development. He emphasized, however, that at no point do the tubules of the Wolffian body provide a passageway between the ovary and ostium. Eventually in the mature mammalian foetus, and in avian foetuses after hatching, the tubules of the Wolffian body lodged between the ovaries and ostium disappear completely.

While the paranephric duct was primarily destined to develop into the female reproductive system, Müller did not think, as we do today, that his newly discovered duct degenerates completely if the foetus develops into a male. As mentioned above, in early embryos before the gonads have begun to differentiaite into either testes or ovaries, both the Müllerian and Wolffian

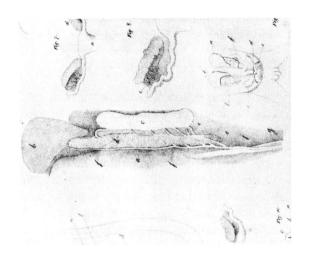

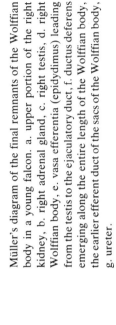

Müller's diagram of the final remnants of the Wolffian body in a young falcon. a. upper portion of the right kidney, b. right adrenal gland, c. right testis, d. right Wolffian body, e. vasa efferentia (epidydimus) leading from the testis to the ejaculatory duct, f. ductus deferens emerging along the entire length of the Wolffian body, the earlier efferent duct of the sacs of the Wolffian body, g. ureter.

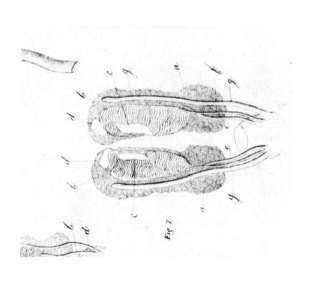

Müller's diagram of the developing urogenital system of a young female avian foetus. a. Kidneys, b. Wolffian body, c. ovaries, d. adrenal glands, e. ureter, f. mesonephric duct, g. Müllerian duct (uterine tube).

ducts are present, emerging from the cloaca. He noted that the Müllerian duct runs over the surface of the Wolffian body almost to the top. Now Müller thought that as the mesonephros shrinks in the male foetus the materially distinct paranephric duct running along its exterior begins to coil like so much excess cording attached to the surface of a deflating baloon. Thus he regarded the remaining portion of the Müllerian duct on the surface of the nephron as the *epididymus*.[149] Its upper end fuses with a process extending from the testes, while the lower end fuses with the Wolffian duct, which has now assumed the function of the *vas deferens* – an extremely ingenious if false solution to a very tricky problem. There is a very clear reason for Müller's insistence on this transformation, however. As a teleologist he could not envision the emergence of a structure which had no function. That the paranephric duct could be present to no avail in potentially male foetuses was incomprehensible in his view.

In addition to providing a classic example of how the genetic or embryological method was to be utilized in solving a complex question regarding the source and function of certain organic structures, Müller's monograph beautifully illustrated one of the guiding assumptions of the teleomechanist program: The functional whole has priority – even in a causal sense – over its constituent parts. To a mind of Müller's persuasion nothing better illustrates this claim than the functional transformation of different elements of the Wolffian body, which at one stage in foetal development is a functional kidney, into parts of a completely different system, the male reproductive system. In Müller's interpretation this wonderful goal-directed process is even further enhanced by the inclusion of a structure in the male genital tract (the epididymus) that at one stage was potentially a completely different structure (the Fallopian tube). Müller and others of his generation, such as von Baer, did not regard it simply difficult to imagine how such sequences of events could be explicated within a strict causal-mechanical framework. It was, in their view, impossible. Such an intimate and finely-tuned orchestration of events required a more 'purposive' mode of causation.

CHAPTER 3

TELEOMECHANISM AND THE CELL THEORY

Teleomechanism as originally conceived in the works of the Göttingen school was a comprehensive program advancing simultaneously on several research fronts, including physiology, systematic zoology, and later embryology. For men such as Kielmeyer who had originally formulated the itinerary of future lines of research these were not to become separate research specialties. Such a notion was far from their minds. Rather these research areas were viewed as closely related aspects of a comprehensive science of life, which was referred to as *Biologie, allgemeine Zoologie*, or *Morphologie*. In each stage in the development of their program, the teleomechanists considered their task to be the construction of general descriptive laws in each of these several areas and their synthesis into a unified system. Only once the systematic interdependence of these laws could be demonstrated in terms of a single unifying principle did they think that a causal explanation could be attempted: as Humboldt expressed it, only then would *Naturbeschreibung* give way to *Naturgeschichte*. In this sense the analogy drawn frequently in the works of early nineteenth century German scientists between the results of Kepler and those of Newton is illuminating. For just as the universal law of gravitation was the principle that systematically united Kepler's laws, Galileo's law of falling bodies and a host of others, so we might consider the systems of laws united by the four Types as analogs in the biological realm to Kepler's laws in the physical realm. What both vital materialists and developmental morphologists hoped – we have seen it often expressed in their writings – was a Newton of natural history, who would discover that more general principle that could unite these into a natural system. While the cell theory is usually celebrated as the foundation for an atomistic reductionist approach to life, it may come as a surprise to learn that teleomechanists regarded the new theory as consistent with their programatic aim of uniting the theoretical concerns of physiology and zoology in a concrete manner. For the cell theory offered the means of uniting the functional and morphological concerns essential to the early versions of the program.

From their inception both vital materialism and developmental morphology rested on the doctrine that the source of the structures and harmoniously interrelated functions asociated with each type of organism must be grounded materially in the unity of a set of *Keime* and *Anlagen*. Comparative anatomy and physiology, paleontological research, and biographies of species, but above all *Entwickelungsgeschichte* offered the best means of revealing the laws organizing the elements of form and function; for a causal account of the relationships binding various forms of organization would be one that demonstrated their *development* from a common fund. A central unresolved problem in this tradition, however, was that of deciding *where* these *Keime* and *Anlagen* reside and in what their nature specifically consisted. Were they actual material particles with a specific locus or were they 'dispositions' generally distributed throughout the organism as a whole? Although the works of Kant and Blumenbach, and certainly those of Reil had seemed to lean toward a localization of these dispositions, especially in the generative fluids, nowhere did they give an explicit discussion of how that localization was to be conceived. Was the source of structural development a set of dispositions in the germinal fluid? Or was its source a set of primordial structures, perhaps like crystals, the arrangement and order among which provided the pattern and plan of the organism? On either hypothesis, how did the first products of the germinal fluid become organized and structured into germinal materials and germ layers? Was the vital force the directing agent in this process, a kind of 'central force' organizing the germ much like the force of gravitation structures the solar system into an 'organic' whole? Or could order and arrangement of the primitive structures be made to carry the entire burden of the argument? Clearly at issue in deciding between these alternatives was the crucial question of what role the notion of a special vital force or *Bildungstrieb* was going to have in physiology.

From the beginning of the debate, which first surfaced in the work of Schwann, the main lines of an acceptable solution to these questions moved strongly in the direction of a more strict and complete mechanism than that espoused by the vital materialists. While rejecting vital forces, the new solution did not dispense with the principle of teleology, however. Indeed, just like their forebears in the vital materialist tradition, the mechanists of the 1840s and 1850s, whose work I will be discussing in this chapter and the one to follow, argued that a teleological framework must always underpin the search for a mechanical explanation in the life sciences. If vital forces had lost respectability, the problem they were intended to solve had not. To the mechanists of the 1840s the problems of function seemed even more destined to fall before advances in organic chemistry and physics than they had to the vital materialists, but these mechanists were not so confident as to believe that mechanical modes of explanation could begin to unravel the questions of order and auto-regulation among the materials in the organized body, the two most significant formal elements that seemed to determine physiological function. In adopting a teleomechanical framework for biology the generation of the 1840s attempted to assume a consistently agnostic position with respect to the cause of order in the phenomena of life. The vital forces of the previous generation became for them – to use an anachronistic phrase but one nonetheless descriptive of the problem – 'Maxwell demons' with no place in biological science. Investigations into the cause of order were to be dispensed with as beyond the limits of legitimate science. Henceforth the only subject of a properly scientific nature was the investigation of the mechanisms employed in *maintaining* biological order.

In the context of this broader debate emerging in the 1840s, the cell theory proved to be a great watershed for the teleomechanist research program. On the one hand it seemed to succeed in revealing the universal law of development that had been the dream of two generations of researchers. But in identifying the seat of development with a specific set of structures the ever-present question of the limits of mechanical explanations in biology stepped once again to the foreground.

It was certainly no accident that the inventors of the cell theory, Schleiden and Schwann, were reared in the perspective of teleomechanism. As students of Johannes Müller and Heinrich Friedrich Link they were in daily contact with the tradition of biological research that originated in the 1790s in the lecture halls of Blumenbach, Kielmeyer and Reil. The writings Schwann, Carl Vogt and Rudolph Virchow will demonstrate that teleomechanism provided the background of issues and the conceptual framework within which the cell theory took shape. In extending the standard portrait of events surrounding the discovery and reception of the cell theory, the analysis presented here will show that the developmental morphologists, particularly von Baer, Johannes Müller,. Rudolph Wagner and their immediate circle contributed substantially to the cell theory. These men welcomed the theory and were quick to apply it. Moreover, they accepted without significant reservation the mechanistic approach to cell development emphasized in the writings of the younger generation.

There were two reasons for the agreement between these two groups. Ultimately both groups were in accord on fundamental assumptions and methods. Both were convinced that a teleological framework must underpin the investigation of the phenomena of life. Secondly, both were in agreement with the notion that a mechanical explanation must be pursued as far as possible, even to the most fundamental structural and functional unit. Until the work begun by Liebig in the early 1840s and culminating in Helmholtz's derivation of the conservation of energy in 1847, even the most detailed structural and descriptive developmental analysis of the cell possible could not decide the issue of whether physiology demanded a special vital force. The vital materialists and developmental morphologists could accept the empirical findings concerning cells of their students and younger colleagues without calling their own assumptions into question.

Our examination in this chapter will show that the role of the cell theory in the transformation of German physiology was not to deal teleomechanism a mortal blow. Rather it exposed the outlines of a broader strategy then surfacing in the early 1840s for solving the major problem that had always stood in the background of the teleomechanist program; namely, how to carry through in a consistent fashion a non-vitalistic, mechanical approach to life constructed within a teleological framework.

VON BAER AND THE GESTALTUNGSKRAFT IN THE OVUM: PRELUDE TO THE CELL THEORY

In his examination of the progress of physiology since Haller, Döllinger had remarked that one of the most important areas in need of examination was not only the development of the foetus, but the processes leading to its actual formation. Because those processes lie hidden in the darkest interior of the organism, Döllinger despaired of ever being able to draw adequate inferences concerning them based on empirical evidence. This problem was taken up by his student, von Baer, resulting in the discovery of the mammalian ovum and the publication of the monograph that assured his place in the pantheon of the history of biology, *Über die Bildung des Eies der Saugethiere und des Menschen* (Leipzig, 1827). Von Baer's interest in this problem was consistent with his conception of morphology and with his program of following the development of form from its first beginnings and throughout each stage of development. Von Baer's discovery is of importance for our story, however, not only for its contribution to embryology but because it served as a paradigm case of the application of the principles of developmental morphology, including its problematic use of vital forces. We have already seen that von Baer regarded the essential burden of an argument intending to establish connections between related forms that it demonstrate a sequence of transformations related through material continuity. This method was exhibited clearly and with spectacular results in his paper of 1827. Equally important, were von Baer's observations and inferences on the formative process of the mammalian ovum, which he viewed as guided by a vital force. While they had reformulated several aspects of the earlier version of the program, residues of the vital materialists' notion of vital force lingered on in the form of an emergent organizing force in the theories of the developmental morphologists. Von Baer's paper became the point of origin for a groundswell of opposition to these continuing vitalistic features in the developmental morphologists' approach to the phenomena of life. But there was an element of irony in the ensuing debate. For the pattern that von Baer had detected in the formation of the ovum was adopted by Schwann as the basic model for all cellular development. Some of the distinctive features in von Baer's description of this formative process will be the focus of the present discussion, for they provide the context within which the emergence of the cell theory is to be seen.

One of the principal conclusions reached by von Baer is that the Graafian follicle is the true ovarium while the egg itself is the 'foetal egg'. The mammals have, therefore, an egg within an egg, or if one prefers, an egg of the second power.[1] Here we see some hint that the formative process always follows the same laws in the same types of animals, each new structure is in some since the repetition of the same basic developmental plan. The developmental process of the Graafian follicle from its initial beginnings to its rupture and the release of its contents is, therefore, for von Baer a clue to development in general.

Von Baer followed Döllinger and Burdach in treating the tissues of the ovary as a matrix, which he called the *Stroma*. Its function was to secrete the fundamental organic material , the *Grundstoff*. The follicle according to von Baer is formed as a swelling within the ovarian stroma. Within the stroma the thecal membrane or shell of the follicle is formed and its interior, or nucleus [Kern] as von Baer called it, consists of a granulated membrane, the *membrane granulosa*, filled with an organic fluid:

> The fluid contained in the follicle is composed, just like all organic fluids except for waste products, form a *Fluidum nativum*, whereby I mean the *Grundschleim* of Nees [von Essenbeck], and the particlus [Körnchen] contained therein.[2]

He went on to note that in addition to these particles there were often small particles like oil droplets. These two types of variously shaped particles join together forming larger particles similar to the yolk granules in hens' eggs. The primitive fluid in which these particles are suspended is sticky and transparent with the consistency of lymph. When it is heated or mixed with tartaric acid, it is, according to von Baer, indistinguishable from egg white. Thus, we see the *Fluida nativa* contained in the follicle is indeed conceived by him as the *Grundstoff* he had spoken of earlier in his lectures on anthropology.

Within the native fluid floats a small disc, the *Discus proligerus* (corona), and the *Cumulus*, both of which von Baer characterized by their white color and halo-like appearance under the microscope. He viewed these structures as constituted from particles of the Fluida nativa pressed tightly together. Finally, within the Discus and Cumulus sits the ovum [Eichen]. The ovum is surrounded by a transparent peripheral membrane, and in the middle stands a tiny, darkly colored spherical mass, the nucleus of the ovum. The nucleus, however, is not a single solid mass:

In order to determine whether the dark central mass is really a solid spherelet (for it has this appearance), I have observed nuclei under the microscope from dogs and from a cow, which were either accidentally or purposively removed from their ova; and I concluded that the spherelet is not solid, but rather its particles form a large periphery around a small empty space.[3]

From this description it is clear that von Baer conceived the structure of the Graafian follicle and that of the ovum to be a repetition of the same formative process. The ovum is in a sense, the nucleus of the follicle. He did not, however, regard either one of these structures to be prior to the other in the temporal sequence of development.

I do not know if the human eye will ever be able to decide whether the ovum or the Graafian follicle is present first. But I regard it as fully certain that in the more immature follicles an ovum with a Discus proligerus is already present.[4]

True to the teleological principles of his science von Baer viewed the parts as mutually cause and effect of one another. The ovum was not formed by the mechanical aggregation of parts around a primary particle analogous to the process of crystallization. Although rooted in the Fluida nativa secreted by the ovaries, the ovum itself was generated from forces obeying developmental laws other than those present in inorganic processes.

While he did not regard any one part of the ovum and follicle as temporally prior to another, it is clear nonetheless that von Baer thought the nucleus of the ovum had an especially important role in the generative process. All organic development seemed to obey the law of being initiated in the center and striving toward the periphery:

All development is based in the fact that the formative process moves from the center to the periphery.[5]

Von Baer here stated the principle law of developmental morphology the notion that life is an internal center of activity, a *zweckmässig Trieb*, manifesting itself in the external realization of structure. Von Baer saw this principle actively present in every developmental stage of nature, not only at the level of oogenesis but, as we have already seen, at the level of systematics as well. In discussing his view earlier we noted centers of organization from which ever more complex animals radiated, the most structurally complex standing on the periphery. Organic nature follows the same laws of morphogenesis at all levels.

Von Baer expanded upon this notion further in discussing the fate of the egg after leaving the follicle. In proceeding to the uterus, he argued, the egg undergoes a further process of maturation which is aided by the intake of fluids. The particles in the ovum press ever more toward the periphery forming a cortical membrane and leaving a hollow center. Furthermore, these particles secrete a substance which forms an encapsulating membrane. The particles remain attached to the inner surface of this membrane.[6] "This membrane is the vitelline membrane and the particles referred to correspond to the yolk particles of the bird's egg".[17] In an extremely remarkable passage, von Baer continues:

After it has brought the particles to the periphery, the formative force [Gestaltungskraft], which acts from the center to the periphery, forms a peripheral thickening in each particle. In this way each particle is tansformed into a circle, in which small particles are formed around a transparent center.

The metamorphosis of the egg corresponds, therefore, completely to that of the Graafian follicle. Just as in the follicle the majority of the particles form a peripheral layer on the surface of which a smooth memebrane is deposited, so the same process occurs in the formation of the vitelline membrane. Furthermore, in the Graafian follicle there is a *Stratum proligerum* with a cumulus present, which corresponds to the developmental history of the blastoderm; namely a portion of the particles in the ovum form a hill under the vitelline membrane, which is gradually transformed into a disc. The cortical membrane corresponds to the theca.[8]

Of special significance in this passage is not only the strong emphasis von Baer placed on the continuity of the center-periphery pattern of development, but also the stress laid on the correspondence of the structures in the egg and follicle. Of further interest was the reference, at the end of the passage, to the formation of the germinal disc in the ovum and the extremely interesting linkage of vital force and the formation of material *Anlagen*. Von Baer conceived the first steps in oogenesis to consist on the formation of certain small particles under the direction of a force emanating from the center of the nucleus. Those particles were then transported to the periphery where they were transformed into disc-like regions of the germinal material. Though he did not mention it in this context, it followed that those were the Anlagen from which the germ would be generated after fertilization.

Von Baer provides further insight into his vital materialist conception of the formative process in the last section of the monograph. There he intended to demonstrate that the *Vessicula Purkinje*, which Purkinje had described in bird eggs in a paper published in 1825,[9] was in point of fact the structure in the avian, amphibian and invertebrate egg analogous to the mammalian ovum. In establishing this claim von Baer pointed to the position and structural similarity of the Purkinje vessel to that of the mammalian ovum. Of particular significance in von Baer's view was that the vessicula Purkinje seemed also to be the center of the same developmental pattern he had observed in the mammaliam ovum. In all the eggs of various animals he observed, von Baer noted that the vessicula Purkinje strives to move from the center of the egg to the periphery. This was especially obvious, he noted, in the case of frog eggs, but it was equally true of all others. The only difference was in the amount of time the process required. This led him to formulate the following hypothesis:

Allow me to formulate an hypothesis concerning those matters which subsequent research may refute or raise to the status of a dogma.

I believe that the Purkinje vessel is the active part of the egg, through which the female principle exercises its power [Kraft], just as the male principle is located in the semen. The fleck, which is present in the hen's egg as long as it is in the ovary, does not deserve to be called the germinal disc, but it is its precursor. It is somewhat sunken in the yolk and it has indeterminate contours. If I am not mistaken, in all eggs prior to the development of the blastoderm there is a granular layer present. It has the shape of a disc or a more or less curved plate. But this always assumes the position of the Discus proligerus in the mammalian ovum, which we have discussed above. ... The Discus or Stratum proligerum may contribute to the formation of the blastoderm in all eggs.[10]

In discussing the morphological turn given to the vital materialist program by Döllinger and von Baer, I emphasized that central to the methodology developed by the embryological school was the notion that laws of organization would only emerge once a radically empirical base had been laid in which each stage of life was directly observed from the very beginning to the completion of the structuring process. Furthermore in tracing those developmental patterns it was essential to establish the concrete material ancestors of each structure and to follow their transition through intermediary structures. One especially important expression of this methodological principle, which I have discussed in the preceding chapter, was the insistence that homologies and hence interconnected groups of organisms be capable of being traced back to a morphogenesis of the same regions in the germ. In the monograph on the mammalian ovum, von Baer had succeeded through the application of this methodological principle in arriving at the very beginning of the structural process itself. He had found the seat of the formative force: It lay in the nucleus of the ovum. It was not itself a material particle, but rather a center of activity giving rise to a sequence of structures. Most significant among those early structures was the germinal disc, a disc of granular particles, each one of which had been formed and guided to its proper place by the *Gestaltungskraft*, working in a primitive organic matter. In the course of these observations von Baer had established what he took to be one of the most important laws of morphogenesis; that morphogenesis always proceeds from the center to the periphery. He proposed the hypothesis that this developmental law and its manifestation in the sequence of events and structures in the formation of the ovum might be universally valid. As we shall see, Theodor Schwann succeeded in raising this hypothesis to the status of a dogma, indeed one of the most important dogmas of biology. For he not only established it as true for the formation of the ovum but showed that all organic structures are connected by the same fundamental law of development.

The discussion in von Baer's treatise on the mammalian ovum brought to completion a line of argument begun by Blumenbach. For Blumenbach too the *Bildungstrieb* was the generative force associated with the whole. The best illustration of this concern is provided by Blumenbach's emphasis on teratology as a source of insight into the laws of biological oranization. What interested him so much about these phenomena was their demonstration of the fact that even when it fails, the formative force always strives to bring forth a whole, complete organism as perfectly constituted as possible within the limitations placed upon it. This is why he named that force the Bildungs*trieb*, for it was a purposive agent unlike 'blind' mechanical forces.

It is tempting to regard von Baer's concern with a *Gestaltungskraft* in his work on the ovum as a moment of weakness in which a bit of speculation forced its way past the otherwise careful censor of von Baer's suppressed adolescent association with speculative Romantics. Perhaps, it might be argued, von Baer would have been more cautious had he been more cognizant of the latest developments in organic chemistry, for the position he adopted was in essence identical to that advocated by Kielmeyer some twenty years earlier, and in the meantime considerable progress had been made in organic chemistry. But it is evident from von Baer's published writings of the

1820s, particularly his lectures on anthropology, that he followed developments in organic chemistry with great interest. In fact his position was in full accord with the best scientific opinion of the day.

An excellent guide to the climate of opinion in the late 1820s and early 1830s regarding chemical processes in a physiological setting is provided by the work of Ernst Heinrich Weber. Weber, then a physiologist at Leipzig, was well-known as a careful experimenter,[11] and he was pioneering in the development of techniques for making quantitative physical measurements of vital functions.[12] As his work on the physiology of blood circulation demonstrates, he was strongly committed to employing the methods and principles of physics in physiology wherever it was possible.[13] In 1830 few could claim to be more directly concerned with constructing a mechanistic physiology than Weber, but he held exactly the same view as von Baer regarding the need for a *Gestaltungskraft* in physiology and for the primacy of the whole over its parts.

Weber expressed his view in 1830 in the revised edition on Hildebrandt's *Lehrbuch der Physiologie des Menschen*, a four volume work which had first appeared in 1789-92. The choice of Hildebrandt's text is particularly relevant for our discussion, for Hildebrandt had been a student of Blumenbach and had, like Girtanner, made one of the first attempts to generalize the theory of the *Bildungstrieb* into a unified theoretical foundation of zoology.[14] Weber's revised edition of Hildebrandt's work attracted widespread attention among developmental morphologists. Johannes Müller cited it in his *Handbuch der Physiologie* as containing the clearest definition of the term *Lebenskraft* and an explication of the relation between organic and inorganic forces.[15] Weber's view of the difference between crystallization and the operation of chemical affinities in the formation of organic compounds set forth in clear terms the issues that were to occupy center stage in evaluating models for the formation and development of the cell.

Weber, like the other physiologists in the research tradition we have examined, did not argue for a material difference between organic and inorganic substances. Organic materials, in his view, are typically composed from materials that are capable of oxydation, and characteristically they are composed of at least three such substances. Organic chemical affinities are ternary or quaternary in contrast to inorganic modes of chemical composition. The major difference between organic and inorganic substances, therefore, "consists in the difference in the relationships binding the chemical elements together":[16]

A connection of 3, 4 or more substances directly with one another – that is to say, such a connection between them so that each element is directly and with equal proximity bound to all the other elements – appears, as the works of Foucroy and Berzelius have demonstrated, only to emerge in organized bodies under the influence of life.[17]

In order to visualize this type of chemical affinity, and its difference from ordinary chemical affinities, Weber suggested viewing binary affinities in terms of a series of (successively larger) boxes nested one within the other. Organic affinities could then be represented by the intersection of four boxes in a corner, while the four boxes thus united were nested within a larger one.

Weber attempted to support this theory by calling upon five sets of data. Among these he included the observation that organic substances are incapable of artificial synthesis from inorganic elements. In addition, basing himself on the work of Berzelius, he argued that inorganic elements are combined in terms of whole number ratios, whereas organic compounds do not always combine in terms of simple numerical proportions but often enter into fractional equivalents.[18] Finally he argued that if binary chemical affinities were the only type of affinity permitted, the explanation of the enormous diversity of organic forms from such a small number of elements would be impossible. Ternary and quaternary combinations for organic compounds would solve this problem.

In order to demonstrate that no ordinary mechanical process could account for the generation of organic structures by 'enhancing' binary affinities, Weber undertook a comparison of the formation of organic structures and crystallization. All crystals having the same chemical properties, he noted, are always composed of smaller unit parts having the same shape. From those smallest parts laminae are formed. These second-order structures are capable of being assembled into various shapes, but they are always united in terms of a characteristic angle and belong to a determinate class. Organic structures on the other hand formed from the same chemical materials, such as bones, are capable of numerous different shapes; and no two bones of the same type are ever in fact identical. Conversely, while chemically disparate mixtures rarely have the same crystal form it is frequently the case that organic structures of the same type are composed of different elements. As an example he argued that embryonic cartillage possesses different chemical properties from fully ossified bones, although the external shape remains the same:

To be sure this does not imply that the chemical composition of matter in organized bodies is absolutely without influence on their shape ... but because only chemically identical parts are capable of uniting to form a crystal and thereby are compelled to adopt a determinate shape, crystallization must be more dependent upon the chemical composition of matter than is the structure of organic matter.[19]

A further difference between crystals and organic structure pointed to the necessity of an organizing force superior to but not separate from the elementary parts as the basis for the organism. As he has already noted, although the elementary parts of crystals are identical, they are capable of uniting to form shapes of such variety that it is difficult to recognize them as members of the same class. The cause of these variations is normally a function of external conditions. Accordingly in the case of crystals, from invariant shapes and positions of the elemental parts a multiplicity of structure and shape is possible. The exact opposite, however, holds true for organic bodies according to Weber. In organic bodies the shape and position of the large organs is not subject to variation, whereas the smaller organs composing them are subject to enormous variation. As an example he cited the variations in the branching of veins in the arm from one person to another:

From this it must be concluded that the formative force in organic bodies is able to give the large parts their determinate position and shape even when the smaller parts from which they are constructed have a variable shape and positional arrangement; and consequently the shape of the whole organs cannot depend upon the affinity of the smallest parts for one another, which they exercise in virtue of their inherent qualities, and which causes them to assume determinate positions with respect to one another as appears to be the case in crystallization. On the contrary the formative activity must be determined by laws which are connected with the interrelationships with respect to size, form and position of the larger parts of the organized body; that is independently of the interrelationships of the smallest parts.

Complex crystals are constructed from individual *parts*; organisms, however, are constructed from the *whole*.[20]

Weber went on to distinguish organisms further by the fact that their structures are formed by invagination rather than the aggregation and apposition of elementary parts which implied the presence of a formative force prior to and independent of the forces generated through the interaction of the parts.

In these passages from Weber's work, passages which were cited widely by other teleomechanists, we seen an explicit formulation of the *Lebenskraft* as a constitutive principle of organization. Like von Baer, Weber treated the vital force as having a determinate locus, but he too regarded it as incapable of being explained in terms of the elements themselves. The writings of Theodor Schwann and Carl Vogt on the cell theory provide some of the first reactions to this interpretation. In their view the initial elementary parts themselves carry all the information needed for the later organization of the whole. It was thus not necessary to assume the existence of a directive force associated with the germ as a whole which gives rise to the origin and differentiation of its parts.

THE TELEOLOGICAL FRAMEWORK OF SCHWANN'S THEORY

Schwann's presentation of the cell theory is typically hailed as starting from the complete rejection of vitalism – by which is understood a rejection of the vitalism of Schelling, Oken and the Naturphilosophen – and placing biological research on a reductionistic framework of physics and chemistry. This interpretation must be qualified. For the young physiologists and zoologists being trained by Müller and Wagner the approach to nature advocated by the *romantische Naturphilosophen* was simply not a matter for serious consideration. Schwann was not arguing against the *Naturphilosophen*. He was criticizing certain aspects of the frameworks of vital materialism and developmental morphology, particularly the use of specific directive, vital forces exemplified best by von Baer's *Gestaltungskraft*. But as I will demonstrate, he was by no means rejecting the framework of teleomechanism underlying the approach of von Baer and Müller. There were many strengths of this program, and its early practitioners need not have committed it to a defense of vital forces.

Schwann argued that there are two frameworks within which to conceive organic phenomena:

> The first view is that a special 'force' lays at the basis of each class of organ which forms them in accordance with an immaterial idea, and arranges the molecules in the manner necessary for achieving the purpose intended by this idea ...[21]

While the reference to vital force as an "idea" suggests the Naturphilosophen, as the context unfolds, it becomes clear that Schwann's chosen opponent is really von Baer. Two aspects of von Baer's work received special scrutiny by Schwann, and they formed the thread of argument throughout the *Mikroskopische Untersuchungen*. I have noted that in his *Entwickelungsgeschichte* von Baer presented evidence which he regarded as establishing the priority of the whole over the parts. In his view the animal that is to be, the end, or *telos* of development actually directs the entire process. Schwann objected to this conception of teleology, for it could never be formalized within a mechanical framework. Secondly, in his treatment of the formation of the mammalian ovum von Baer attempted to identify the biological whole with the *Gestaltungskraft* at the center of the ovum. This force was really the 'whole' organism *in potentia*, for it arranged the organic particles of the germinal disc, and the order and arrangment of the organic materials in the germinal disc was the material basis of the future organism.

Schwann's principal objection focused specifically on this conception of vital force. It was unacceptable within a mechanical framework. In his view, the term 'vital force' was to be regarded merely as metaphorical language. Development is not organized and initiated by a force concentrated in a single point, as von Baer had claimed. In Schwann's view, the 'force' is not at all localized, but rather it is the effect of a particular order and arrangement of material particles, and the effect consists in specific sorts of patterned motion resulting in a definite pattern of structures. At first glance Schwann's objection may appear trivial, for he too employs the language of vital force. He speaks, for instance, of the 'metabolic force' associated with the cell. But Schwann conceives this force differently from von Baer. To see the contrast in their conceptions it is useful to make an analogy once again to Newton and gravitation. In applying the inverse square law acting between point masses to large objects such as the earth and planets, Newton demonstrated that the combined effect of an indefinite number of particles forming a sphere and attracting one another inversely as the square of the distance between them could be treated as though the entire mass were located at a single ideal point, the center of mass of the system of particles. Von Baer conceived the vital force similarly. It was not independent of the matter constituting the organic body. In fact, as we have seen, he treated it was arising directly out of the arrangement of matter in the organic context. And like Newton, he regarded the force resulting from the combined effect of these materials to be a center-periphery pattern in the formation of the ovum. Moreover, von Baer laid great importance on the fact that while the center of the ovum was the dynamic center of initial vital activity, it was not occupied by a material particle. But there was a major difference between the Newtonian model and von Baer's vital force. In Newton's demonstration the force resulting from the combined effect of the point masses interacting according to the inverse square law is itself an inverse square force. For von Baer, on the other hand – and this point is crucial to the viewpoint of other teleomechanists such as Blumenbach, Reil, and Kielmeyer – the resulting force or *Trieb* is of a completely different character from its constituents. It is a 'force' in von Baer's view because it obeys certain laws, but it is a force of a different sort. The laws it obeys are *teleonomic* rather than strictly mechanical. Von Baer's viewpoint and that of the early teleomechanists is best described, therefore, as 'emergent vitalism'.

The framework Schwann proposed agrees in all particulars with that of developmental morphology except in regard to the postulation of emergent vital forces. For Schwann too the key to understanding vital phenomena is the order and arrangement of material in the organic context, but for him there was no special force resulting from that arrangement. The phenomena connected with life were not the results of an emergent quantum leap into a different framework of laws and physical processes. For Schwann the science of biology consisted solely in the study of order and arrangement of materials in an organic setting acting according to normal physical laws. In his view, life is not something resulting from order and arrangement; nor is it different from order and arrangement. Life consists *in* the order and arrangement of particles of matter *tout court*.

In establishing the framework for his criticism of the vital materialists and developmental morphologists, Schwann attempted to draw a sharp dichotomy between their approach and the one he was advocating. The dichotomy he attempted to draw was much stronger than that actually separating the two viewpoints. He argued that there are really only two strategies for dealing with the phenomena of life; vitalism and physical reductionism. Having identified the viewpoint of the vital materialists as naively vitalistic, he went on to give the false impression that the position he advocated was strongly reductionistic:

The other view is that the elementary forces of the organism are in fundamental agreement with the forces of inorganic nature to the extent that they operate blindly according to the laws of necessity independently of any particular purpose; in short that they are forces which are presupposed with the existence of matter, just like physical forces.[22]

We have seen, however, that a *third* framework is possible for investigating the phenomena of life in addition to vitalism and physical reductionism; namely, that which assumes certain levels of organization as given, for which no mechanical explanation can be advanced, but within the domain established by this framework research is to be conducted in accordance with the principles of mechanical explanation. This is the teleological-mechanical framework set forth by Kant, Blumenbach, and Reil, and it is based on the notion that the phenomena of life are the special results of vital forces which are inseparable from a particular order and arrangement among mechanical processes, an order which cannot itself be explained as the result of a mechanical process. As we have seen Kielmeyer, von Baer, Ernst Weber, and Müller departed from the core doctrines of this framework by attributing a further direct causal role to the vital effects resulting from the functional arrangement of the organic elements. This was an unnecessary addition to the teleomechanical framework set forth in the works of the vital materialists, but the developmental morphologists had made it a central feature of their approach to the phenomena of organization.

The framework adopted by Schwann is not physical reductionism. Already in the last sentence of his formulation of the alternative to vitalism we see that he had something close to the teleomechanical viewpoint in mind, for he imagined a framework that assumes the existence of another set of forces operating just like ordinary physical forces but inseparably bound to the matter of organization. He did not speak of a *reduction* of organization to inorganic physico-chemical forces. That the framework Schwann had in mind is the same teleological framework at the basis of vital materialism and developmental morphology is made explicit in the continuation of the previous passage, which has in the past frequently been overlooked:

It could be assumed that the structuring forces of organisms do not appear in the inorganic realm, because this or that particular combination of molecules through which they obtain expression is not present in the inorganic realm, and yet these organic forces would not be essentially different from chemical and physical forces. Purposive organization in nature is not to be denied, even a high degree of purposiveness in each individual organism; but according to this view the reason for it is not because each organism is produced by a force operating in according with some purpose, rather the reason for this purposive organization is to be sought in the creation of matter itself together with its blind forces by a rational Being.[23]

In this view purposive organization must be assumed even for the inorganic realm. Thus, Schwann went on to write, the fact that physical and chemical forces of matter operating in accordance with blind necessity should produce a solar system, an organized whole of dynamically interacting material particles, can only be explained by assuming that a rational being created matter with such forces that by following the laws of mere mechanism a purposive organization, in this case the solar system, would result. Thus even at the highest levels of theory construction in physics certain teleological principles must be assumed for which no further explanation can be demanded; but that need not imply a failure of mechanical models of explanation. Rather it merely recognized the limits of a mechanical explanation: *That* material particles act upon one another in accordance with the principles of mechanism is not to be denied; reason demands it. *How* and *why* they came to be so constituted to act in this way rather than in some other manner equally consonant with the principle of mechanism is not a question that can be answered. At this stage reason must assume a purposive organization at the basis of inorganic nature.

The same principle applies to the organic realm:

> Purposive organization in the organic realm is only a different degree of the same purposiveness manifested in the inorganic realm; and an explanation based on the assumption that organisms, just like all phenomena in the inorganic realm, arise through the blind action of forces created simultaneously with matter itself cannot be refuted. Reason, however, demands an explanation for this purposive organization; but it must ultimately be satisfied with the assumption that matter, together with its indwelling forces, owes its existence to a rational Being. Once created and preserved in their integrity, these forces can then bring forth combinations exhibiting purposive organizations acting only in accordance with the laws of blind necessity. If this Rational Power acts only in a preservative manner after the creation and never as immediately active in the world itself, then we can abstract from it completely in the realm of natural science.[24]

When we compare the preceding formulation of the role of teleological principles in the natural sciences with that set forth by Kant, we see that Schwann too was well aquainted with the teleomechanical core of vital materialism.[25] Like Kant, he emphasized that how organized forms are first created is impossible to unravel even assuming that they emerge somehow from physico-chemical forces. But that does not hinder the construction of a biological science based on the principles of mechanism which examines the laws in terms of which organisms function, even though the first causes of organization itself do not enter into the discussion.

The view of the relationship between inorganic forces and biological organization at the basis of Schwann's cell theory bears an unmistakeable family resemblance to the approach sketched in the writings of Kant, Reil and Blumenbach. But Schwann was deeply concerned to establish the limits of teleological thinking in biology, and in his opinion there had been misuses of the principle of teleology by the vital materialists and developmental morphologists. At issue was the problem of weighing the relationship of the whole of animal organization to its parts. The essential position from which no vital materialist departed was that the parts are only intelligible as units functionally related to others in the integrated organism. Thus a four-chambered heart demanded lungs and determines a particular organizational type among organs of motion. But there were some teleomechanists, such as von Baer for instance, who took this notion of teleology a step further by adding that the whole is *ontologically* prior to and determinative of its parts, that the whole *directs* the organization of its parts. Schwann wanted to exclude any sort of directive agent from discussion of biological organization. This move was, as we shall see, thoroughly consistent with the Kantian foundations of the teleomechanist program. In attacking this problem, Schwann proceeded in an extremely ironic fashion. For Schwann's cell theory is modeled almost exactly on von Baer's model for the development of the fertilized egg. In demonstrating that the same fundamental pattern of development underlies every organic formation, Schwann sought to develop a framework that did not require the directive agency of the whole over the parts.

Schwann attacked this issue by posing the question of whether the forces of organization are to be conceived as existing only in the totality of the animal form, that is whether the end product, the functioning organism as a whole, determines the manifestations of the vital phenomena, or whether the organizing principles of the whole are to be conceived as contained in each of its elementary parts, the cells. The former alternative had, of course, been preferred by von Baer, who urged that the whole is prior both developmentally and functionally to its parts. Schwann objected to this Aristotelian version of immanent teleology. He proposed an alternative model, consistent with that formulated by Kant and Blumenbach, but based on a different conception of the role of the 'whole' in determining animal form. As we have seen, Kant and Blumenbach stressed the genetic continuity of structure and function as originating from a single source, the *Keime* and *Anlagen* of the generative fluid. Differentiation of structure and function must already lie 'dormant' in each elementary part. Just as each race receives the full complement of *Keime* and *Anlagen* constituting the generative stock of the species, a different set of which lies dormant in each race, so must the various structural elements of the entire organism be present *in potentia* in each of its elementary parts. That this latter alternative must be correct is illustrated, according to Schwann, directly in the organic phenomena themselves. If it is the case, he says, that any elementary part can live independently of the whole, then the forces of organization must not depend upon the whole for their manifestation. Such is the case with the fertilized egg. According to Schwann, it is a cell possessed of an independent, self-regulating life. Further evidence for the primacy of the cells as the locus of the principles of organization is provided by the fact that, according to his extensive comparative studies of cell development, all cells develop in exactly the same general pattern:

Since all cells grow according to the same laws, the cause of development cannot in one case lie in the individual cell and in another be based in the organism as a whole. Furthermore since certain individual cells, namely the fertilized egg, which also follow the general pattern of growth of all cells, can develop completely independently, we must assign an independent life to the cells; that is to say, the combination of molecules present in an individual cell is sufficient to generate the force whereby it is able to attract new molecules. The basis for metabolism and growth does not lie, therefore, in the organism as a whole, but rather in its individual elementary particles, the cells.[26]

This discussion points to one of the characteristic features of Schwann's cell theory; namely that he conceived the fertilized egg as a single macro cell. By examining the various stages of cell development he described and comparing them with contemporaneous treatments of the development of the germ, the student of Schwann's work obtains the clear impression that the accounts of the early stages of embryonic development, particularly the 'first period' described by Pander and von Baer, shaped Schwann's perception of general cell development. Although, there are several distinct patterns of cell development they are, in Schwann's view, all specialized cases of the pattern of development of the blastoderm.

This impression is supported by Schwann's presentation of cell formation in the *Mikroskopische Untersuchungen*. The first stage of cell development occurs in the cystoplasm of an already formed cell. There a group of cell granules are formed through a process analogous to crystallization (we will discuss the differences between this process and crystallization later). These granules, so-called *Körperchen*, come together to form the *Kernkörperchen*, the nucleolus. Next, around the outside of the nucleolus a layer of small granules is formed. This layer, indistinguishable from the nucleolus at first begins to expand through intussusception so that eventually a membraneous layer separates sharply from the nucleolus, forming the nucleus [Kern] of the future cell. This same process is repeated with the outer surface of the nucleus serving as the locus for the formation of the cell membrane, only in this case the process is very rapid. A consequence of the formation of the different membrane layers by intussusception is that the nucleus remains attached to the cell wall at one point and the space between the cell wall and nucleus is filled with fluid.[27] Similarly the nucleolus remains attached to the wall of the nucleus. In Schwann's view there are no cells without nuclei. All cells follow fundamentally the same process of development. The only difference is that in most cases after the formation of a surrounding membrane the cell nucleus (or nucleolus) does not continue to develop. In some cases, however, development is not arrested but rather continues until all the nuclear material is exhausted. The result is a large, somewhat thick membrane filled with clear cellular fluid. The blastoderm is the paradigm example of the limit of cellular development:

The development of many nuclei into hollow vessels and the difficulty often in distinguishing such hollow cell nuclei from cells themselves must lead to the suspicion that the nucleus is not fundamentally different from the cell itself; that a normal nucleated cell is nothing other than a cell that grows around another cell, the nucleus; and that between the two the only difference is that the inner cell develops more slowly and less completely after the formation of the outer cell.[28]

This description is almost word-for-word identical to von Baer's description of the formation of the ovum within the stroma: a series of 'eggs within eggs'. According to Schwann's view, therefore, the blastoderm is the most perfect and completely developed cell; and conversely all nucleated cells are only arrested stages in the developmental pattern of the blastoderm.

We see several familiar themes in Schwann's portrait of cell development. Foremost, of course, is the notion that the same pattern underlies all modification of structure and that later developments are variations of a 'fundamental organ'. The source of these developmental patterns and the related structures to which they give rise is originally the material contents of the cellular fluid from which the cell nucleus and related structures originate. The 'Entwickelungsgrad' of this material substrate determines cell type and function. Equally pronounced is von Baer's beloved center–periphery pattern.

Not only do some of the central ideas of Schwann's theory bear strong resemblances to the theories of the developmental morphologists but his treatment of the localization of the 'Keime and Anlagen' was modeled from contemporary embryological theory as well. Pander for example had argued that the germinal disc [Hahnentritt] directs the entire development of the embryo and that it consists of two parts, a round disc, which he called the blastoderm [Keimhaut], and beneath it a small lump of matter, which he called the nucleus [Kern] of the germinal disc:

With the formation of the blastoderm [Keimhaut] the entire development of the embryo is founded, ... for whatever noteworthy events may occur in the future, they are never anything other than a metamorphosis of this membrane and its tissues endowed with the inexhaustible fullness of the *Bildungstrieb*.[29]

In the hands of Purkinje and Rudolph Wagner, Blumenbach's successor at Göttingen, this same idea was carried a step further. The yolk sac, Wagner had argued in his *Prodromus*, contains the Purkinje vessel attached normally to its wall, and within the Purkinje vessel small, granular particles are united to form the 'Wagnerian flecks'. Schwann observed the same pattern in the germinal disc. The disc, he observed, consists of round cell-like vessels each containing a small lump of dark particles.[30]

Since it appeared to Schwann that everywhere the same developmental pattern occurs, and that its initial phases and subsequent pattern is determined by the contents of the nucleolus, the question that ultimately concerned him was how to conceive the formation and subsequent functioning of this cell material as well as a possible mechanism explaining differentiation of cell type. The solution of these problems demanded a consideration of the cell from two perspectives; first its 'plastic' or formative processes and second the mechanism for its metabolic function. In the final analysis everything turned on the answer to the question of whether the organism derives its formative and functional force from the manner in which its atomic, elementary particles are linked into molecules.[31]

It is well known that in addressing these issues Schwann compared the process of cell formation to crystallization, but it has not been sufficiently emphasized that for him crystallization was only a useful metaphor for placing the process within a mechanical framework. Cell formation is not crystallization, he emphasized, for crystallization proceeds by apposition while cell formation proceeds by intussusception. No chemical process is known, he argued, which works in precisely this way. Nevertheless very little was known about the chemical affinities involved in crystallization, and with greater knowledge of these processes some means might be discovered for chemically generating crystalline growth by intussusception. What he wanted to emphasize was that *if* the laws of chemical affinity could somehow explain growth by intussusception, *then* cell development and function could be conceived as a process of crystallization. In the absence of such a demonstration, however, crystallization continued to serve as a useful metaphor. Qualified in this sense the crystallization model could not produce a source of friction with teleomechanists. Kielmeyer had spoken of special organic laws of affinity parallel to those for crystallization; and Reil had compared the processes of growth and nutrition to crystallization without implying that one could be *reduced* to the other.

That the entire process of cell formation and function depends upon a special set of organic chemical affinities was beyond question for Schwann. To be sure, the *Kernkörperchen*, the granular materials forming the nucleolus are "distilled out of the concentrated fluids of the cytoblastema by a kind of 'Herauskristallisieren'".[32] But Schwann emphasized that this formative process does not proceed independently of the cellular metabolic process, by which the cell attracts appropriate materials out of the cytoblastema through a special sort of elective chemical affinity. It is not the case that the cell forms *and then* begins to function; from the earliest stages of their formation, cells are carrying on their specialied functions. Hence the *form* of the chemical affinities and the functions they lead to mutually influence one another:

> The attractive force of the cell acts with a certain selectiveness. Not all the substances contained in the surrounding cytoblastema are attracted, rather only certain ones; namely some that are partially similar to the cell substance itself (Assimilation) and others partially different. Through assimilation the individual layers develop; on the other hand in the formation of a new layer different substances are attracted, for the granules, nucleus, and cell membrane consist of chemically different substances. ... The expression of this plastic force pressupposes, however, another capability. The cytoblastema in which the cells develop contains to be sure the elements of the material from which the cell is composed, *but in different combinations*; the cytoblastema is not simply a solution of cellular substance, rather it contains other definite organic substances in solution. The cells, therefore, must not only possess the ability merely to attract substances out of the cytoblastema; they must also have the ability to transform chemically the material in the cytoblastema.[33]

For Schwann, fermentation offers the paradigm example illustrating the necessity of assuming a structural-functional unity of the nuclear granules and their independence from the materials of the cytoblastema out of which the mature cell is ultimately constructed. After establishing that yeast is cellular in nature, he describes experiments demonstrating that several hours after yeast cells are placed in an organic solution alcohol appears on the outer surface of the cell.[34] Clearly the cell has transformed the material of the 'experimental cytoblastema' according to its own structural-functional identity. What this example further illustrates is that the unity of *Keime* and *Anlagen* from which the primitive nucleus is formed cannot come from the material of the cytoblastema itself. They must come somehow from without, fully constituted in this case in the form of the yeast cell. A further reduction is not possible.

Assuming a certain degree of organization as a precondition for cellular development, what form might that organization take in order to determine later differences between structures and functions? By what mechanisms are *Keime* and *Anlagen* stored? In answering these questions Schwann united the perspective offered by the crystallization metaphor with resources also utilized by the founders of developmental morphology. We have seen, for example, that von Baer sought a mechanism for the initial manifestation of type in the positional arrangement of particles along the central axis of the egg, and he regarded histological differentiation within the germ layers as possibly due to differences emerging from positional arrangement of the organic molecules in the germinal substance and the resulting "lines of force". Schwann's crystallization model led to similar conclusions:

Since all cells possess the metabolic force, it is more probable that the basis for it lies in the particular positional arrangement of the molecules, which is most likely the same in all cells, rather than being due to the particular manner of chemical synthesis of the molecules themselves, which is very different in different cells. ... The qualitative differences in metabolic function are probably due to differences in chemical constitution.[35]

In answer to the question of how it is possible for the material contents of the cell to be essentially different from the material of the cytoblastema Schwann took recourse to *qualitative* differences in the products of different levels of organization; Difference in the mode of organization produces differences in the material product:

One could maintain that the cell membrane alters the substances coming in contact with it, but that the products of these alterations are different because the cellular substance producing the alteration is different from the external cytoblastema. But then the question emerges how the cell contents are different from the cytoblastema. ... I believe that the explanation must be not only that the cellular material is able to transform chemically the imbibed material in contact with it but it also must posses the ability to separate this material so that certain substances remain on the inner surface of the cell membrane and others remain on the outside. ... But there is nothing bold in this assumption, for it is a fact that in the galvanic cell exactly this sort of chemical dissolution into separated, different materials occurs. Perhaps it might be concluded from this characteristic of metabolic activity in cells that a particular position of the axis of the atoms constituting the cell membrane play a special role in bringing about this phenomenon.[36]

For example, only by assuming some such organizational difference in the membrane structure of kidney cells, Schwann concluded, could the synthesis of urea be explained. Through alterations in the positional arrangement of atoms forming cell membranes, differences in chemical affinity must result, leading in turn to functional differentiation. Each cell membrane is like a small galvanic cell, a tiny chemical factory transforming imbibed substances into chemically distinct, spatially separated substances and resulting in a difference between the extra and intra-cellelar environment.

GENERIC PREFORMATIONISM RE-INTERPRETED: CARL VOGT AND THE CELL THEORY

Schwann's cell theory was immediately received as a fundamental contribution by both physiologists and zoologists. One of the most interesting examples illustrating its perceived role in zoology is provided by a work of Carl Vogt entitled *Historie naturelle des poissons de l'eau duce* (1838-1842). Vogt had a deep understanding of vital materialism, and like Schwann he rejected it. Until forced to flee Germany in 1835 as a result of his involvement in several radical political intrigues, Vogt had been a student of Liebig in Giessen.[37] Working from morning to night six days a week for a year at Liebig's side, Vogt not only mastered the techniques of analyzing organic compounds Liebig was pioneering in developing, he also absorbed Liebig's stringent mechanistic approach to vital phenomena.[38] Lack of laboratory facilities in Bern comparable to Liebig's lab in Giessen forced Vogt to break off his study of chemistry. Fortunately, in 1836 Gabriel Valentin, a microscopical anatomist who had worked with Purkinje in Breslau, joined the faculty in Bern. Having completed his degree in 1839 and in need of employment, Vogt succeeded in attracting the attention of Louis Agassiz, who was engaged in a massive project on the natural history of living and fossil fishes. Vogt worked closely with Agassiz for the next five years in Neuchâtel, joining Agassiz and others in many expeditions to Swiss glaciers.

Vogt's assignment in Neuchâtel was to write the embryology of the fresh water fishes for Agassiz's monumental *Recherches sûr des poissons fossiles*. Circumstances surrounding the publication of later volumes of that work ultimately led to a break between Agassiz and Vogt. Vogt and another assistant Desor claimed that Agassiz had signed his name to work done by them.[39] The *Histoire naturelle des poissons de l'eau douce* bore the name of its rightful author, however. It was one of the most exacting descriptive and carefully experimental pieces of embryological research in the early nineteenth century.

The version of the cell theory accepted by Vogt differs in one essential respect from that proposed by Schwann, the most generally accepted version of the cell theory from the work of Schwann to the formulation of it given by Rudolph Virchow in 1858 was the so-called Hourglass model'. According to this model, formulated by Schwann, cell development begins in the cytobastema. This fluid was, as we have seen, of a very special sort, for in it were contained the *potenntialities* for particular kinds of organized elements. Thus a specially organized fluid had as its product a specialized organiced structure. These first elementary organic forms were called the *Elementarkörperchen* by Schwann, and every investigator of cellular development up to Virchow claimed to see these elementary structures floating in the inter-cellular fluid, although some dissenters argued that they were to be found only in the intra-cellular fluid. As Virchow later demonstrated convincingly, these were in part optical illusions stemming from the use of too much sunlight resulting in dispersions having the appearance of small spherelets.[40] Other sources for these observations could be traced to earlier methods of making slide preparations and to the assumption that fatty substances and other granular materials floating in the cellular fluids were themselves cellular substance in its primitive state of formation. According to the hourglass model these granular particles came together by a sort of mutual attraction forming larger conglomerates, ultimately leading to the so-called *Kernkörperchen* or nucleolus. Afterwards a further layer of paricles was assembled around the surface of the nucleolus and through intussusception, this layer expanded forming the outer cell membrane.

While most accounts of the development of the cell theory during this early stage pass briefly over the discussion of these granular particles, they are of singular importance for our present discussion of the transformation of teleomechanism. One of the central threads in the development of this research tradition was the attempt to discover the mechanics in terms of which the *Keime* and Anlagen, the potentialities for determinate structure and function present in the germinal fluids, passed over into localized structural components. The discovery of the germ layers and the mammalian ovum were the first important steps in this direction.

The granular materials of the cytoblastema could be viewed as a furtner step in this direction. But in the early attempts at localizing the phenomena of life a transcendental element remained; namely the assumption of an organized set of potencies in a material substrate which were incapable of being identified with any particular structural elements or chemical constituents. This held equally for Schwann's cell theory no less than for the germ layer theory of Pander and von Baer. In Schwann's case it had the consequence that, in spite of his efforts to exclude it, elements of the notion of a directive teleology remained in his approach to biological phenomena. One of the characteristic features of the further development of the cell theory after Schwann was the success in removing this difficulty. It occurred through increased emphasis on the impossibility of spontaneous generation and the simultaneous linkage of the phenomena of life to the pre-existence of structure rather than to hypothetical potencies. The *Keime* and *Anlagen* of the earlier tradition, in short, took on an ever increasing actualistic, materialistic interpretation. This point becomes plain when we compare the reception and further development of the cell theory by members of the older generation, such as Döllinger and Müller, with parallel attempts of younger men to develop the theory. Whereas the notion of non-physical, organizational potencies continued to be a strong element in discussions of the older generation, younger men such as Carl Vogt and Rudolph Virchow identified these potencies with material structures capable of direct observation.

Instead of allowing new cells to develop freely in the cytoblastema, Vogt argued that new cells only emerge from the nuclei of fully developed cells. In his view the nucleolus is already a fully formed cell and its nucleus is constituted by granular material which he called 'tâches germinatives', i.e. the *Elementarkörperchen* of Schwann. This version of the cell theory, which might best be characterized as a sort of preformation theory on the cellular level, was consistent with the result of Rudolph Wagner, and Vogt supported it by his own observations as well as those on the embryology of the rabbit reported by Martin Barry in the *Philosophical Transactions* for 1840.[41] From the standpoint of earlier attempts to establish a mechanistic theory of organization Vogt was raising an extremely important issue; for in spite of efforts to envision a mechanical model accounting for differential organization, the problem always remained of accounting for the beginnings of this organizational differentiation in the *Kernkörperchen*. Schwann himself, we have seen, had taken resort to a special sort of chemical affinity; but then the 'potentiality' for these functional-structural units must already lie somehow in the cytoblastema undisturbed by its own organic chemistry. It was more consistent with the teleological-mechanical framework required by physiology – and less objectionable from a metaphysical standpoint – to assume that those tiny elements of organization lie already formed in each cell, and that they are the foundations for all later cellular development.

Vogt developed this point of view by examining first the stages in the formation of the egg in the ovaries, the changes initiated by fertilization, and then the formation of the germ. The fresh water salmon were especially useful for providing detailed embryological observations and opportunities for experiment. The eggs develop normally in winter, so that the lower temperatures cause the embryo to progress more slowly, which offers better possibilities for distinguishing various stages and their relationships to later stages. In order to study the role of fertilization, Vogt fertilized the eggs artificially in his laboratory.

Vogt distinguished three major structual elements of the egg as it forms in the ovaries; the *vitellus* surrounded by the vitellary membrane, the germinal vesicle, and the so-called *tâches germinatives*. Within the vitellary fluid Vogt distinguished numerous small granules, which had the general shape and color of droplets of oil or fat. Later as the egg matured these droplets coalesced to form small fatty particles, which had the appearance and characteristics of albuminous matter. This matter, Vogt believed, is the material from which the embryo will be nourished and not as Schwann suspected the basis for new nuclei. Within the germinal vesicle Vogt distinghuished normally about a dozen small granular objects which ultimately became grouped closely together, though they appeared not to coalesce into a single body. These he called the *tâches germinatives*, and as we shall see they had an extremely important future role to play:

> Here is the way I conceive the formation of the eggs: in a cavity of the ovary a cell forms, the germinative vesicle or vesicula Purkinjei, and after it has attained a certain size a second cell, the vitellary membrane is observed to form around it, thus containing the germinal vesicle. At the same time a considerable number of young cells, the *tâches germinatives*, are observed to form in the interior of the germinal vesicle. These cells always remain small and are grouped together.[42]

As he later pointed out, the egg has thus the structure of a nucleated cell.

Vogt devoted extreme care to the determination of the process of fertilization and the contribution of the sperm. In spite of the fact that his technique for artificially fertilizing the eggs enabled him to observe all the events connected with fertilization, he did not feel he could conclude anything with certainty. However, he observed that under certain circumstances unfertilized eggs would exhibit some signs of early, often irregular development. He concluded, therefore, that the sperm makes no material contribution to development, but that it serves as a *directive force*, an organizer bringing the egg to maturation.[43]

Although he was unable to provide direct observational support for the claim, Vogt believed that the vesicula Purkinjei of the 'immature' egg becomes the germinal disc of the 'mature' egg. "Just as the germinal vesicle rested in the middle of the vitellary fluid surrounded by granules that later become fat droplets, now the germinal material is located in the center of our 'oily disc', the germinal disc". Furthermore, according to his own observations as well as those of Barry on the rabbit embryo, after fertilization the germinal disc is observed to contain numerous small, primitive cells. These cells have the exact appearance of the *tâches germinatives* of the unfertilized egg:

> We are permitted to conclude, therefore, that the cells of the embryonic germ develop from the *tâches germinatives*, and consequently that the *tâches germinatives* are in reality the true primitive embryonic cells, and that in the fishes at least they form the rudiments of the embryo itself.[44]

Thus the small particles or 'germinal flecks' which appear in the formation of the egg were the rudiments of structure for the future embryo. Each fleck was in reality a tiny cell, having a different structural and functional constitution. Fertilization 'awakened' those cells into development, the probable function of the sperm being to guide them into the proper positional arrangement within the germinal disc.

Having adopted this generic preformationisit scheme clothed in contemporary model of cellular development, Vogt's framework demanded that he go one step further than Schwann in refuting the holistic teleological conceptions of development held by von Baer and other developmental morphologists. Since the whole is the guiding principle for the development of the parts and since the type is laid down first as the pattern for the whole, it was typical for von Baer and those adapting his approach to the beginnings of biological organization to prefer division by infolding or gastrulation as the principle mechanism for embryonic development. Vogt's generic preformation scheme, however, demanded that gastrulation and all subsequent infolding be a consequence of differential rates of cell development from the primitive stock:

If we consider now the cells of the germ and their relations among themselves and to later structures it is evident that the changes, and most notably the emergence of hills and folds, are due to the growth of these original cells and not due to some exterior cause: The reason is that in the beginning only tiny, imperfectly developed cells exist in the germ; gradually, however, they expand, but not necessarily in a uniform manner, but rather resulting in small rises in certain points. ... As a result of this continuous growth the hills multiply, superimposing one upon the other which causes the principle fold gradually to recede and ultimately disappear as a result of the cellular development progressing from the base to the surface of the germ ... the folds are therefore not the precursors for the formation of the embryonic cells. On the contrary they are the primary phase of cellular life to which the embryo owes its entire existence.[45]

Vogt thus proposed an elegant and simple (unfortunately too simple) solution to the problem of embryonic development. In the beginning the embryo is contained in potentia in the maternal egg. The germinal flecks are the *Keime* and *Anlagen* for the future organism. Once suitably arranged through fertilization, these *tâches germinatives* form the germ plasm of the future organism. Each tâche is the nucleus of a cell having a different structural-functional constitution. Growth and multiplication of these primitive embryonic cells gives rise to the structural and functional organization of the mature organism.

Although Vogt's embryological studies focussed on the familiar theme of *internal* principles of organization, his work also took cognizance of another familiar theme in the zoological writings of the teleomechanist tradition; namely the relationship of the organism to its environment. The quality of the water, he observed, had an enormous influence on the development of the germ. Eggs of species of trout that normally spawn in rivers around Neuchâtel do not reach maturity in the waters of the lake; similarly the eggs of the *Palée*, which spawns in the lake do not flourish in the streams. Even more locally specific factors seems to play a role. Thus the eggs of the Brochet living in marshes around Neuchâtel and which normally spawn later than the Brochet in Lake Neuchâtel, dot not flourish in the water of the lake, "even though there is no specific difference between these two fish".[46] Variations in temperature and especially exposure to light were often factors affecting the development of the organism. Moreover, it appeared that constellations of external and internal factors could combine to alter the rate of development of some organ systems, retarding one and accelerating another. This was especially apparent in the development of the circulatory system, the first vestiges of which demonstrated a close correlation with the quality of light received by the embryo.[47] In the hand of someone given more to speculation, such as Ernest Haeckel, such evidence would be used as a source for breaking up the germ plasm into genealogically related species. Such implications were no doubt apparent to Vogt as well.[48]

RECEPTION OF THE CELL THEORY BY THE OLDER GENERATION: DÖLLINGER AND JOHANNES MÜLLER

That the interpretation of cellular development and its implications for future research being explored in the works of young men such as Schwann and Vogt were perceived as consistent with the aims of the program first set forth in the early years of the 19th century can be gathered from its reception by some of the participants in that early project. Ignaz Döllinger's view of the cell theory is perhaps exemplary of other members of the founder generation. In the last years of his life Döllinger began to reshape his lectures on physiology in light of developments that had occurred since 1805 when he had first assembled his thoughts on theoretical zoology. The work was not completed by Döllinger himself, but a former student and colleague, M. Erdel, assembled the remaining materials needed for its completion from Döllinger's *Nachlass* and published a volume entitled *Grundzüge der Physiologie der Entwickelung des Zell- Knocken- und Blutsystems* (1842). This volume offers a clear picture of Döllinger's perception of the relationship between the earlier versions of the program and the new shape it was being given by the cell theory.

Here alongside the emphasis on developmental history as the key to discovering the laws of organic form together with the theory of the type and the correlative rejection of Meckel's recapitulation argument, Döllinger discussed the material basis for the differentiation of form and structure in the germ. Difference in structure and function, he wrote, presupposes a difference in the physical characteristics of a material basis. These differences may not emerge all at once but may only appear in the course of development; moreover, since the multiplicity of organic bodies is constructed from a relatively small number of elements, the material condition of form must reside in a complex mosaic of interrelations. Minute alterations in chemical affinity must lead eventually to structural and functional differentiation. But the beginnings of this diversity lay already fully formed in the generative substance:

These beginnings out of which the body and its functions are constructed are at first only laid in formative structures [formelle Bildungen] like seeds [Keimen] more in a state of potentiality than one of actuality, so that no trace of the characteristics of the tissues, that fineness of tissue elements which will later distinguish the individual structures, is at first noticeable.[49]

Although indissociable from a material substrate, Döllinger still regarded the 'potentialities' characteristic of organization as somehow distinct from matter itself.

These 'seeds of potentiality' were localized generally in the germinal disc and awakened to action by the generative fluids of the parents.[50] The initial manifestation of this generative stock was the blastoderm, which consisted of a granulous mass of material. It was, wrote Döllinger, the so-called '*Urgewebe*' of the future organism. All structures emerge from the further development of these small granules, the former *Keime* of the germinal disc.[51] The *Körner* or granules themselves became the cells of the new organism. Unlike Vogt, however, Döllinger did not conceive the contents of the cells as different. Each contained a portion of the germinal material, the *Keimstoff*, of the original germinal disc. Each cell contained in·short, a replica of the original *Keimstoff*:

The cellular system [Zellensystem] is the organ which is continually renewing and conserving the *Keimstoff*. It is like the germinal disc itself in that it contains the germinal material and binds it. It is the inner root of the entire developmental process, the foundation of all organization. ... In the cell system, therefore, two similar processes take place; first the reception of the germinal material, and second its transmission to other, differentiated structures.[52]

This manner of conceiving the relationship between cells emphasized the continuity of the germinal material, the essential unity of each structural element with the fertilized egg:

> In this manner the process of successive development from the germ is repeated in each particular structure of the mature body just as it happened first in the egg. The cell is thus an internal germinal organ [Keimorgan] which determines the progress of life; it is at the same time the egg of the individual through which the supply of germinal material is maintained and transmitted to the individual structures.[53]

The assumption of a continuity of germinal material in each cell brought with it, of course, the problem of accounting for differentiation of cell structure and function. Döllinger's solution to this problem made use of a model long familiar to teleomechanists. As we have seen, according to this model, the structure of the organism is conditioned by two closely related factors; the internal *Keime* and *Anlagen* of the generative stock, and the environmental circumstances affecting their manifestation in particular adaptive combinations. Through contact with the external world "a division of organs occurs which corresponds to the different influences, and an interrelation between the internal development of life and the external conditions is affected so that through this exchange each organic system is adapted to each situation in accordance with the characteristics of the system itself in such a way that the strongest [kräftigste] internal harmony is preserved".[54] Similarly to the manner in which Kant had accounted for the phenomena of race by assuming a common generative stock for each individual which is manifested in a limited set of particular adaptive combinations according to external circumstances, Döllinger assumed the genetic continuity of individual cells, and he attempted to explain cell differentiation in terms of adaptation to environmental stimuli.

Among the developmental morphologists the most important response to the cell theory was made by Johannes Müller. This aspect of Müller's work has been, until very recently, entirely neglected.[55] In my opinion a thorough revision of the present understanding of Müller's view of physiology is in order. To historians who have characterized Müller as a vitalist unable to free himself from the influences of Naturphilosophie, it must come as a terrible shock to learn that Müller not only understood the work of his students, Schleiden and Schwann (some have claimed he could not have!), but that Müller regarded the cell theory as consistent both with his most recent research in pathological anatomy and with the general theoretical framework of teleomechanism. Two major sources offer insight into Müller's reaction to the cell theory: a treatise entitled *Über den feineren Bau und die Formen der krankhaften Geschwülste* (Berlin, 1838), and the changes introduced into his

Handbuch der Physiologie des Menschen in 1839 and 1844.

The swiftness with which Müller was prepared to take up the cell theory and apply it technically to specific medical problems, which he supported by a wealth of exacting microscopical researches, certainly indicates that Müller did not feel himself slipping into the alien world of a reductionist physiology with which he could not cope. Schleiden's treatise appeared in Müller's *Archiv* late in 1837, and Schwann's generalized theory appeared in Froriep's *Notizen* in January of 1838. Müller's *Über die krankhaften Geschwülste* appeared a few months later. In his work Müller pointed out that as early as 1836, in three papers delivered in Berlin, he had indicated the apparent cellular–like nature of several cancerous growths.[56] The theory set forth by Schleiden and especially by Schwann had obvious implications for pathological anatomy, and Müller tested it on his own pathological specimens. The results totally confirmed the theory:

In numerous growths in which I had not previously found cells, I now found them by stronger magnifications. ... Therefore Schwann's observations are confirmed by these pathological structures, just as expected.[57]

Where previously no clear method seemed available for ordering the various pathological specimens he had been accumulating for years, Müller now seized upon cell-type and cellular development as a powerful theoretical foundation for ordering the phenomena; and through it, combined with improved chemical analysis, he expected an explanation for cancerous growths with far-reaching implications for medical practice.

Müller's findings are worth considering in detail, for they reveal important aspects of the inner workings of developmental morphology and its potential for growth. Müller observed that the same developmental pattern ascribed by Schwann to healthy cells is also evident in pathological development: The cell always emerges from the nucleus, "which then surrounds the nucleus like an hourglass". The nucleus normally remains attached to the cell wall. Cells develop in two ways: most typically they develop from tiny free nuclei (Kernchen) within the mother cell, the cell wall of the mother cell either dissolving or bursting, thus allowing the daughter cells to exist independently. But cells may also form *outside* of a completely developed cell. In this case, best typified in Müller's view by epithelial cells, the cell wall of the mother cell either dissolves or bursts allowing the embryonic nuclei to escape. In the extracellular fluid they then begin to initiate the developmental sequence resulting in the formation of the cell wall enclosing the nucleus. It may also be the case, Müller argued, however, that the nuclei form freely in the extracellular fluid.[58]

Müller's reason for preferring an extra-cellular origin of the nucleus becomes apparent in light of his theory of disease. He seems not to have thought that a cancerous cell growth is in any way a deterioration of healthy tissues or due to the degeneration of healthy cells. "The first appearance of cancerous degeneration", according to Müller, "does not consist in the transformation of elements [Formenelemente] of the cancer within the tissue of the organ".[59]

The original cells [Keimzellen] of a carcinoma, for example, arise not only from already existing fibres, but rather independently from a *seminium morbi*, which develops between the tissues of the organ.[60]

In Müller's view the diseased cells are not really foreign organic structures. The *seminium morbi* refered to here was not a 'germ' coming from without. "The positive character of the structure of the carcinoma does not demonstrate anything heterologous or foreign to the healthy organism; the formative elements are in part the same as those that emerge in a fully developed organism, while in other cases they come forth only in the foetal condition".[61] As we shall see presently, the basis for this view is in large part traceable to the fact that for Müller different chemical constituents are capable of producing the same structure; and similarly the same chemical elements are capable of combining organically to form different structures.[62]

Müller attributed the cause of the *seminium morbi* to a particular localized 'disposition' of the extra-cellular fluid.[63] This disposition resulted in the formation of 'krankhafte' cell nuclei and ultimately cancerous tumors. The source of the 'disposition' itself was, in Müller's view, not mysterious; it must somehow originate in a breakdown of the command of the integrated, whole organism over its parts. In this condition cells could develop, but they would not develop as integrated parts subject to the command of the whole organism; rather, left to themselves as independent units, they would either arrest development in a foetal condition or overstep the limits assigned by their relationship to the whole:

The differences between pathological and healthy cartilaginous developments consists principally in the continuation of embryonic cell formation. In numerous other tumors the same observation can be easily made. It is not the form of the elementary parts that distinguishes diseased structures. The problem lies in part in the formation of normally primitive structures where they are not necessary and do not contribute to the purposiveness of the whole, and partially in the incomplete development of these tissues, which usually only reach a particular stage of development which is transient in healthy life. This is the mode of operation of diseased vegetative life. In the development of sound primitive cartilage, however, the monadic life of the cells is controlled by the *Lebensprincip* of the entire individual; it reaches its limit, the cells coagulate and the interstitial, unclear fibrous mass of the cartilage emerges between the cavities of the germinal cells. In the Enchondroma on the other hand the regulated life of the part no longer attains a particular limit, and it slowly continues to increase in size. The cell walls in this case do not thicken normally; the cartilage remains in its embryonic condition and this embryonic structure is continually repeated.[64]

Lest we be tempted to regard this theory of disease as an attempt to interpret the cell theory within a naturphilosophic framework, it is instructive to consider the manner in which Müller introduced the cell theory into the framework of his own general theory of organization in the *Handbuch der Physiologie*. A comparison of the third and fourth editions of that work demonstrates that Müller was able to incorporate the cell theory into his view without modification. The cell theory makes its first appearance in the second volume of the third edition (1837-1840) of Müller's *Handbuch der Physiologie des Menschen* where it is included as the basis for the theory of generation and cited as the fundamental unifying principle for physiology. Schwann's work was completed after the new edition of the first volume had already gone to press in 1837, and accordingly the general discussion of biological organization in the opening section of that volume does not mention the cell theory. But of particular interest is the fact that in reworking the discussion on biological organization in order to incorporate the cell theory for the fourth edition (1844), Müller did not modify the structure of the theory presented in the first volume in any essential way.

One of the characteristic features of Schwann's theory was its attempt to employ the model of crystal formation, suitably modified for the 'organic' context, as a heuristic guide for visualizing cell development within a mechanical framework. As we have also seen, the model of cellular development that particularly attracted Müller was the possibility of nucleus formation in a 'suitably disposed' extra-cellular fluid. A comparison of the third and fourth edition of the *Handbuch* confirms that the emphasis on the model of crystallization did not in any way disturb Müller, for the 'crystallization' required by Schwann was a special sort depending upon 'organic affinities' incapable of being constituted out of regular chemical affinities. It was this point, essential to his perspective as a developmental morphologist, that Müller had always insisted upon: organization cannot be explained out of the simple aggregation of chemical elements.

In the third and all earlier editions of the *Handbuch*, Müller insisted that although many organic solutions are capable of crystallization through the agency of heat, chemical reagents, or galvanic currents, they do not normally crystallize while the organism is functioning. This did not imply, however, that the normal functioning of the organism proceeded in terms of laws other than those regulating normal chemical reactions. On the contrary, "The laws of attraction of matter through dissolution and mixture, as well as the laws of equal distribution of fluid mixture, are applicable to the animal fluids".[65] But this did not lead to the conclusion that the formation of organic structures could be effected through standard chemical processes of aggregating molecules. To be sure microscopic examination of tissues seemed to reveal that all structures are composed of aggregates of molecules. The clearest example, he noted, was the blastoderm. Microscopic analyses of the embryo might be interpreted by an uncautious mind to imply that the blastoderm consists of 'molecules' of a particular size and that it develops by attracting small 'molecules' from the yolk and combining them in aggregates. This was contradicted, however, by the fact that

all formations within the blastoderm itself proceed through the dissolution and transformation of these aggregate particles into matter so fine that elementary particles of this substance cannot be detected, and they must in any case be much smaller than the aggregate molecules of the blastoderm. ... Ehrenberg's discovery that monads 1/2000 of a line in length still have organs composed of parts makes highly improbable the theory that organisms are constructed out of rows of spherelets [Kügelchen], since each of these little molecules is greater than 1/2000 of a line. The construction of tissues out of molecules is, due to the difficulty of identifying molecules with the microscope, a questionable hypothesis. In any case, however, the organic molecules are only the smallest forms in which organic matter appears, never the atoms of organic structure. ... organic matter never arises out of an accidental combination of physical elements.[66]

In this passage Müller was arguing against *two* schools of biological thought which seem totally unrelated; namely Naturphilosophie and mechanistic reductionism. In a twist that appears curious to us today, the developmental morphologists always accused the reductionists of introducing a dogmatic, monistic philosophy of nature which was the mirror image of Schelling's absolutism; the materialists had really just "turned Schelling on his head". That was, for example, Müller's line of attack against Magendie; and, as we shall see it was von Baer's line of attack on Darwin. Idealism and reductionism were just obverse sides of the same coin, which Müller had characterized as 'falsche' or 'dogmatische Naturphilosophie' in his inaugural lecture of 1824. But in this early edition of his *Handbuch* Müller was arguing against one of the most celebrated doctrines of the Naturphilosophen.

In discussing the differences between the approach of the Naturphilosophen and that developed by Döllinger, von Baer and others, I pointed in particular to the different conceptions of morphology held by the two schools. Oken, Spix, and Carus, like their counterparts in France led by Geoffroy Saint-Hilaire held what, following E.S. Russell, I have described as a 'transcendental morphology'. The morphotypes of this school were sets of geometrical shapes, and organic connections were traced by geometrical transformations of these fundamental units. Oken's and Carus' vertebral theory of the skull are classic examples of the transcendentalist approach. This style, which led to the introduction of much mathematical mysticism in Oken's Naturphilosophie contrasts sharply with the functionalist conception employed by teleomechanists such as Kielmeyer, who stressed *forces*, interrelated *processes* and eventually developmental pattern as the basis for conceiving the type. Now, according to my view, just as the developmental morphologists were prepared to incorporate the cell theory into their program, so too were the trasnscendentalists prepared to absorb it into theirs. But for them its significance lay in its *spherical shape*. Oken had mused on the geometrical properties of the sphere as the ultimate *Grundform* of organic structure in the opening sections of his earliest system of Naturphilosophie. Similarly Carus' diagrammatic presentation of the *Urtyp* employed an array of spherelets suitably arranged. It came quite naturally to them, therefore, to see the cell as the geometrical building block for their animal morphology. Oken and Milne-Edwards, for example, were among those who had argued that an atomic organic element, the Kügelchen, is the fundamental building block of all organized bodies in both the plant and animal realms.[67]

Müller, by contrast was, in the passage cited above, defending the position that the mode of arrangement of the parts is the effect of processes governed ultimately by the life of the whole. One could not assume the existence of a structural-functional unit, the same for all organisms, which when combined in various ways leds to different sorts of organs. For Müller the relationship of cell to organism is the precise reverse of this 'atomistic' view. In his view the cell is the agent operating in the service of the whole, and its structure is determined by the functions that it must carry out in harmony with others to insure the whole. This argument could also be marshalled into an attack on the reductionists, however; and as we shall see, it was in fact employed in precisely this fashion by Müller's student Carl Reichert a decade later in his criticism of Carl Ludwig.

In the edition of 1844 the discussion on 'molecules' was repeated, only now it was supported by the conceptual framework of the cell theory. After a brief description of basic cellular structure, Müller continued:

The cells are also, according to Schwann's discovery, the elements of all complex structures in animals. ... Aggregation of spherelets [Kügelchen] is, on the other hand, never the cause of the formation of fibres or any tissues whatsoever. The discovery of Ehrenberg (etc.) ...[68]

The context in both editions makes it clear that by 'cells' Müller meant organized structures dependent on life processes defined by the whole, not 'organic atoms' which would serve as the elementary building blocks for all organisms through different modes of aggregation. In Müller's view there are no special organic 'molecules' out of which animal structure can be built. To be sure, the initial stages of cell development, the formation of the nucleolus, for instance, could be generated through a process analogous to precipitation; but it only occurred in an organic solution 'purposively' organized to generate this structural unit. The cell is, in Müller's view the structural manifestation of potencies already present in a previously organized solution. The production of the organic solution in turn presupposed the presence of the same type of organized structure as a precursor: It was required to generate the proper order among the chemical constituents. Moreover, the context in which Müller presents the cell theory makes it clear that the cell, once constructed, would function in terms of the laws of chemical combination regulated by the *Lebenskraft* expressing the life of the whole. In Kantian language, the cell is a chemical machine, but one that is purposively organized.

This attempt to interpret the cell as the mechanism through which a purposively organized whole, a living individual organism, maintains its organization and transmits it to others was perfectly consistent with the general theoretical framework of earlier editions of the *Handbuch*. Just as in the earlier editions so in the fourth edition Müller's theory of animal organization was based squarely on Kant's conception of teleomechanism:

Kant says that the cause of the manner of existence of each part of a living body is contained in the whole, while in inert matter the cause is contained in each part itself. As a result of this characteristic it is clear why a part of an organic whole is normally incapable of surviving apart from it and why the organic body appears to be an individual incapable of division.[69]

In this same context Müller, like Kant, also argued that the ultimate cause of organization must remain beyond the limits of a scientific explanation. In the third edition, supporting this view by Schwann's work on fermentation and spontaneous generation, Müller argued that an explanation of living phenomena must always begin from the assumption of an original state of organization. Müller stressed emphatically, however, that organic phenomena cannot be viewed as seperate from matter, that "in actuality organic phenomena are peculiar only to certain combinations of elements, and even organic matter can be broken down into inorganic combinations as soon as the cause of organization, the *Lebenskraft*, ceases".[70]

If 'Lebenskraft' did not designate the property of a special particle or force capable of existing separately from an organized body, neither did Müller think that it should be conceived as the result of a harmonious, integrated interaction of elementary parts; and though mention of it was not made, this was precisely the position advocated by Schwann. Müller was attempting to head off the construction of a reductionistic physiology by incorporating the cell theory into his own perspective of developmental morphology. By 1844 the issues surrounding reductionism were beginning to heat up right within the walls of Müller's own lab!

In Müller view, possible misconseption of 'organization' had been encouraged by Reil's treatment of the *Lebenskraft* and its adoption by others such as Rudolphi. Swann and others might have been misled by the ambiguity in the writings of the older teleomechanists. While it is true, as Reil had argued, that particular forms are associated with particular 'Mischungen' or combinations of elements, nevertheless said Müller, the problem of identifying a cause of the harmony and integration which could explain how form came to be linked to particular mixtures in the first place and how certain mixtures lead to particular forms of organization remains unsolved, and it must remain forever unsolved:

Life is therefore not simply a consequence of the harmony and interaction of the parts, rather it begins to express itself in the germinal material as a force of imponderable matter which goes into the construction of the parts themselves and imparts the qualities of organic combination to them.[71]

In Müller's view, which is identical to that of Kielmeyer and von Baer, the integrated 'harmony' demonstrated by organic forms is dependent upon a 'force' which acts throughout the entire animal and is identical with the animal whole. Even if it does not have an ontological priority over the material elements of organization, this force is logically prior to the harmonized individual parts, and like the *Bildungstrieb* it is first created in the development of the germ. "The germinal disc is the whole *in potentia*, possessed with the essential and specific force of the later animal".[72]

From the theoretical standpoint adopted in the *Handbuch*, therefore, we see that Müller's principal objection to any attempt to explain animal structure and organization through organic chemistry was not that it led to the view that organization is deeply rooted in matter itself, but rather that it necessarily implied a priority of the part over the whole; this, in short, was to deny the *zweckmässig* organization of animals. This is why in all editions of the *Handbuch* Müller argued that organic compounds require ternary and quaternary affinity relations while inorganic compounds require only binary relations of affinity. These 'quarternary' compounds were believed incapable of being synthesized artifically; they required the presence of a previously organized body for their constitution. To be sure Wöhler had synthesized urea, but urea, Müller argued, stands at the most external border of organic matter. As an excreted byproduct it is not capable of generating or sustaining life.[73] The proceses leading to the artificial synthesis of certain organic compounds should not be identified with the sources of animal organization. Only if the artificial synthesis of tissues and organs proved possible could one speak of a 'reductionist' physiology.

When we consider the thoroughly teleological orientation of Müller's theory of organization the question arises, how was he able to regard Schwann's cell theory as consistent with his own conceptual framework? Schwann, we have seen, argued that it was necessary neither to assume the whole as primary to the parts nor to invoke a special organizing force residing in the whole as the cause of the integration of the parts. The answer to his problem is, of course, that in Müller's view Schwann had, to be sure, localized the force regulating the production and integration of animal structure in the cell, but he had not thereby reduced the whole to an assembly of correlated individual parts. Each cell was a replica of the whole, specially modified to carry out its specialized assignment.

This evaluation of Schwann's work is set forth clearly in the second volume of the *Handbuch* in the section on generation. Here Müller argued that every organism must be considered a multiple of an original set of constitutive 'Teilchen'. Thus he argued that young organisms can be expressed by the formula, $abc\ .,..$, where a is the constitutive liver cell, b the germinal nerve cell, c the germinal muscle cell, etc. The mature organism could then be expressed by a^n, b^n, c^n. These individual, specialized cells are incapable of being separated from the integrating life of the whole animal:

However, even the higher organizational forms at maturity must be regarded as virtual multiples of the germ since through growth they are capable of forming a new germ.[74]

In order to conceptualize the possibility that each specialized cell is at the same time a multiple of the original germ, Müller observed that in some lower animals (he does not specify which organisms he has in mind) each part of the germinal material is capable of generating an entire organism, but gradually these parts are transformed into particular structures such as muscle tissues, nerves and 'cellular tissue' of various sorts, which are incapable of an independent existence.[75] It could be, he reasoned, that through some sort of alteration of the original stock of generative material cells of different types are produced and that these cells are in turn capable of reproducing themselves uninterruptedly but are subsequently incapable of generating an entire new organism. Somehow both the process of generating new organisms as well as the continual renewel of specialized structures integrated into a whole organism had to be explicated in terms of the same mechanism; namely cell development:

> The mutiplication of all organized bodies, is therefore, a dual process of growth: on the one hand of the present, fully developed form by means of the multiplication of the parts constituting it, and on the other hand the multiplication of the form of the species in an undeveloped state containing all differentiated structures in an unseparated form. ... Both the multiplication of the cells, which is the mechanism by means of which the whole is assembled, as well as the formation of multiples of undeveloped forms as germinal cells [Urzellen] ... are processes continuously present in the organism from the very beginning.[76]

The internal processes regulating the original constitution of the germinal material, which contained all future structures *in potentia*, therefore, were functionally identical with what Müller had earlier described as the organizing force, the whole which is both one with and prior to the development of the parts.

How this regulative process was conceived by Müller to be related to the cell theory can be gathered from his discussion of the development of fish and birds. According to Müller the unfertilized egg consists of an outer membrane, an interior filled with yolk, within which floats the germinal vesicle or Vesicula Purkinje, and inside this latter structure is the macula. As Müller pointed out, the egg is a cell in Schwann's sense. Before fertilization occurs the germinal vesicle disappears in most animals and is transformed into the germinal disc, the *Keimscheibe*. Directly beneath the germinal disc is a lump of granular masses, the germinal spot or *Hahnentritt* (in the case of the chicken egg). Both the germinal disc and the germinal spot consist of cells, according to Müller.[77] The future embryo is contained in this germinal material.

After fertilization cell nuclei begin to form in the yolk and migrate toward the germinal disc, where they become attached to it. The germinal disc directs all future development of the embryo in that it gives rise to the basic germ layers. This requires a transformation of the yolk cells into cells of the various germ layers. Thus all structures ultimately have the same cellular origin in the yolk cells and are transformed first to their more specialized germinal materials by the germinal disc and ultimately into specialized cell-types. The first of these layers which formed on the germinal disc was the blastoderm. After its initial formation it multiplied and enclosed the yolk. The blastoderm, of course, also consists of cells.

Only the cells of the germinal layers [Keimanlagen], and the yolk sac have a direct part in the development of the embryo. The cells of the yolk material can never be shown to pass directly into the structure of the chick embryo. The nucleus of the germinal sport, which is to be regarded as the first aggregate layer of cells of the yolk sac in the canal, cannot be separated from the germ layer [Keimanlage] in the early state of development. After the first layer [Anlage] of the embryo has been formed, however, it separates and quickly disappears.

It is characteristic for the first period of embryonic development that entire structures arise and develop in the cells of the germ layers, and subsequently those neighboring cells of the yolk sac are deposited and the layers are then enlarged through the assimilation of new, properly suited cells. Nourishment is never transported to these cells. They emerge solely through the development [Ausbildung] of *Anlagen* at the expense of their rich supply of granular material. These contents gradually disappear, while at the same time numerous young cells become visible in differentiated structures.[78]

After the formation of the last of the germinal layer, the mucous layer [Schleimhaut], the germinal disc [Keimhügel] has served its function of directing the formation of the specialized cellular basis of the organism. "Previously designed [geeignet] completely to separate out the various layers [Anlagen] of the embryo from the aggregated, predisposed [disponibeln] cells of the yolk sac, it now becomes superfluous to further development, although it is still visible for awhile beneath the embryo as a white mass of cells ...".[79]

Comparing the treatment of genration in the *Handbuch* with the theories discussed earlier, it emerges that Müller was attempting a synthesis of the various elements of the embryological model developed by von Baer and Pander with the rich physiological mechanisms offered by Schwann's cell theory. Moreover, Müller's discussion of generation was in essence a full elaboration of theories proposed by the teleomechanists in the 1790s rendered up to date with the advances of the 1830s and early 1840s.

Müller himself characterized the theory of generation emerging from his discussion as a theory of generic preformationism, exactly as Kant and Blumenbach had envisioned it, only supplemented by the mechanism of cell development. The germ, itself a cell he argued, contains the form of all future generations encapsulated within it. This was a 'geistige' version of the old preformation theory.[80] Since the production of germinal material and the replacement of cells were continuous processes in the organism there was no need to assume the actual presence of the future organisms in miniature. The mechanisms of segregation of germinal material and cell development insured that the potential for future generations of new individuals would be preserved. In keeping with Kant's generic preformationism, Müller assumed that the "Anlagen for particular structures are contained in the egg",[81] and similarly to Reil, he argued that the sperm too must contribute to the future organism, that the future organism only emerges from a direct, material contact of the egg and sperm.[82] Spallanzani's artificial fertilization of frog eggs demonstrated the impossiblility of fertilization through an *Aura seminalis*, and the production of races and hybrid forms demonstrated both a paternal and maternal contribution to the germ.[83] Nevertheless, although *contact* of sperm and egg were required, the sperm did not contribute material to the future embryo: "the sperm is so organized that it is ... in the form of a vital, fluid *Incitament*".[84]

Müller also expanded the viewpoint of the teleomechanists in discussing the implications of generic preformationism for the explanation of the variation of animal forms. Here his views correspond with those elaborated by von Baer. Species are not transformed into one another in Müller's view.[85] The causes of the variations of species are partially internal and partly external. Each species has a particular '*Variationskreis*' built into it which permits adaptation to particular environmental conditions. Each individual carries within it, independently of all external influences, the potential to produce a member of the '*Variationskreis*', insofar as each individual does not just reproduce offspring identical to itself but rather is governed by the law of the species in reproduction.[86] Through prolonged exposure to particular environmental conditions and altered means of subsistence, the potential for variation can be awakened, leading to the production of different forms.[87] Müller also assumed that species of the same genus and closely related genera could interbreed, but that behavioral traits and instinct usually prevented this form occurring.[88]

In the work of Johannes Müller we see the clearest elaboration of the principles of teleomechanism for empirical research in physiology. What von Baer and Rathke had accomplished for zoology, Müller had now succeeded in expanding to physiology. His work represented the most systematic development of ideas at the basis of the teleomechanist research program for physiology since Kant and Reil. The teleologico-mechanical framework is unmistakeable in Müller's work. As we have seen, according to the teleomechanists organization is the primary starting point for biological research for which no mechanical explanation can be advanced. The task of the physiologist is to discover the mechanisms in terms of which organized beings function, preserve their organization and pass it on the future generations. Those mechanisms, while not differing in form from other physical and chemical agents, are to be understood only in reference to their role in maintaining a functional whole; for they are the means employed in maintaining the purposive organization of the whole. Because these mechanisms can never be viewed independently of their function within a particular organizational setting, it is necessary that the whole be assumed as the primary starting point for a causal explanation of function. Something must integrate these mechanisms into a functional individual organism, and this is the whole, which Müller, like Blumenbach, Kielmeyer, and von Baer expressed by the term *Lebenskraft*. This vital force, like the force of gravitation, does not exist independently of the material constituents, of which it is at one and the same time the *expression and the organizing principle*. That is to say, in Müller's language, the form of the whole is peculiar to a particular material constitution but it is not on that account the *effect* of an integrated harmony of parts which could exist independently. The concept of the whole in biology is one of the logical priority in establishing a framework of intelligibility. The parts only have their existence as parts of a whole. This is without question vitalism, but it is material vitalism.

Once this framework has been adopted the task for physiological research becomes one of discovering what mechanisms, what material processes are employed by an organized whole in maintaining itself as such. In this context Müller's work demonstrates the synthetic culmination of a single line of development beginning with Kant and Blumenbach passing through Kielmeyer to von Baer and Müller. Their point of view was compatible with the cell theory. According to the views sketched first by Kant and Blumenbach, purposive organization is invested in a set of *Keime* and *Anlagen*. By what mechanism did these Keime and Anlagen get developed into an individual? Pander had first suggested the localization of the organized material substrate in the germinal disc, with the nucleus of material directly beneath the disc serving as the directive agency for the production of germ layers with different potencies for special types of structures. Meckel, for example, had imagined that the further mechanism for development lay in the different constitution of areas of the germ layers. The work of Purkinje, Wagner, von Baer, and finally Schwann provided the next step in localizing the mechanical processes and establishing the chain of continuity from the organized but structureless beginnings of the germ with its *Keime* and *Anlagen* to the differentiated regions of the germ. Cell development and differentiation of cell structure provided that continuity of mechanical interaction within a teleological framework leading from the structural organization of the egg, 'awakened' by the catalytic action of the sperm, to the differentiated cells of the germ layers.

There were still many questions to be answered by future research within the framework set forth by Müller. Thus, according to him the germinal vesicle of the egg becomes the germinal disc, and the macula becomes the granulous lump of matter directly beneath it in the mature egg. By what mechanisms did these developments occur? Similarly, the 'nucleus' of matter beneath the germinal disc was the directive agency guiding the differentiation of cells into the germ layers which formed successively on the disc – just as Schwann had suggested that the nucleus and cell wall are successively formed as a precipitate around the nucleolus. By what mechanism did this process of differentiation take place? Those were logical extensions of the problem-context established by the framework of teleomechanism. For as Kant had written, the mechanico-teleological principle does not imply that a mechanical explanation should not be pursued as far as possible. It only asserts that at the highest levels, in the explanation of the source of organization itself, the principles of mechanism are not applicable.

CHAPTER 4

THE FUNCTIONAL MORPHOLOGISTS

During the 1830s developments in organic and physiological chemistry were taking place which considerably altered perceptions of some of the central issues in previous debates over biological organization. As we have seen, teleomechanists had two principal lines of argument to offer in defense of their holist approach to physiology. On the one hand it had been argued by no less a figure than Johannes Müller that *in vivo* experimentation could not yield reliable results. Intervention supposedly disrupted the systematic interconnection with the whole of the organ under study, and its normal functioning depended essentially upon this vital connection. The most convincing support in defense of this position, however, concerned the damage inflicted to the organism through the crudeness of surgical technique and instrumentation. Although success in this area was limited, improvements had definitely been registered, particularly by the French.[1] Magendie's classic work on absorption, his investigation of the lacteal and lymphatic vessels were notable demonstrations of how *in vivo* experiments could resolve questions of organ function.[2] In the early 1840s further advances in surgical technique, such as the fistula developed by Beaumont and Blondlot, which was exploited by Claude Bernard as early as 1843 in his own investigations of digestion, provided further evidence that through refinement of technique valid inferences concerning physiologic processes could be made through direct experiments within living animals.

The second and perhaps most important consideration for German physiologists of the early nineteenth century was related to the nature of chemical processes taking place in the animal body. As we have seen, the view defended by Kielmeyer and Weber was that chemical affinities in the physiological context stood under the higher guidance of a 'dominating force'. While belief in special organic laws of chemical affinity was gradually abandoned, the view that chemical experimentation in the laboratory could not duplicate reactions within the animal body was still regarded as compelling.

Organic chemists developed two sorts of technique which eventually disarmed these criticisms. An extremely important field of work was exemplified in the laboratory analyses undertaken by Liebig and Wöhler.

They had pioneered in the development of techniques for controlling oxidation reactions and analyzing organic compounds. Liebig and Wöhler foresaw great reform in physiology based on these techniques of organic analysis. Their approach was to make inferences concerning physiological transformations on the basis of the elementary organic analyses carried out in the laboratory. It was assumed by Liebig that the pathways of material exchange established by his analytical techniques were duplicated in the organism.

A second approach, designed to overcome some of the objections to inferences based on straight laboratory analyses, attempted to follow the chemical process directly in the organism. A pioneer study in this method was Tiedemann and Gmelin's work on digestion.[3] Their approach was to analyse the digestive juices in the laboratory, exploiting the chemical procedures developed by Berzelius and others for identifying the components of the digestive juices as well as the constituents of the aliment. They then compared these results with the contents of the digestive tracts of animals maintained on various diets. This type of approach produced landmark discoveries in the chemistry of digestion. Their experiments on the conversion of starch to sugar, for example, produced direct evidence that a substance had undergone the same specific chemical transformation that had also been demonstrated in the laboratory.

The questions explored by Tiedemann, as we have seen, were generated directly within the vital materialist tradition. But he was not the only teleomechanist to pursue such work. Among the researchers taking up this line of investigation during the 1830s, the most important for our purposes was Johannes Müller and his student-assistant, Theodor Schwann. The chemical studies of digestion undertaken by Müller and Schwann culminating in Schwann's independent isolation of pepsin and the first development of ideas concerning the role of catalysts in metabolism were major contributions to the transformation of physiology. These technical achievements set the stage for a reconsideration of the question of biological organization.

As we have seen, by the mid-1830s Von Baer and Johannes Müller had reshaped and extended the original vital materialist tradition into a wide-ranging program of research into the mechanisms of vital phenomena based on a holistic, teleological conception of organic form governed by special vital forces. Their program, which I have termed 'developmental morphology', had its roots in the theoretical writings of Kant, Blumenbach, and Kielmeyer, and it led to some of the central positive achievement of zoology and physiology in the early 19th century including the cell theory. But in Johannes

Müller, Karl Ernst von Baer, Heinrich Rathke, and Rudolph Wagner it found its last representatives. To be sure, elements of their approach continued to play a major role in the conceptual organization of the life sciences in the second half of the century, as we shall see; but during the 1840s vital materialism and its younger sibling, developmental morphology were extensively modified and reformulated. The cell theory provided the forum in which a broader set of issues concerning the limits of physical methods in biology was about to surface for the first time. The debate over the cell theory, however, revealed only the outlines of discussions well in progress by 1840. It was merely the tip of a very big iceberg concerning the question of whether special vital forces could be dismissed from physiology and whether the problem of order in biological systems could be re-interpreted in a completely non-vitalistic mechanical framework. Schwann and Vogt merely expressed the strong conviction that this shift could be made. We get no clear picture, however, of how they thought it would in fact be effected. Nor do we see any treatment of the theoretical foundations upon which a mechanical framework could be constructed which acknowledged the primacy of order, arrangement, and sequence among the material components of organic systems without invoking a vital force. The persons who attempted to provide those necessary underpinnings were Justus von Liebig, Hermann Lotze, Carl Bergmann, and Rudolph Leuckart. Two factors provided essential background for this final transformation of the teleomechanist program: the internal logic of the research tradition itself, and progress in organic chemistry which the vital materialists and developmental morphologists had all encouraged and to which they had even contributed. So long as developments in organic chemistry seemed to require special organic forces, *Lebenskräfte*, the old conceptual foundations seemed to offer a consistent, progressive basis for research. All of this began to change in the 1840s, forcing a re-evaluation of the conceptual foundations of the system.

ORDER, ARRANGEMENT, AND THE CATALYTIC FORCE: JUSTUS VON LIEBIG'S TREATMENT OF THE LEBENSKRAFT

The writings of the vital materialists in the period between 1790 and 1830 are all characterized by a persistent ambiguity in the meaning of the term *Lebenskraft*. While Blumenbach was most cautious in attributing to it a primarily regulative significance, others were not always so careful. Reil's ambivalence was typical. On the one hand, he treated the *Lebenskraft* as the *effect* of the peculiar order and arrangement of organic and inorganic elements in animal substances. For him 'form', conceived as the order among constituent elements, was the source of function. When he used it in this sense, *Lebenskraft* was not taken to be the constitutive cause of this order. It was a term expressing a causal complex incapable of further analysis.

A similar approach to the relationship of form and function had been taken by Johannes Müller in his theory of specific sense energies. Müller assumed that the special effect of which a particular organ was capable was not the result of a vital force residing in it, but rather the effect of the organization of the constituents of the organ. The specific physiological effect of the functional, whole organ was to be interpreted completely in terms of the normal physico-chemical laws governing the action of its parts. This approach clearly incorporated both teleological and mechanical modes of explanation: It was mechanical in that the specific functioning of the organ was to be explained not as the result of a vital force but in terms of the forces of physics and organic chemistry; it was teleological in the sense that, according to Müller, the same sorts of physico-chemical causation that account for the functioning of the organ are not capable of explaining the source of its organization.

This treatment was consistent with Kant's prescriptions for the purposive use of teleological judgement. Kant had emphasized that ultimate biological explanations require a special form of causality in which each part is so ordered with respect to others that it is both cause and effect at the same time. Since parts related to a whole in this way transcend the legitimate use of the category of causality, the order and arrangement of parts constituting a whole had to be assumed as given and had to be taken directly from experience. How successful the biologists could be in discovering the most primitive unity of order and arrangement establishing interconnections between various forms in organic nature was then a matter of constructing the most powerful empirical method of observation. But Reil and Müller were not always consistent in preserving this methodological sense of the term *Lebenskraft*. They sometimes treated it as a cause of organization.[4]

This constitutive usage was even more strongly pronounced in the work of Kielmeyer, Hildebrandt and E.H. Weber, who all attributed to the *Lebenskraft* the power of generating certain elements essential to the animal body not present in inorganic nature. While he did not attribute to it a creative power, von Baer's treatment of the *Lebenskraft* was also ambivalent. On the one hand, he regarded it as expressed in a certain order and relationship among materially related parts. This usage was especially visible in his embryology, but it was also present in his early notion of a material

Grundstoff. As we have seen, this special organic material was characterized by the order among its parts. On the other hand, in his discussion of the mammalian ovum von Baer on occasion employed a notion of Lebenskraft that was constitutive and directive. This was especially evident in his reference to a *Gestaltungskraft* directing the formation of the ovum from its center of activity in the *Fluida nativa*. In such contexts von Baer employed a special organizing force responsible for actively arranging physico-chemical forces into a functional animal.

In the 1830s these amibiguities in the conception of the *Lebenskraft* became a source of concern, a primary result of advances in organic and physiological chemistry. A principal advance in establishing the meaning and limits of the notion of *Lebenskraft* was made by Berzelius and Liebig. Their work reinstated the importance of conceiving *Lebenskraft* as the expression of a complex interrelation of material parts incapable of further analysis but inseparable from the order and arrangement of matter. *Lebenskraft* was to be understood as the expression of this state of affairs rather than its sustaining cause, and the object of physiology was understood to consist in the investigation of the lawlike effects of this state of organization.

Although Berzelius had developed similar ideas earlier, he laid special emphasis upon the way in which the *Lebenskraft* was properly to be conceived in his *Jahresbericht über die Fortschritte der physischen Chemie* in 1836.[5] Berzelius noted that until recently it had only been possible to conceive chemical reactions as resulting from the separation and union of different substances in terms of stronger or weaker affinities, which were ultimately electrical in nature. The investigation of animal substances had proved resistant to analyses based on simple chemical affinity, however. In terms of such models it was impossible to understand, for example, how in an organic body a single fluid could serve as the raw material for a variety of substances, how for instance in a continuous network of vessels, blood could be taken up and successfully broken down into milk, gall, urine, etc., without the apparent addition of any other fluids. "It was clear from this that something was transpiring here for which the processes of inorganic nature had not yet given a clue".[6]

The long sought clue to these reactions was revealed through several discoveries made by Thenard, Edmund and Humphrey Davy, Döbereine, Dulong and most importantly the Göttingen chemist Mitscherlich. Through their work it was established that some substances have the capacity for initiating a chemical reaction among other substances without entering into the reaction as a constituent of the resulting product. These substances were designated 'catalysts', and Berzelius argued that in virtue of the special capacity with which they are endowed, a capacity so completely different from the normal properties of chemical reaction, they should be regarded as manifesting a special *catalytic force*:

> If I call this a new force, I do not thereby wish to be interpreted as though it were my view that this force is to be explained as a capacity independent of the electrochemical relationships of matter; on the contrary I can only imagine that it is a special sort of expression of those relationships. As long as the mode of interconnection of these electrochemical relations remains hidden from us, however, we can treat this as a special force in itself, just as it lightens our task when we designate it by a single name. I will therefore ... call it the 'catalytic force' of bodies, and the reaction initiated by such bodies I call 'catalysis'...[7]

Berzelius went on to note that this new force obviously had very important consequences for physiology and animal chemistry. For it could serve as a model for interpreting how differently constituted fluids can be precipitated from a common ground substance. Each organ has a specific constitution, a special arrangement of parts related to one another by means of electrochemical forces. Acting in unison as a single complex, the organ could be viewed as analogous to a substance capable of exercising a catalytic force.[8] In contact with the 'ground substance', the organ, through its catalytic force, could re-arrange the components of that substance, thereby initiating a chemical reaction without itself entering into the reaction as a constituent.

Berzelius' discussion of the catalytic force had obvious parallels to the conception of the *Bildungstrieb* in the works of Blumenbach and Reil. It was regarded as a force whose effects could be investigated without having to inquire further into its origins. Also of special significance was Berzelius' conception of this force as really designating a special mode of order among electrochemical forces. What resulted was a sort of emergent or second order 'force' which always acts as a unit. This idea was, of course, closely related to another notion central to Berzelius' organic chemistry, namely the notion of special organic atoms or isomers. These 'atoms' were not simply substances incapable of further reduction, but in organic reactions they acted as units. They were distinguished by a specific structural arrangement and order among inorganic elements. This specific order and arrangement was responsible for their properties.

A feature of Berzelius' catalytic force significant for later developments was the fact that it worked by contact. This provided significant support for those who wanted to argue that the development of the embryo is initiated by contact between the germinal material and the sperm rather than chemical mixture of the two substances. Thus each stage of the generative process could be regarded as dependent on a previous state of organization.

The approach to questions of organization and the conception of vital force developed by Berzelius was amplified by Justus von Liebig in his work, *Die organische Chemie in ihrer Anwendung auf Physiologie und Pathologie*. Liebig was opposed to the use of specific vital forces in physiology. Instead of explaining muscle action as the result of chemical exchanges under the guidance of the force of irritability as Kielmeyer had advocated, for instance, Liebig proposed that muscle action be explained completely in terms of chemical and physical forces. Similar concerns motivated his controversial stand on fermentation. Whereas others considered alcoholic fermentation to be the result of a vital force in yeast cells, Liebig wished to explicate the phenomenon completely in terms of chemical processes. Even if fermentation were to be demonstrated as inseparably connected to the life process of yeast cells, that process was to be understood in terms of chemical exchanges.[9] The same issue motivated Liebig's debate with Berzelius over the catalytic force. While Liebig was very receptive to Berzelius' emphasis on order and arrangement in describing the origin of the catalytic force, there was a danger in Berzelius' approach of re-introducing vital forces into organic chemistry and giving up the hard-won advances made by organic chemistry since the first decade of the nineteenth century.[10] Specifically, Liebig wanted to avoid the possibility that catalysts and enzymes might work somehow inexplicably through the exercise of the 'catalytic force'. He wanted to insure that the notion of order and arrangement emphasized so effectively by Berzelius in defining the source of the catalytic force was carried over consistently into explaining its mode of action. It would not do simply to say that the catalytic force 'catalyzes' a specific reaction, but rather Liebig demanded that, since the catalytic force resided in the order and arrangement of the atoms in the catalyst, its effect must be manifested in the transmission of specific ordered *movements* to the substance with which it was in contact, resulting in a specific re-arrangement of that substance. The focus was to be upon the mechanism of the transmission of motion. In fact this was characteristic of the reform Liebig wanted to make throughout physiology: wherever possible, to replace the action of vital forces with a continuous series of material exchanges. In a very real sense, Liebig's goal was parallel to that of the Cartesian physiologists of the seventeenth century who had sought to replace Galenic 'faculties' with mechanical descriptions.

Liebig was an outspoken proponent of the *Lebenskraft* in physiology, but he was very careful not to attribute to it either a special constitutive, active, or directive role:

> In the ovum, in the seed of plants, we recognize a remarkable activity, a cause of the increase in mass, of the replacement of spent material, a force in the condition of rest. By means of external conditions, through fertilization, through the presence of air and moisture, the condition of static equilibrium of this force is removed. In going over into motion it expresses itself in a series of structures, which are quite different from geometrical forms of the sort we find in crystalizing minerals even though they are sometimes enclosed by straight lines. This force is called *Lebenskraft*.[11]

Liebig acknowledged that in plants and animals a completely different sort of causal relation is obtained from that found in inorganic processes such as crystalization. He therefore imagined that at the basis of organic processes a special force is present. But this force was incapable of initiating anything on its own. The expression of which it was capable could only be generated in conjunction with the interaction of external factors. The formulation of *Lebenskraft* in the passage above hints at the notion of a sort of mechanical potential that can only be set in motion by external causes. These external factors act as constraints on the motion the potential is capable of generating. In fact, as Liebig conceived it, the physical processes of material exchange, particularly those connected with oxidation, are the necessary conditions for the manifestation of the potential locked up in the organization of the germ.

> Respiration is the falling weight, the coiled spring, which keeps the clockwork in motion.[12]

> Whatever the role of electrical or magnetic disturbances in the functioning of the organs, the final cause of all these activities is a process of material exchange which can be expressed as the conversion of components of food into oxygen compounds taking place in a definite period of time.[13]

Implicit in this line of reasoning, was a formulation of the *Lebenskraft* as an ordered or pre-arranged set of motions, a mechanical potential in the germ capable of being activated by oxidation. Linking this idea with the line of thought developed by Berzelius, this potential must be thought as related to the arrangement of parts in the germ. In a later passage Liebig illustrated this notion of the *Lebenskraft* through an analogy to the voltaic cell: By closing the circuit the particular sort of motion potentially present in the arrangement of the parts of the apparatus is actually manifested.[14]

Liebig attempted to incorporate two essential features in his conception of the *Lebenskraft*. On the one hand he regarded vital phenomena as the expression of a special sort of force unlike other physical forces; they could not be reduced pure and simple to electricity, magnetism, or chemical interaction. On this point he was quite explicit:

The cause of the *Lebenskraft* and the phenomena of life is neither chemical force, electricity, nor magnetism. It is a force that possesses the most universal characteristics of all causes of motion, form and structural change of matter, and yet it is a different force because it is associated with manifestations that none of the other forces carry in and of themselves.[15]

Nevertheless Liebig did not want to go so far as to argue that any mode of understanding organic phenomena was possible other than that provided by chemistry and physics. The problem was to unite the best features of the vitalist and materialist frameworks. His strategy for resolving the apparent contradiction between those two viewpoints was to treat the *Lebenskraft* as a form of motion which is manifested as a group of interconnected chemical, electrical and magnetic phenomena. These different forces were assumed to be particular manifestations of this general form of motion, and they were called forth by factors conditioning and limiting its expression in particular contexts. Without further qualification, of course, this notion threatened to take on a metaphysical status incapable of further empirical analysis. In order to guard against this, Liebig emphasized that the *Lebenskraft* was to be conceived solely in terms of the order and arrangement of natural forces, and that its only mode of appearance was through the material interconnections of those forces.[16] This implied that the *Lebenskraft* had to be analyzed within the same conceptual framework as all other forces, namely in terms of the general principles of motion.

Liebig turned to mechanics for assistance in constructing a generalized treatment of motion. He observed that the relationship of force to matter can be expressed in two different ways, either as a function of distance or as a function of time. To illustrate this idea he noted that at every moment a force of attraction exists between a stone and the earth. This force is incapable of being manifested, however, when a table is placed under the stone, but, he asserted, the force is still present nonetheless. "The measure of this force is what we call the weight of the stone".[17] If the table is removed the stone falls, of course, but in falling it attains another type of force by virtue of its motion which was not formerly present. This is evident through the impact it makes when allowed to fall for some interval of time. Liebig argued that this force, which he called the *Kraftmoment*, must be defined in terms of the effect it is capable of generating, and this he said is measured by the product of mass times velocity [Geschwindigkeit] (mv).[18] The second way of discussing the relationship of force to matter is to treat it as a function of distance. "Thus a man who raises a thirty pound stone two hundred feet has expended twice as much *Kraft* as a man who lifts a thirty pound stone one hundred feet". Liebig called force treated as a function of the distance the *Bewegungsmoment* and he defined it as the force to be overcome multiplied by the distance.[19]

Liebig was clearly laboring with an impoverished knowledge of mechanics in these sections of his *Organische Chemie*, but the application he wanted to make of these concepts within the context of physiology was profoundly original, and as we shall see, it laid the groundwork for a rigorous reformation of the vital materialist program by Lotze, Bergmann and Leuckart. For what Liebig suggested in the pages following these definitions is truly astounding:

> In mechanics *Kraftmomente* and *Bewegungsmomente* are, therefore, expressions or measures for the effects of forces which are related either to the velocity attained in a given period of time or to a given distance; in this general form they are capable of being applied to the effects of all other causes of change in motion, shape or composition ...[20]

In this general form Liebig was suggesting that all forces in the organic body – chemical, magnetic and electrical forces alike – be expressed in terms of momentum and work. If we correct his notion for the measure of impact by replacing it with *vis viva*, we see that in these passages Liebig was struggling to formulate a conception of the *Lebenskraft* in terms similar to the framework that Helmholtz would later provide with the conservation of energy.

That Liebig had something of this sort in mind is clear not only from his insistence that all forces be expressed in terms of the same mechanical measure, but even more importantly from the manner in which he attempted to relate the two different modes of conceptualizing force to one another. He argued first that the total quantity of work, or the *Bewegungsmoment*, capable of being generated is dependent upon the order and arrangement of the parts of the system. Next he argued that the *Bewegungsmoment* or potential to do work "present in a living part of the body is capable of being used in order to impart motion to other matter at rest".[21] Once activated by an external agent, Liebig conceived this potential as being manifested in a definite quantity of momentum, *Kraftsmoment*, and this quantity of motion could then be used to activate other processes of material exchange:

> The motion of matter obtained through any sort of cause cannot in itself be destroyed; it may, to be sure, no longer be perceptible, but even if it is hindered by some opposition, its effect cannot be destroyed.[22]

Once a given potential for motion had been released, that quantity of motion could be preserved in another latent form. This latent form itself was conceived by Liebig to lie in a different arrangement of material parts with respect to one another:

Just as the expression of the chemical force (the *Kraftsmoment* of a chemical compound) appears as dependent on a particular order of contact among the elementary particles, so experience shows that the phenomena of life are inseparable from matter, that the manifestation of the *Lebenskraft* in a living part of the body is conditioned by a particular form of the vehicle [Träger] and through the particular manner in which its elementary parts are arranged; if we change the form, composition, or order, all the phenomena of life disappear.

Nothing hinders us from considering the *Lebenskraft* as a special quality which is attributed to certain materials, and is perceptible when their elementary particles are brought together in a particular form.[23]

The preceding passage reveals Liebig's intention of reformulating the *Lebenskraft* in a manner consistent with the notion of force in physics. According to this notion, *Lebenskraft* is a special effect of the relationship of natural forces and *not* a hyperphysical directive agent superimposed on them. Rather than attempting to explain how these forces became so arranged, Liebig preferred to examine the effect and the means employed for maintaining it. He designated these effects by a single term, *Lebenskraft*. But these effects could only be manifested through phenomena of material exchange, *Stoffwechsel*, which, he argued, is first initiated and is always accompanied by the binding of oxygen. The process of oxidation trips the lever, so to speak, that gives rise to a systematically interconnected set of related chemical, electrical and ultimately morphological phenomena, and in his view, these phenomena are tightly interrelated through laws capable of being expressed in a common mathematical language. For Liebig the *Lebenskraft* could not be conceived as existing independently of this set of quantitative interrelations:

Material exchange, the manifestation of mechanical power and the assimilation of oxygen, stand so closely related to one another in the animal body, that one may assume that the quantity of motion, the amount of organic material converted [des umgesetzten, belebten Stoffes], stands in a definite relationship to the amount of oxygen assimilated and used by the animal body in a given interval of time. For a determinate quantity of motion, for a particular portion of *Lebenskraft* expended as mechanical force, an equivalent of chemical force attains expression [gelangt ein Aequivalent von chemischer Kraft zur Äusserung]. The quantity of oxygen that has been assimilated by the organ is equivalent to the quantity of *Lebenskraft* lost, and in the same measure an equal portion of the matter is expelled from the organ in the form of an oxygen compound.[24]

For Liebig the *Lebenskraft* was a kind of potential energy connected with the organization and arrangement of material parts but capable of assuming certain concrete material forms of expression in the structure of the organism itself. Here Liebig imagined that the quantity of motion released by the process of material exchange had not been lost, but merely transformed into a latent state. It is consistent with his formulation, moreover, that in its new latent state the quantity of motion might exist in a form no longer useful for the functioning of the organism.

However original these ideas may have been with Liebig,[25] there is no doubt that with his sponsorship they received immediate resonance among physiologists. This is particularly evident in the writings of Liebig's colleagues in Giessen, Theodor Ludwig Wilhelm Bischoff and Rudolph Leuckart. Both of these men attempted, for instance, to use Liebig's ideas in support of the contact theory of fertilization; Bischoff in an article for *Müller's Archiv* in 1847 and Leuckart in his extremely important article, 'Zeugung' for Rudolph Wagner's *Handwörterbuch der Physiologie*. As previously with Kant, now with Liebig "ging die Morgenröte auf".

Bischoff's account of Liebig's ideas is extremely illuminating for the historical transformation of vital materialism we are now considering, for he laid emphasis on two aspects of Liebig's theory; first on the notion of catalytic forces and secondly on the importance of understanding all physiological processes in terms of a general theory of motion. Liebig's ideas, wrote Bischoff, permit an explanation of the riddle of generation:

Such an explanation I now believe can be found in the deep, and in my view, infinitely important ideas which Liebig has developed concerning chemical action through mere contact, about which I have had the pleasure of discussing and being instructed by my friend and honored colleague.[26]

It was the genius of Liebig, according to Bischoff, to have shown how catalytic agents could provide "the internal level for the transfer of motion" within the animal body.[27] The most significant aspect of Liebig's ideas on catalytic forces according to Bischoff was the notion that the remarkable effect of which these agents are capable resides ultimately in a particular order and arrangement, in the form of the elementary parts. It is in virtue of this arrangement that they are capable of converting one type of motion or *Bewegungsmoment* into another. They are in a sense the chemical cogwheels of the animal machine.[28]

Bischoff explained that Liebig's concept of the action of catalytic force was closely connected to a general theory of motion. Although he had himself given an account of Liebig's theory in the *Jahresbericht* on the progress in organic chemistry for the year 1843, Bischoff felt that the main idea behind Liebig's theory was still not well understood:

I must call attention to a second idea for which Liebig is responsible, one which will certainly be the subject of more research in the future, although it has attracted little attention up to now. I refer to the point of view in terms of which the phenomena of motion in organic bodies and elementary particles is to be considered. I scarcely need mention that for many persons today it suffices to attribute these motions to a particular motive force or irritability which belongs to these moving parts, or that they are the direct expression of the *Lebenskraft* acting in a particular direction. When I first commented on Liebig's views, ... I congratulated physiology upon finally opening the perspective that would free it from its previous mystical schema. Taking our lead from analogous phenomena in inorganic nature, we must consider these effects in the organic context as effects of material exchanges to which variations in the manifestation of force are also connected. The motion thus initiated is nothing other than the effect or the measure of material variation in form and constitution; and wherever we see motion begin, we must assume a displacement of matter.[29]

Building upon these general principles of Liebig, Bischoff constructed a theory of the process of fertilization. A material mixture of egg and sperm need not be assumed, he argued. The egg must be considered an organized whole. The developmental potential of the organized germ materials is activated by contact with the sperm, which sets the developmental process in motion through its 'catalytic force'.[30]

While Bischoff's assessment of the widespread lack of appreciation and immediate reception of Liebig's ideas was well founded, there were others who were sensitive to the issues raised by Liebig and who advocated a similar approach to the interpretation of vital phenomena. Foremost among them was Hermann Lotze, who in 1842, the same year in which Liebig's *Organische Chemie* appeared, published an essay entitled *Lebenskraft* for the first volume of Rudolph Wagner's *Handwörterbuch der Physiologie*.

THE LIMITS OF TELEOLOGICAL THINKING IN BIOLOGY: HERMANN LOTZE'S CRITIQUE OF THE LEBENSKRAFT

It was not Lotze's intention to reject the Kantian foundations of the vital materialist tradition. On the contrary it is clear that he intended to correct a fundamental error that had crept into the conceptual foundations of the program. The main line of Lotze's attack against the vital materialists, among whom he singled out Reil, Kielmeyer, Humboldt, and Johannes Müller for special consideration, was that they had mistakenly introduced a false notion of 'force' into their conception of *Lebenskraft*. He did not deny that biological organization must be approached within a teleological-mechanical framework. On the contrary he argued that both the teleological and the mechanical frameworks must be employed together even in the investigation of inorganic phenomena. Rather than asserting that recourse to teleological principles must be taken when mechanical modes of causation fail, he argued that:

Both principles are equally general, and they are to be employed equally for every object of investigation. Not just life but even every inorganic occurrence must be investigated in terms of the notion of a hidden purposive organization. Not just the final cause of a phenomenon but equally the proximate cause demands a continuous, at no point interrupted instrumentation of causes in order that the purpose can be realized.[31]

What Lotze wanted to emphasize is that the telological framework thus conceived does not offer an explanation, rather it serves as a regulative principle of inquiry into the interrelated sets of conditions and means in terms of which an explanation can be given. This point, it should be recalled, had been emphasized by Blumenbach, but the next generation of physiologists he inspired did not always manage to observe it.

Accordingly we can never demand from 'purposive organization' the same thing that we expect from a causal explanation. 'Purposive organization' can never be the basis of anything real, rather it can only be a *command* which requires a certain formal order in the composition of the real so that out of the causal interrelations of those instrumental means that actual content emerges as a later result.[32]

According to this point of view the concept of purposive organization as a regulative principle serves only to promote the investigation of all the means, which, when properly ordered and operating according to laws of mechanical necessity, have as a predetermined consequence the formation and functioning of an organized body. It is important to emphasize in addition, however, that the starting point for the investigation into the instruments of functional organization is the organized body itself taken from experience as given. It is this fundamental datum which renders the framework teleological:

Purposive organization obtains thereby power over the course of events only in that it is already present as a germ [Keim] in the disposition of causes; without being supported in this way, however, it is in no way to be regarded as an independent reality capable of driving the causes to their realization, nor is it capable of modifying them in accordance with its own contents.[33]

The teleomechanists from Reil to von Baer and Johannes Müller had made the mistake of treating forces as if they were objects of experience. Forces, Lotze emphasized, are not objects of experience; they are supplements of reason. "Things do not act because they have forces, rather they appear to have forces whenever they effect something". The mistaken notion that forces are causes leads immediately to the further mistaken attempt to identify them with some particular material substrate or even to treat them as independent beings existing in and for themselves just like ordinary material objects. Were it the case, Lotze observed, that the fundamental forces of nature were somehow independent agents, they would necessarily call forth a *perpetuum mobile* of interrelated motion. "Organic bodies are not such free mechanisms in nature, however; they are more similar to artificial machines in that their motions require continuous replenishment and renewed impetus".[34] Forces are not independent agents nor can they be transmitted from one body to another as Johannes Müller had argued in his theory of generation. Thus, according to this view, the only thing that can be transmitted from one body to another is a change in a pre-existing state of motion or rest.[35]

When applied consistently this notion implied that two types of forces must be present within organic bodies. Every part must exercise the forces proper to it as an individual; but each part must also enter into dynamic relations. Each of the gears in a clock, for example, must have its own mechanical properties, but the effects which each is capable of generating are only possible so long as it is an integrated part of the entire clockwork:

Thus all parts of the animal body in addition to the properties which they possess by virtue of their material composition also have *vital* properties; that is, mechanical properties which are attributable to them only so long as they are in connection with the other parts. ... Life belongs to the whole but it is in the strictest sense a combination of inorganic processes.[36]

Here Lotze emphasized that once the term '*Kraft*' is understood in its proper sense, namely as the measure of change of a state of motion or rest, then the term '*Lebenskraft*' must be stripped of any implication as a cause of organization. This does not mean that there are not vital properties, but only that they must be interpreted as partial effects resulting from an integration of mechanical processes into a systematic whole. On the other hand, Lotze refrained from claiming that the whole was somehow the sum of its parts:

the Pantheist error is to be completely rejected; as though it were the case that organization could be the automatic product of accidentally arranged substances.[37]

The parts could be understood without reference to the functional integration of the whole; but the systematic organization of the whole was not to be understood in terms of a force independent of the parts, imparting organization to them. What this implied was the following extremely important definition:

Biological organization is, therefore, nothing other than a particular direction and combination of pure mechanical processes corresponding to a natural purpose. The study of organization can only consist therefore in the investigation of the particular ways in which nature combines those processes and how in contrast to artificial devices she unites a multiplicity of divergent series of phenomena into complex atomic events.[38]

This definition emphasized the central theme of Lotze's article: Biological organization is to be understood in terms of the *order* of mechanical processes associated with the phenomena of life. Life does not exist independently of the mechanical processes through which it is manifest. But equally important, life is not simply a mechanical process. It is inseparable from an order among mechanical prcesses. How mechanical processes came to be arranged in the manner exhibited by living beings can never be answered, for it lies beyond a mechanical conception of cause and can only be understood in terms of the anthropomorphic analogy of 'natural purpose'. Ultimately, however, the same teleological conception must regulate all inquiry into nature. Thus Lotze pointed out that *why* there are 56 elements cannot be explained; similarly *why* oxygen has the property of being sour, or why the physico-chemical processes initiated in the retina in the presence of particular stimuli give rise to the sensation 'blue', for example, simply cannot be explained. The experience of 'blue' is the result of an organized body, and that experience must both *precede* and guide the investigation of the mechanism through which it is generated. To attribute that organization to a special force disregards the necessary limits placed on mechanical explanations. In so doing a regulative principle of reason is hypostasized, and at the same time a contradictory use of one of the notions fundamental for experience, namely 'force', is introduced.

Although the question of the origin of biological organization lay beyond the limits of a possible scientific account, the problematic posed by this teleological-mechanical conception of life would have, according to Lotze, important consequences for future research in both·the physical and life sciences. The leading questions for the life sciences in accordance with the definition of biological organization given above must be that of determining how a mechanism must be constructed in order that it maintain itself in equilibrium in the midst of continually varying external conditions. "In order to solve this problem a mind of the magnitude of Laplace is required in order to determine the sum of mathematical possibilities. ... For this question demands as a solution the principles of material exchange Stoffwechsel in the most general sense . ... The general form of the difficulty is this: to establish the first phase of a regulatory motion at the moment when it first enters after some distsurbance has been introduced."[39]

Since we find on the basis of experience that the exchange of matter in animal bodies is utilized in order to regulate such disturbances, we are permitted to regard this as the midpoint of the organic machine, around which all other processes of the animal economy are to be attached.[40]

The first step toward resolving this question of the interrelation of material processes of exchange as a prelude to understanding the mechanisms employed by organized bodies was taken by Hermann von Helmholtz, the student of Johannes Müller thoroughly schooled in Kantian philosophy and the biological theories of the teleomechanists.

FUNCTIONAL MORPHOLOGY IN THE WORK OF BERGMANN AND LEUCKART

By the early to mid-1840s, as a result of the critical reflections of men such as Liebig, Lotze and Schwann, it was becoming increasingly clear that the teleomechanist program as originally conceived could no longer stand and that in particular the notion of the *Lebenskraft* as one of its guiding ideas would have to be abandoned. But the criticism waged by these men did not result in the demise of the program. It led rather to a critical re-examination of the approach needed to preserve the teleological framework, which was believed to be essential for understanding the phenomena of life, without assigning a special force as ths cause of organization. The clearest and most successful attempt at constructing a transformed vital materialism along the lines suggested by Lotze and Liebig was contained in the works of Carl Bergmann and Rudolph Leuckart, particularly in their monumental synthetic collaborative work entitled *Übersicht des Tierreichs* of 1852.

The conception of biological organization and its relationship to inorganic processes at the basis of Bergmann's and Leuckart's work is fundamentally that developed by Liebig and Lotze. The agreement between their points of view was fostered no doubt by the close personal contacts between the four men. Bergmann and Leuckart were student colleagues at Göttingen. Lotze was a member of the faculty there. Bergmann was afterward appointed to the medical faculty at Göttingen, where he continued his association with Lotze. Leuckart, on the other hand, went off to Giessen, where he became the colleague of Bischoff and most importantly, Liebig.[41] That Bergmann and Leuckart adopted Liebig's approach to organization is evident from their discussion of the proper way to interpret the special nature of vital phenomena:

The earlier notion that organic compounds must be understood simply from the quantitative relationships of their component elements had the unfortunate consequence that differences in the properties of chemical compounds that appear to be composed from the same quantities of the same elements were too readily ascribed to a *Lebenskraft* which is capable of producing different effects within the same material substrate; whereas in the case of identical chemical components the possibility is now near to hand that here just as in inorganic nature the cause of the properties of a substance may reside in the manner of grouping the atoms; and therefore a different grouping may result in an essentially different chemical nature.[42]

The discovery of catalytic agents and organic radicals, which seemed to owe their unusual effect to the structural arrangement of atoms rather than to any specific quantitative feature of the component elements, seemed to eliminate the role previously ascribed to the *Lebenskraft*.

Drawing upon advances in the understanding of organic compounds made during the 1840s, Bergmann and Leuckart, like Lotze, no longer saw a need to postulate a special force as the cause of organic affinities. Whereas vital materialists had asserted a kind of quantum leap between the binary compounds of the inorganic realm and the ternary and quaternary compounds of organic substances, Bergmann and Leuckart saw no reason why the process leading to the dissolution of quaternary compounds into binary affinities could not be reversible. Moreover, recent research, by Liebig in particular, had indicated that, contrary to the view of the vital materialists, not all living bodies are capable of synthesizing organic compounds from inorganic substances: "Animals do not possess this capability at all. They receive organic substances in their sources of nourishment and are only capable of transforming, assimilating or breaking down these substances. Plants are the true chemical laboratories in which ternary affinities are formed from carbon dioxide and water, and with the addition of ammonia (and probably nitrogen) substances structured in terms of quaternary affinities are formed".[43]

If a special force, a *Lebenskraft*, could not be attributed to the animal in virtue of which it forms the organic substances necessary for its structural organization, neither was there a necessity for assuming a special force which maintains the animal in its state of organization. All chemical bonds are maintained only so long as external conditions do not admit the action of stronger affinities. Organic bonds are no different in this regard. In fact, animal bodies depend upon the breakdown of organic substances, and the processes of material exchange – transformation, assimilation and breakdown – are continuous in the animal body. In normal conditions, however, limits are set to the degree to which these processes can proceed by various structures and mechanisms within the animal economy itself. Once these 'protective' or 'zweckmässig' interrelations are interferred with, the physico-chemical processes, which are always taking place in the organism and upon which its existence depends, continue beyond the limits imposed upon them. The object of physiology is to investigate the limits and conditions, both internal and external, governing the processes of material exchange that define the life of the animal body. To attribute the maintenance of the *zweckmässig* state of organization to a *Lebenskraft* effectively puts an end to the investigation of the structural and systemic arrangements that condition and limit the operation of these chemical and physical agents, and this, they argued, was contrary to the spirit of scientific inquiry. But the concept of the *Lebenskraft* as the cause of the maintenance of organization could still be used in a metaphorical sense, they argued, so long as one considered the object of physiology to be the investigation of the natural laws in terms of which this effect is achieved through the structure of matter and the systemic relations governing material exchanges. This was in the spirit of Blumenbach's and Kant's original formulation of the teleomechanist program. What Bergmann and Leuckart objected to was the attempt to use the notion of a *Lebenskraft* as a means for drawing an arbitrary limit to the domain in which a mechanical explanation can be pursued:

We assume, therefore, that within plant life conditions are offered which permit the affinities between carbon dioxide, oxygen, and nitrogen to come together to form organic compounds. In plants and animals organic matter is everywhere chemically transformed in regular patterns, because the conditions for its breakdown and transformation are provided in the proper sequential order. At the same time this process is a requirement of the plan of the organism within which the organic matter is formed and for which it is purposeful [zweckmässig].[44]

This statement emphasized that although explanations of structure and function in biology were to be mechanical in nature, the necessity of exploring teleological explanations was not thereby eliminated. The plan of organization, the functional whole in traditional teleomechanist terminology, not only depends upon those mechanical processes for its existence, but equally important it provides the set of conditions which permit chemical and physical processes to run a determinate course. The problem with contributions to the tradition by men such as Kielmeyer and Johannes Müller was not that they had invoked a teleological framework of explanation in addition to the mechanical, for physiology required both; rather they had misconceived the way to integrate these two types of explanation.

According to Bergmann and Leuckart, the mechanical and the teleological frameworks are two separate perspectives from which to analyze an organized body. The mechanical framework seeks to unravel the complex of effective causes from which an organized body emerges. The teleological framework poses the question of the purpose or 'zweck' of that structure and its particular mode of organization. When taken together as mutually supportive both are equally valid. Both must be employed to understand the organic body fully. Neither by itself provides a sufficient explanation of form and function. To illustrate these points, Bergmann and Leuckart supposed, as a hypothetical example, that a zoologist familiar with vertebrates but having never seen a bird is suddenly confronted with the skeleton of one. From the similarity of the skeletal structure and particularly from the similarity between the system of bones in the wings and that of the extremeties of the mammals, "he would conclude that the causal complex that produced this animal must be quite similar to those which produced the mammals and reptiles. ... A completely independent problem, however, would be posed by the determination of the purpose of this structure, which might be discovered if one had a wing with a feather before him or the musculature of the wing".[45]

The problem pinpointed by Bergmann and Leuckart runs deeper than the fact that the complex of effective causes from which the organism emerges is so complicated that recourse must be taken to a teleological framework as an interpretive guide. Like Kant they argued that the reason why a mechanical framework alone is incapable of providing a foundation for the life sciences is that the analysis of causal relations in the organic realm must proceed differently from those treated in the inorganic realm. A teleological framework would be unnecessary, "if in the organic realm the relation of cause to effect were always such that the effect [Wirkung] of one thing always resulted in the effective action [Bewirkung], of another; and vice versa that the effective action of the one were capable of being reduced to a simple action of another, just as an impact is transmitted through a linear series of equal spheres. The simplicity of this relationship does not exist in the organic realm however".[46] In the organic realm everything is so intertwined that, while resulting from certain conditions, each thing is itself in turn the condition of something else. "So that the question of the purpose [Zweck] of one thing and the effective cause of another frequently fall together; ... as is the case in the question of the purpose of the heart and certain structures of the vascular system and the question of the effective action of the circular system".[47]

One of the most important features of their work is the emphasis on the rewards to be gained by pursuing the search for the 'complex of effective causes' on the one hand in tandem with pursuing the demands of functional organization on the other. The functional organization of the animal is *conditioned*, in their view, by its material constitution. Thus they accepted as axiomatic that biochemical analysis would ultimately provide a clue to understanding organization:

If so many able researchers did not balk at the notion that the differences of the germinal materials of animal species must be the cause of the difference between these species themselves, ... then biology would certainly possess a richer fund of material for comparative oology than is now the case.

On this point we must admit our ignorance. Only for the major groups of animals are we able to offer even a few characteristics of the eggs, but the points in which eggs of animals of the same family or species agree and disagree with one another has not up to now been the object of any extensive research. We do not doubt, however, that herein is to be recognized one of the important problems for chemistry and microscopical anatomy. But it will probably require much more time, and completely new methods of research will have to be invented before anything satisfactory can be achieved in this area.[48]

While Bergmann and Leuckart argued that the effective cause of animal form was to be sought in its material constitution, they did not view animal form to be completely *determined* by its constitutive material alone. The genetic material of the organism provides, in their view, the condition for the expression of form. Form itself, however, is determined by a broad complex of effective causes, a tightly interconnected set of external constraints placed on this material potential. Access to this broader causal complex is only made possible in their view, through the use of functional analysis. Bergmann and Leuckart attempted to reconstruct von Baer's developmental morphology by placing it on strictly functionalist foundations.

An excellent illustration of how the mechanical and teleological frameworks were conceived by Bergmann and Leuckart to be mutually supportive in attacking a specific problem is provided by their examination of general plans of animal organization and their relationship to weight distribution. They began with the assumption that the primary determinant of organizational plan is the "physiological relation of the animal to its environment".[49] In this context locomotion is not only the most important vital function but it also involves the solution of the most difficult problems concerning the plan of structure and organization. "No other end [Zweck] involves such a large number of organs and to no other function is such a large proportion of the mass of the body devoted".[50] In order to overcome its own inertia and the resistance of the surrounding medium the animal body must not only have an appropriate system of locomotive organs and a body as compact as possible; it must also have the most advantageous distribution of the individual organs in the mass of the body. "Any other distribution would result in a useless expenditure of energy rendering motion more difficult or even impossible".[51] An extremely important physical principle dictates the organization of the animal body: the principle of least waste of energy [das Prinzip der grössten Kraftersparniss]. This principle, the basis of an organic physics, determines through mechanical necessity the structure and arrangement of certain organs.[52]

The principle, accordingly to Bergmann and Leuckart, explains why admidst the myriad of animal shapes three fundamental shapes – the sphere, disc and cylinder – constantly occur. In illustration of this claim they assume a hypothetical aquatic animal, spherical in shape, homogeneous, and having the same specific gravity as water. For such a body the center of gravity and center of rotation would be identical and occupy the midpoint of the central axis. This animal would be capable of moving in all directions with equal ease. Now alter these ideal conditions to correspond to those affecting real animals, whose masses are not homogeneous and whose specific gravities differ from the surrounding medium. The arrangement of organs most advantageous for purposes of mobility is that one in which the center of gravity and the center of rotation coincide as nearly as possible in the middle of the central axis, for this arrangement would insure equilibrium. There are such organisms in nature Bergmann and Leuckart point out – the sea urchins.

Corresponding to this equal distribution of weight, there must be an appropriate grouping and arrangement of the internal and external organs. "To this end [Zweck] it is probably simplest to divide the various structures for performing specific tasks, the sizes of which are always determined by the needs of the organism, into a number of equal pieces and group them at equal distances around the midpoint of the organism".[53] Of course one organ resists such a division, namely the intestinal tract. Accordingly it must be placed along the central axis, the poles of which now become the openings of the animal:

Around this axis lie multiples of the remaining organs in accordance with the requirement of weight distribution. Each organ is divided into a number of pieces fixed at equal distances from each other. The right and left halves of the body as well as the upper and lower halves, in short any arbitrary section of the central axis exhibits an equal weight. In such an arrangement we encounter symmetry from all sides, the so-called radiate structure.[54]

Thus the presence of symmetries in organic forms need not be idealized as attempted by the Naturphilosophen; rather they are capable of an exposition in terms of mechanical principles operating within limits prescribed by functional necessity. This, in essence, was functional morphology.

Bergmann and Leuckart went on to show that similar considerations would establish the possibility of radiate structure of internal organs within cylindrical or disc-shaped bodies. But radiate organization has as a consequence the splitting up of the locomotive forces, and this is not without influence on the speed with which a radiate can move. This is the reason why radiates live primarily in circumstances in which the preservation of life is possible with little or no necessity for movement, namely in water. "For whatever reason, as soon as a land animal possesses a cylindrical body, it can no longer retain a radiate structure. The organs of locomotion must then expend a significantly more considerable amount of work [Leistung], and the whole intensity of the force generated must be utilized for a very specific mode and direction of motion: the body can no longer remain a radiate body".[55]

Bergmann and Leuckart next applied the same principles of analysis to cylindrically shaped organisms. In this case the center of gravity must lie beneath the center of rotation along the axis and an equal distribution of weight must obtain between the anterior and posterior halves. In the case of land animals, physiological considerations render the multiplication of organs impossible as a means for establishing static equilibrium. In order to determine the most effective [zweckmässig] arrangement of different organs of unequal weight and shape, therefore, they observed that the expenditure of force required to move any weight is directly proportional to the weight and inversely proportional to the square of the distance from the center of rotation. If this physical law is coupled with the principle of least expenditure of force governing physiological organization, "then the heaviest organs must be as near the center of rotation as spatial and physiological considerations permit. If those organs are cylindrical in shape, they will be along the central axis, if spherical, they tend toward the midpoint of the axis".[56] By further analysis of the mechanical necessities connected with the most advantageous distribution of weight in the vertebrate body, Bergmann and Leuckart attempted to explain all the various symmetries connected with animal forms.

Although Bergmann and Leuckart's considerations of the determinants of the plan of organization depended essentially upon the application of mechanical principles, these principles were applied within a fundamentally teleological framework. The position developed in their work is that however successful a mechanical-causal analysis which attempts to reveal the physico-chemical conditions behind the generation of form, it can never succeed in providing a full understanding of organization. In order to understand why various structures, capable of performing determinant functions, get organized as they do, the physiologist must take as his starting point the animal that is to be, the end or *Zweck* of organization itself. The explanation of how such a functional whole gets assembled and why it is organized in one way rather than another always proceeds by way of careful application of physical laws, but in a very real sense the whole determines the organization of its parts. In so doing it never violates physical laws. On the contrary, its very existence depends on their most efficient possible use in given sets of circumstances. The end or *Zweck* determines organization, not by altering physical laws, as Keilmeyer had seemed to require, but by establishing the parameters of possible physical solutions. The 'functional whole', 'end' or *Zweck* determinative of biological organization was not to be conceived as due to an Idea existing somehow independently of material processes; nor was it to be regarded the effect of a special force, the *Lebenskraft*, which subordinates inorganic physical processes to a special rule. 'Functional wholes', in their view are in the strictest sense the expression of mechanical necessity. But as the result of the relationship of means to an end, they are necessities of a different order.

This application of mechanical principles as a guide to understanding functional organization is applied throughout the *Übersicht des Thierreiches*. Another instructive example is offered by the organization of the skeletal and muscular systems.[57] At first glance it may appear that the typical manner in which bones and muscles are attached makes a most inefficient use of the principle of the lever. The muscles that extend the arm, for example, are attached very close to the elbow whereas the law of the lever would seem to demand that they be attached as far from it as possible. Such an attachment would not be efficient within the animal machine, however: the material to be used as a lever in this case sets the conditions of the problem. In order to extend the arm by some other more 'efficient' attachment of the muscle, the muscle would have to contract a greater distance. The physiology of muscle is such, however, that it becomes weaker the more it has to constrict. In this case, then, by attaching the extensor muscle close to the elbow the inefficiency of using a 'short lever' is compensated by the advantage in power accruing to the muscle. Or conversely, the most advantageous expenditure of the mechanical energy capable of being delivered by the muscle requires the use of a short lever. Similar considerations help to explain why the part of a bone nearest a joint is always provided with an enlargement. If the muscle were to lie plane upon the two bones and were there no prominences at the joint, an enormous amount of energy would have to be expended in moving the extremity. The prominence provides better leverage. Of course from considerations of leverage, the best angle between muscle and bone is that in which muscle constriction is perpendicular to the bone; but once again this arrangement would require a greater muscle contraction than that associated with the normal attachment of bone and muscle. Given the materials and the conditions of their employment, the manner in which bone and muscle are attached as well as their positional arrangement makes the most efficient use of the lever principle within the animal economy.

At each ievel of their consideration of animal form, Bergmann and Leuckart focused on the problem of motion within the general framework provided by mechanics. Animals in their view are various ingenious and extremely refined solutions to general problems of motion. At the most fundamental level the theory of organization must concern itself with the analysis of processes of material exchange, *Stoffwechsel*, and as we have seen, Bergmann and Leuckart shared the view of Liebig and Lotze that this involved the most universal form of the laws of motion as such. *Stoffwechsel*, however, requires the assimilation of matter and the interconversion of forces of various types – chemical, electrical, and mechanical – and absolutely central to this conversion process is heat. Berzelius and Liebig both thought that what was so very special about catalytic agents and enzymes is that somewhow through the arrangement of their parts they could initiate the release of heat, which could set chemical reactions into motion.[58] The entire strategy of life is conditioned by the manner in which the process of material exchange takes place in the animal body; in particular whether – to use Liebig's phrase – the process of binding oxygen proceeds rapidly or slowly.

Adopting this point of view Bergmann and Leuckart speculated that the animal realm might be capable of being analyzed in terms of certain general plans of material exchange, that is certain purposive or *zweckmässig* combinations of material processes, which remain organized and only function within a definite range of temperatures. For each general plan there would be numerous potential developmental grades corresponding to different types of animals; or, to remain consistent with the dynamicist schema of teleomechanism, different forms of animals could be viewed as the stable solutions to the problem of fitting a plan of material exchange to the environment. An immediate consequence of this idea was that if an animal is to be based on a particular scheme of material exchange, it must also be outfitted with a corresponding set of locomotive and sensory organs which enable it to capture the appropriate food supply.

> While in the locomotive organs of the birds we encounter one of the highest efforts of nature to construct under very difficult circumstances a light and powerful mechanism by giving up all superfluous weight and by using numerous other aids, and while the multiplicity of structures designed for the most different purposes found in the mammals is a source of wonderment, we find in the reptiles a remarkable deterioration of the capacity for locomotion and the other structures connected with it. We can relate this to one of our earlier discussions through a very simple consideration. The preservation [Erhaltung] of animal species presupposes among other things a proper balance between needs and capacities [Bedürfnisse and Fähigkeiten]. The reptiles do not have the very great need universally present in the higher vertebrates, namely the need to maintain a regular internal temperature. This requirement is satisfied by means of a very energetic process of material exchange and it is reflected furthermore in the more highly developed digestive and respiratory organs of the mammals and birds. In the same manner we find this requirement expressed once again in the organs of locomotion, upon which the acquisition of food depends. If a reptile were to be given the food requirements of a mammal or bird, *which are deduced from the need for heat*, without also perfecting its organs of locomotion, it would necessarily go under.[59]

Of course the problem might also be understood the other way around in the manner of Cuvier. That is, dependence on a particular food supply and being outfitted with certain organs of locomotion could be considered, from a purely functionalist point of view, as laying down the requirements for an appropriate system of material exchange. For Bergmann and Leuckart, however, this put the cart before the horse. In their view the system of material exchange established the parameters for other functional and adaptive solutions.

For instance, the type of locomotive system an animal has depends, of course, essentially on the process of material exchange in the muscles, and this can in turn be a conditioning factor for numerous other aspects of its organization. Thus, in the birds flight demands not only a very finely tuned musculature and skeletal system, but it also implies that the reproductive system must be organized in such a way that the young develop outside of the mother's body and moreover that the eggs be expelled from the body as early as possible.[60] Heat requirements also enable the zoologist to understand some of the interesting features of animal behavior. Large birds, for instance, do not construct their nests in the same detailed and careful fashion as small birds. In fact Bergmann and Leuckart claimed that the attention to detail in nest building is directly related to the size of the animal.[61] The reason for this is that physical considerations relating to volume and surface area imply that the small eggs of the smaller birds lose much more heat to the environment than do the eggs of large birds. Thus while many large birds simply place their eggs in a depression in the earth, the wrens build well protected warm nests and the kinglet forms its nest from moss in the shape of a large sphere.

Thus a central theme running throughout Bergmann and Leuckart's *Übersicht des Thierreiches* is the notion that the organization of the animal body is conditioned by the generation of heat through processes of material exchange. On occasion this notion is advanced as the supreme regulative principle governing the production of forms within what Bergmann and Leuckart called the 'total causal complex'. Exploration of exactly such an idea was the subject of a paper Bergmann had written just prior to embarking upon the grand project of the *Übersicht*. It was entitled *Über die Verhältnisse der Warmeökonomie der Thiere zu ihre Grösse.*[62]

In this work Bergmann assumed as given that the generation of heat in animals rests upon physico-chemical processes of material exchange connected with respiration. The exact manner in which heat is generated and regulated in the body remained a problem for future research, but Bergmann assumed as axiomatic that a mechanical explanation for those phenomena would one day be given. In the absence of a complete explanation for heat production and regulation, however, it was still possible to examine the role of animal heat as a regulative principle in the organization and distribution of animal forms. In order to investigate this problem, certain simplifying assumptions had to be made. First he assumed that each unit mass of animal respiratory substance is capable of generating a certain maximum quantity of heat. Accordingly, in an idealized setting where all other differences between animals and even variations within the same animal at different times are ignored, total body volume could serve as a measure of heat production. Second he limited his investigation to homiotherms. From the point of view of understanding heat production as a regulative principle of form the warm blooded animals were of primary significance, for they have the ability to maintain a certain internal body temperature under changing external circumstances:

It is obvious that the organizational plan of the homiotherms is to be considered the higher. Through a subtle combination of organic activities nature has provided them with capacities which are lacking in other animals and with which much necessity is tightly bound.[63]

Of premier importance from a purely physical and thermodynamic point of view is the ratio of the surface area of the animal to its volume, for this ratio places conditions on the degree to which an animal can raise its temperature above that of the environment. Since volume increases as the cube while surface area increases with the square, and since the surface is the main source of heat loss the environment through conduction, it follows that simply by virtue of its greater mass a large animal must breathe less per unit volume than a small animal. Accordingly, birds respire more rapidly than mammals, they are also smaller, and they must consume greater quantities of food in proportion to their body weight.[64]

It is thus settled, that the larger animals are, the less heat they produce in proportion to their size in order to obtain a certain elevation of temperature over that of their surroundings.[65]

On first glance the law, in its general form, seemed to be obviously false, for in any given geographical location there are animals of various sizes living alongside one another. The law might seem to imply that in any given medium of temperature, animals of the same size ought to appear. What the law had provided, however, was an ideal case in which animal substance with the maximum capacity for heat generation was coupled with an ideal organic insulating system. If the law is true, therefore, it calls attention to the very important fact that particular systems of generating animal heat are functionally linked to specific organic insulating systems, and that both are expressed in the relative size of the particular animal form. The different size of animals living in the same region is thus a result of their different plans of organization.

The law, Bergmann noted, must therefore have some very important consequences for the organization of animals and their manner of life, and accordingly it can serve as a guiding principle in the investigation of organization. It calls attention to four closely related sets of variables. On the one side are surface area and volume; on the other are the organic means employed to insulate the body against heat loss, such as hair, feathers, fat, etc., and the particular organic structures that serve to generate heat, i.e. respiratory and digestive organs. If volume and surface area are given, then the two other factors must be determined accordingly; and conversely, the systems of heat production and insulation must determine the size of the animal body. As these two systems vary, so must animal size. If Bergmann's law was correct, it provided him with a heuristic guide to the analysis of warm blooded animals, for:

From the difference in size actually present in nature, we could then recognize according to the converse of the rule just how extensively nature is able to modify these two factors.[66]

Thus, while Bergmann's rule concerns directly the relative size of animals and their relation to the environment, it is indirectly a clue to a host of related organic systems and the degree to which these systems are capable of being modified.

Before the full impact of this approach could be evaluated, evidence had to be gathered which indicated that it was in fact a useful guideline. To provide this evidence was Bergmann's goal in the remainder of the *Wärmeökonomie*. In order to gain some insight into the validity of the law in the case of real animals, which are not perfect heat generators with ideal insulation, it was necessary to examine some specific animals which approximate the ideal state as nearly as possible. Bergmann therefore posed the question:

In applying these means, through which such a variety of creation is possible, has nature ever been exhausted? Is it possible that in any particular warmblooded animal nature has reached the largest or smallest form which the plan of the homiotherms was capable of achieving? Or has she ever approximated this extreme?[67]

Two points of methodological significance were apparent in Bergmann's paper. First it is clear throughout the work that Bergmann was constructing an ideal model from physical principles and using it as a guide to the analysis of actual forms. We have seen this same approach followed in the analysis of plans of organization in terms of the principle of least action in the *Übersicht*. Secondly, and more importantly, it is clear that with this approach Bergmann sought a means of analyzing the problem that had always been central to the work of von Baer and other developmental morphologists, namely of establishing the potential developmental grades of a particular plan of organization. Bergmann sought to resolve this problem in a manner which avoids the difficulties connected with the assumption of a *Lebenskraft*. At issue had always been the problem of accounting for the different degree of complexity of forms based on the same plan. The manner in which the developmental morphologists had solved this problem was to invoke a developmental force or *Lebenskraft* as a source of activity that strove to express itself as an ascending complexity of form. Bergmann, by contrast, had denied the *Lebenskraft* any active, constitutive role in zoology and replaced it by interrelated systems of material exchange capable of being expressed in several stable configurations. In a very important sense, *Wärmeökonomie* in Bergmann's system had replaced *Lebenskraft* as the regulative principle for understanding the animal realm.

As a first test case for the question of whether nature has produced any forms approximating the ideal conditions stated in the law, the whales seemed to be an obvious choice. According to Bergmann's rule the largest animal forms ought to be present in the coldest environments. The whales seem to satisfy the law quite well, for not only are they the largest mammals, but they live in an aqueous environment which is the best conductor of heat, and they also frequent the polar regions. Bergmann did not regard them as really suitable for examining the upper limits of his proposed law, however. Precisely because they live in water, almost no concern has to be taken by nature for supporting their large body. Water presses around it from all sides providing the support that muscles and bones would have to provide on land. Furthermore, living in water, this animal does not have to exert much effort in generating motion through muscular action. In spite of its apparent agreement with the law, therefore, the whale was not a good choice for examining the upper limits of its applicability, for several secondary factors intervene in such a manner as not to demand the 'exhaustion' of all the systems interrelated within the organizational plan of the homiotherms.

The hummingbirds, on the other hand, appeared, in Bergmann's view, to satisfy all the conditions for producing the smallest form possible within the homiothermal plan of organization. "The smallest of the homiotherms, the hummingbirds, come very close to these limits".[68]

If that is indeed the case, then we can infer that for the realization of the enormous number of much smaller animals that nature has produced, a portion of the perfection [Vollkommenheit] of the homiotherms has had to be sacrificed and that the organizational plan of the polikilotherms was necessary for their production.[69]

Several factors supported the claim that the (Kolibris) hummingbirds express the lower limit of homiothermal organization. First they live in warm climates and the smallest forms among them live in the tropics. Secondly these birds are extremely active, the motion of their wings being one of the swiftest motions in nature, and they continue this activity for long periods of time. This was especially important to Bergmann, for he regarded flight as an extremely difficult activity, requiring far more expenditure of energy than locomotion on either land or in water. With this enormous muscular activity, the upper limit of heat generation in animals is reached. While he felt confident that the hummingbirds generate the most heat per unit volume of any animals, Bergmann could advance no conclusive argument proving that their exterior surfaces are also the best heat insulators. He believed, however, that further research would firmly establish this claim.[70]

Having established that the conditions postulated by the law are closely approximated in the smallest homiotherm, Bergmann next went on to evaluate its applicability to middle forms. The law predicts that within a primary group of physico-anatomically related animals, species will differ from one another mainly with respect to size, and that, accordingly, the smaller species will occupy the warmer, the larger forms the cooler part of the range of these forms. As a test of this prediction Bergmann concentrated on the birds. He relied on Naumann's *Naturgeschichte der Vogel Deutschlands* for information on the anatomy, behavior, and geographical distribution of the European birds. In carefully examining Naumann's data Bergmann was able to confirm the rule for several families of European birds.[71]

In the opening sections of his paper, Bergmann had made certain simplifying assumptions in order to establish the rule and its usefulness. He had assumed, namely, that the systems of organs responsible for the production of heat and those responsible for insulating the body were either close to the limits of efficiency attainable in the animal realm, or – as he had assumed in his treatment of the size and of distribution of birds – that they were nearly identical physico-anatomically and that all other physiological considerations were equal. Nature, Bergmann realized, was much more complex, and in order to use the rule as a guide for understanding form, a host of other conditioning factors would also have to be taken into account. In an appendix to the work Bergmann examined several variables that must be considered together in assessing the validity of the rule. He showed, for example, that body size is conditioned by the strength of the skeleton and muscles that must support it. Muscle weight increases with volume and therefore as the cube while muscle strength – as Eduard and Ernst Heinrich Weber had shown – increases as the cross-sectional area and hence as the square. Muscle capacity and skeletal support place limitations of their own upon body size and must be considered in conjunction with *Wärmeökonomie* in the analysis of form.

Nature has devised several ways of overcoming the limitations on size placed by the constraints of the muscular and skeletal systems. These entail reducing the amount of muscle force expended in relation to body size. In large animals, for instance, the elements of the extremities are connected in a manner designed to minimize flexion. By placing the bones of the extremities directly over one another in a column so that the center of gravity of the upper bone lies directly over the surface supporting it, the bones themselves can be longer and heavier, and proportionatly less muscle power is required to support the entire structure. The muscle structure can accordingly be simpler. For this, however, the animals must sacrifice both precision and complication of mobility. Extremities whose elements are positioned at angles to one another displace the center of gravity and require, accordingly, muscles in a more constantly tensed state. The muscles must necessarily be stronger and more complex in structure. The animal will thereby be smaller but capable of more complex motion such as jumping, climbing, etc.[72]

An equally important constraint on form is exercised by the reproductive requirements of the organism. Small animals having many predators must reproduce many more offspring in order to maintain the balance of nature. To this end they are usually provided with large gonads and ovaries and are burdened with the support of many more young. This necessarily taxes the vegetative organs. But as the vegetative organs increase in size and weight in order to meet this demand, so too must the muscular and skeletal systems be more heavily taxed. Compsensation is made by sacrificing structural complications for increase in size.[73] Similarly, spiders are provided with specially constituted digestive organs for producing the materials of their webs and extremities specially constituted for spinning them. These features necessarily place demands on the system concerned with locomotion and ultimately, therefore, on the size of the animal.

Throughout the *Wärmeökonomie*, but most strongly expressed in its appendix, Bergmann was developing the new approach to biological organization based upon the critique of the *Lebenskraft* emerging from the works of Liebig and Lotze. Bergmann conceived each type of animal form to be grounded in a certain potential for generating chemical and mechanical energy of various sorts. This potential is determined by the arrangement of the material constitution of the organism and the processes of material exchange conditioned by these material relations. As his analysis of the complex of interrelated factors determining animal size indicates, Bergmann conceived this fund of potential energy as capable of being expressed in the various organ systems of the animal economy in such a way that any expenditure of energy in the development or complication of one system must be compensated by simplification of another. Bergmann conceived the economy of animal organization as a dynamic system of checks and balances, and its determinants were the laws of mechanics and the general laws of material exchange. While the emphasis on mechanical principles as the guide to functional and organizational constraints in the animal economy was new, the basic idea of Bergmann's dynamic system differed only in its mode of expression from the *Physik des Tierreichs* envisioned by Kielmeyer and that described later by the developmental morphologists. Both Kielmeyer and von Baer had viewed the animal economy as a functionally integrated set of systems grounded in the *Lebenskraft*. They too had regarded these systems as dynamically related to one another in terms of checks and balances, so that a more highly developed set of locomotive and sensory organs necessarily entailed a loss of reproductive and regenerative capacity. Liebig, Lotze, Bergmann and Leuckart had not fundamentally altered the teleomechanical conceptual structure of the earlier versions of the program. But in requiring that the notion of *Kraft* in organic contexts be consistent with that in physics, they had been able to replace the earlier constitutive concept of the *Lebenskraft* with the notion of general laws of motion and mechanical interaction as regulating animal form and function. In transforming one of its heuristic principles they were actually giving birth to the organic physics that had long been the goal of developmental morphology. Rather than postulate special organic forces as the cause of structure and functional control, they viewed the task of the morphologist to consist in finding the laws of action of organized materials and the constraints placed on them by the nature of the substances themselves and by their functional integration with other organic systems. For the program thus transformed the search was not for causes of organization; as true teleomechanists their object was to understand principles of organic design. And for this project physics provided the most effective interpretive guide.

Throughout my consideration of the works of Liebig, Lotze, Bergmann, and Leuckart, in this chapter as well as in my discussion of the work of Schwann in the preceeding one, I have emphasized that these men did not view their work as in fundamental opposition to that of the vital materialists and developmental morphologists. Rather they were in agreement with both schools in their emphasis on experimental method, and they accepted the common postulate of both that physiology must press on as far as possible in understanding the phenomena of life in terms of mechanical models. Moreover, functional morphologists were in fundamental agreement with the sort of questions posed by the vital materialists and the basic objectives of their approach to the animal realm. Where they parted company with their teachers and predecessors was over the question of the necessity of assuming a special set of vital forces in order to carry that program through. The increasing success in analyzing and synthesizing organic compounds after the second quarter of the century called the need for these forces into serious doubt. Rather than explaining phenomena such as muscle action, secretion, or even growth and development in terms of specific vital forces, the prospect was now near to hand of interpreting those phenomena in terms of mechanisms supplied by physics and chemistry. But this elimination of vital forces did not entail the removal of teleological thinking from biology.

The functional morphologists rethought the conceptual foundations of the vital materialist program in order that they might preserve and further expand the fruitful approach to the understanding of form and function developed by von Baer, Müller, Wagner, and E.H. Weber. As they examined the foundations of the older program it became clear that the principal reason behind the vital materialists' and developmental morphologists' defense of the notion of a *Lebenskraft* was the need to account for order within biological systems and the apparent inability of solving this problem by mechanical means alone. Their work on 'catalytic forces' suggested a model that might succeed in understanding problems of material exchange more generally and their own early attempts to formulate a general principle for quantitatively interrelating mechanical forces convinced them that they could reformulate the program in terms of a strict teleo-mechanism which avoided distinct vital forces.

All critically thinking students do not reject the ideas of their teachers. Some attempt to further develop and improve upon the work of their mentors. The functional morphologists were such students. This is particularly true of Bergmann and Leuckart. One of the major characteristics of the research of these two men is that they explored questions directly raised by von Baer and they attempted to re-examine some critical elements of his approach in terms of mechanical principles employed within a teleological and functional framework. Thus von Baer had once posed the question why the radiate type is mostly found in aqueous environments while the vertebrate type is found principally in land animals. As we have seen, Bergmann and Leuckart answered this question in terms of the requirements of motion governed by the principle of least expenditure of energy. I have argued that behind Bergmann's *Wärmeökonomie* stands the question of whether von Baer's earlier concept of 'developmental grades' could be explained in terms of functional demands placed on a common plan of material exchange. Similarly in his important article on generation for Wagner's *Handwörterbuch der Physiologie*, entitled 'Zeugung', Leuckart constantly reiterated the importance of von Baer's ideas for his conception of the entire subject.

In light of the genetic relationship I am claiming between the approach of the developmental morphologists and that of the functional morphologists it is important to have some means of assessing how each side viewed the work of the other. While internal aspects of the sort considered above indicate a concern among functional morphologists for arriving at new solutions to some of the critical problems in the theories of their mentors, it is of some importance to the interpretation being offered here to show that the functional morphologists considered themselves actually to be engaged in revising and extending the older approach. Equally important is the assessment of the older generation of the work of their students. While Johannes Müller and Rudolph Wagner strongly supported the work of their students, Schwann, Lotze, Bergmann, and Leuckart, it is important to know whether they also recognized the work of these younger men as compatible with their own view of physiology and whether they endorsed the new approach.

One source of insight into the way in which the relationship of their respective approaches was viewed by members of the two schools is a correspondence between von Baer and Leuckart. This correspondence is particularly valuable because it touched upon matters deeply central to both conceptual frameworks; namely generation and the transmutability of species. In several places in the correspondence Leuckart explicitly acknowledged his long-standing commitment to the general framework and methods developed by von Baer in approaching these problems. A student who considered himself to be building and improving upon the ideas of his mentors is quite naturally concerned to know how his work is received by them. Leuckart expressed such concerns in a letter to von Baer dated December 1, 1865.

Leuckart had been invited by von Baer to participate in the celebration of his jubileum on August 29, 1864 in St. Petersburg, a great honor indeed. In this letter Leuckart was conveying his belated apology for having been unable to attend. Along with his original letter von Baer had enclosed a copy of a recent paper on paedomorphosis. In this paper he had once again reaffirmed the importance of employing a teleological framework in biology, and he had specifically pointed to Leuckart's article, 'Zeugung', in Wagner's *Handwörterbuch* as paradigmatic of the proper use of teleological thinking in biology generally. Leuckart viewed von Baer's selection of the problem of paedomorphosis as the forum in which his own conception of teleology was praised as extremely significant. For it indicated that von Baer had truly understood that the spirit motivating Leuckart's work was an attempt to augment and further develop the science of morphology as von Baer had first conceived it. Von Baer had not only understood Leuckart's aims, he also approved of them:

What you wrote has deeply absorbed me not simply on account of the factual matters that you report but especially on account of the reflections which you have added to them in your usually brilliant and deeply concerned fashion. If I would not be suspected thereby of estimating my own worth too highly, I would dare say that those words and thoughts were spoken to me directly from the heart. This much I can say, however; that your view of the essence of alternating generations and the entire process of generation in general appears to me to be the only natural one. What you said therein in defense of so-called teleology as I have portrayed it in my article 'Zeugung' encourages [tröstet] me far more than any of the offhand expressions both solicited and unsolicited that I have received.[74]

Leuckart concluded the letter by expressing his deep respect for the great man. His comments also provide indication of the importance of von Baer's work for the young teleomechanists trained at Göttingen by Rudolph Wagner during the 1840s:

I look forward very much to hearing from you again soon. The good fortune to see you in person and to express the boundless respect and honor that I have held for you since my student days has until now been denied me. Your autobiography, which has given me tremendous pleasure, and your portrait in addition to what I have learned about you from others, particularly Rudolph Wagner, and through your works, has had to substitute for a personal impression.[75]

In addition to confirming his high regard for Leuckart's work, von Baer was indirectly expressing approval of Leuckart's functionalist approach to physiology. Von Baer was not only capable of leading others, he was capable of responding to criticism by modifying his theories: He was not an inveterate paradigm-captive. This emerges clearly when we consider the teleological framework of Leuckart's article 'Zeugung', with which von Baer had expressed his agreement. For while von Baer's works were treated as seminal to understanding the whole question of generation in that article, other views held by von Baer and developed explicitly by him in print were directly called into question. Thus, in section four of the article, titled 'Die chemische Vorgänge der Entwickelung', Leuckart wrote:

Earlier it was believed that the interconnected phenomena of development belonging to a self-contained whole must also be deduced from a single active central point. One constructed a proper Craftsman of the animal body, a *Bildungskraft*, which shaped the formless matter immediately in accordance with later needs and gradually unfolded the germ as a finely articulated organism. But the time is past in which one subsumes the phenomena of life under the command of such specific forces. We have come gradually to understand that what one had earlier elevated to the status of a simple and final principle of explanation is only a summational expression for an entire system of conditions and forces, which work together to a common end in accordance with certain physical and chemical laws under a variety of changing circumstances.[76]

This 'older view' directly criticized here, of course, was von Baer's notion of the *Gestaltungskraft* as presented in his treatise on the ovum. Leuckart went on to explain that according to the new way of viewing this problem the genetic substance [Bildungssubstanz] was assumed to consist of heterogeneous material the molecules of which were regularly arranged such that, in consequence of the movement induced by the semen, the generative substance would begin to undergo division, separation, and segregation. There was no mixing of male and female generative substances, rather the semen acts as a catalyst upon the original pre-established pattern of order and arrangement built into the molecules of the generative substance in the ovum. Once set in motion, the germ materials were then assumed to enter into new arrangements as a result of the particular physico-chemical properties of the molecules themselves. According to this scheme, then, a continuous set of external forces and conditions would be acting upon a changing substrate of germ materials:

> Nothing is therefore more certain than that under such circumstances more and more new and different products will gradually be formed, that these products undergo a gradual divergent combination and structure in the germ itself.[77]

A set of preformed organs need not be assumed therefore. Generation could be viewed as a continuous metamorphosis but one pre-determined by the order and arrangement of the initial materials. Assuring his readers that the evidence of organic chemistry amply support this teleo-mechanical model of generation, Leuckart reiterated the core of the doctrine:

> We cannot doubt that the particular morphological changes of the yolk are not only universally and continuously accompanied by a corresponding chemical transformation [Umbildung] of the form-elements but that they are in large measure determined by those transformations.
> *The possibility of these chemical metamorphoses must, of course, already be given in the elementary compositions of the generative substance. ... and therefore we may assume at the outset that the essential elements of the animal body are not only contained without exception in the generative substance, but that they are contained in the correct relationships and interconnections [Verhältnissen] und Verbindungen].*[78]

Similarly to this argument in the *Übersicht*, Leuckart went on to draw as the consequence of this view that all the characteristics of the species are contained in the order and arrangement of the germ substance, and that the preservation of the species was therefore dependent upon the maintenance of that original constitutive arrangement.[79]

In consequence of his views on generation, we might expect Leuckart to have been a strong supporter of the evolution of species, and indeed he was.

But it was not Darwin's theory of evolution that won his allegiance It was von Baer's. Leuckart discussed his view in another letter to von Baer dated April 26, 1866 and once again this letter is no less interesting for its revelation of von Baer's views than for what it shows us concerning the background of Leuckart's scientific work.

In reporting a geological theory of a colleague named Baff, Leuckart was provided the context for discussing his views on Darwin. Baff and Leuckart had developed the hypothesis that the sharper and more eroded eastern shelves of large landmasses is an effect of the earth's rotation. After explaining Baff's theory, Leuckart launched into the following discussion of Darwin, which I quote here in full due to its significance for the thesis I have been advancing in this and the previous chapters concerning the relationships between developmental and functional morphology and the acceptance by both traditions of a limited evolutionary theory:

Whether we morphologists will ever succeed in unraveling the original formative structures [Pregeanstalten] of nature remains questionable.[80] Certainly they will remain hidden from the present generation – and until then Darwinianism will remain hypothetical. I must admit however, that I am in principle completely converted to it. If it is true that we are less capable of recognizing 'species' the more we know about them (and even the 'best species' are mortal, which are represented by one or some few museum exemplars), then in my view it is demonstrated that species conceived as closed unities simply do not exist, that different species, even when they are locally distinct from one another, pass over into one another on the whole and evolve out of one another. It is unfortunate that you have withheld your thoughts on this theory from us Germans.[81] You were once very near to substituting your own name to it instead of that of Darwin. Your judgement would therefore be of double weight.[82]

Leuckart was convinced of the thesis that species evolve, but he was not a convert of Darwinian evolution without qualification. He could not have been without overturning all the principles of the science of morphology as he understood it and to which he and Bergmann had contributed so much. For central to their thought was the notion of functionally conditioned organizational plans. Hence in the passage quoted above, Leuckart spoke of the 'Pregeanstalten' of nature. Transformation within the confines of the plan could be understood in terms of physical principles. Descent from a common set of ancestors could not. While Leuckart was convinced of evolution, it was evolution understood in terms of the principles advanced by von Baer that he had in mind:

I would consider it one of the most beautiful days of my life if it were possible for me to see face to face the Man, who, next to my dear teacher Rudolph Wagner, has had the greatest influence upon me. Gratify me soon with the mission [Botschaft] that will be my true evangelical calling.[83]

It was not the gospel according to Darwin that Leuckart felt himself called to preach but rather the Book of Nature according to Karl Ernst von Baer.

CHAPTER 5

WORLDS IN COLLISION

It is not uncommon to find that even among the students of recognized giants, some choose not to accept the viewpoint of their teachers; and it is not infrequently the case that some of the very brightest students are found among the rebels. Thus Aristotle was the rebellious pupil of Plato. Helmholtz, DuBois-Reymond, Ernst Brücke, and Matthias Schleiden were the rebellious students of Johannes Müller. While some of Müller's students, such as Bischoff, Schwann, and Virchow, accepted the main lines of his approach to physiology and attempted to improve and expand upon it, these men rejected Müller's view of the subject and set themselves up in conscious opposition to their great master.

The deep split in Müller's school was already clearly in evidence in 1848, ten years before his death, and by 1858 it had reached such apparent bitter proportions that some of his loyal students seized the opportunity of eulogizing Müller in order to denounce the others as traitors to the science he had helped to found. In his *Gedachtnissrede* Virchow branded his former colleagues 'Auswuchse' [excrescences], and in his lecture, 'Über Johannes Müller und sein Verhältniss zum jetzigen Standpunkt der Physiologie', Bischoff chastized them as 'Schaum und selbst Schmutz' [scum and filth] that had unfortunately floated to the surface of the ferment initiated in physiology by Müller. Having been fired upon, the irreverent students did not hesitate to return a fusillade of their own. Schleiden's essay, 'Über den Materialismus der neuen deutschen Naturwissenschaft', not only denounced Müller as a Naturphilosoph but concluded with a lengthy and scathing personal attack on Virchow, whom he characterized along with Schelling and Hegel as a mystical dreamer. Though in a much more subdued and less vitriolic fashion, both DuBois-Reymond and Helmholtz expressed their radical break with Müller's vitalistic physiology and underscored the completely original character of their own enterprise.

That was a difficult claim, and as I will demonstrate, the historical evidence does not support it entirely. Just as Aristotle dit not walk in the shadow of Plato for twenty years without absorbing some of the contours of his point of view, neither was Helmholtz completely independent of Müller. Helmholtz's formulation of the conservation of force was not only deeply indebted to the emphasis on sound experimental technique practiced within Müller's school, but it also emerged from pursuit of a line of research directly encouraged by Müller himself. Helmholtz profited greatly from the approach being developed by the teleomechanists, particularly Schwann and Leibig, but he rejected the essential outline of their approach to physiology. For in Helmholtz's view physiology could be reduced ultimately to physics without reference to purposive organization. The term 'zweckmässig' was carefully expunged from his scientific lexicon.

In the previous chapter I argued that Liebig contributed significantly to the reformulation of vital materialism by showing that the Lebenskraft could be divested of its privileged status as cause of order and arrangement and simply be reinterpreted as the combined effect of appropriately arranged physical and chemical forces. According to Liebig's colleague at Giessen, T.L.W. Bischoff, a student and friend of Müller and collaborator with him in writing the famous *Jahresberichte* for physiology in the *Archiv*, it was in large part due to Johannes Müller that the original ideas in Liebig's work came to be so rapidly accepted as suggesting a theoretical framework for future advance in physiology:

Although it must be said that Müller was not as well prepared for chemistry as he was for physics by virtue of his background in mathematics, nevertheless he possessed not only the fullest comprehension of its significance and made use of it, but he engaged in chemical research himself ... I believe it could be demonstrated that Müller was one of the very few who was able to evaluate the complete significance of Liebig's work for physiology and who not only perceived its importance for contemporary work, but rather who grasped its underlying principle and its significance for the future.[1]

Bischoff's claim is amply supported by Müller's writings. Müller, for instance, delayed the publication of the 1844 edition of his *Handbuch* in order to incorporate Liebig's work into it.[2] In 1893 Müller's lab was a beehive of activity assimilating the results of Liebig's work on respiration and nutrition and incorporating it into the *Handbuch der Physiologie des Menschen*. Helmholtz was involved in checking Liebig's work, an opportunity which set him on the path to establishing the conservation of energy. The work opening the path to this great discovery was begun by Helmholtz in a paper on fermentation. In this paper Helmholtz attempted to examine aspects of Liebig's work and to expand upon the earlier results of a collaborative effort of Müller and Schwann on digestion. With this research Helmholtz began to thrust deeply into the heart of the teleomechanist program.

LIFE WITHOUT PURPOSIVE ORGANIZATION; HELMHOLTZ'S PATH TO THE ERHALTUNG DER KRAFT

Just one year after the publication of Liebig's *Thierchemie* and Lotze's defense of teloemechanical frameworks in his article, 'Lebenskraft', Helmholtz began the first of a series of researches that would embolden him not only to reject all forms of vitalistic explanation in physiology but also any use of teleological reasoning as an explanatory strategy.

At first glance we are apt to dismiss the first of these papers as insignificant because it comes to some curiously false conclusions. In light of Helmholtz's later major successes its primary interest may ultimately be in illustrating an attempt to draw results from over-eager experiments assigned by Müller to students in his lab, a young scientist demonstrating that he has earned his spurs in the rigors of experimental technique. When the paper is examined against the background of issues I have been discussing however, it takes on special significance. The paper was titled 'Über das Wesen der Fäulnis und Gährung', and it appeared in Müller's *Archiv für Anatomie und Physiologie* in 1843.

The object of the paper was to demonstrate that putrefaction and fermentation of organic substances are essentially different types of organic processes. The view rapidly in ascendance among Helmholtz's contemporaries was that both were organic processes only capable of being produced by the presence of an organized living body such as a yeast cell, and that neither were capable of transpiring in organic substances hermetically sealed in contact with water previously boiled or air previously heated to 100°C. The principal support for these theses, Helmholtz noted, had been furnished by Liebig and Schwann. But their arguments were not strong enough to support the conclusion in his view. Schwann had clearly demonstrated that fermentation was a chemical transformation dependant on the presence of a living organism, but it did not necessarily follow on the basis of the similarity of the products of putrefactions and those of fermenting substances that the two processes were identical in nature. Helmholtz clearly wanted to take his older colleague to school, providing him a lesson both in logic and the way to isolate all the variables in an argument through proper use of experimental technique. It was a brilliant piece of specious reasoning.

Helmholtz devised two beautiful sets of experiments. In the first set he demonstrated that in sterilized solutions normally capable of generating fermentations or putrefactions, neither the simple presence of oxygen nor chemical reactions generated by electrolysis are sufficient to initiate these processes. This left two hypotheses to be explored. Purefactions and fermentations might be initiated by a substrate, a 'miasma' coming from matter already putrid, or they might be results of organic 'seeds'.[3]

In order to distinguish between these two cases he fashioned a glass receiver with two tubular extensions. Over the end of one of these he attached a fine membraneous bladder, of such fine texture that "the entrance of small solid particles such as the seeds of microscopic organisms is completely hindered, while fluid and gaseous materials are free to pass".[4]

Next Helmholtz placed the substance or fluid to be examined in the retort and heated it to 100°C. After cooling, the end of the apparatus with the membrane covering was placed in contact with a putrid solution or with water. A short time after this contact the material in the retort began to exhibit typical signs of decay, whereas no such signs were apparent so long as the end of the retort was not in touch with the products of the external, decaying substance. For Helmholtz these results were extremely significant. Since he regarded the materials entering the retort through endosmosis not as organic bodies but as fluid or gaseous substances in no way differing from the end-products of normal chemical reactions, the conclusion seemed to be warranted that putrefaction is a chemical process capable of being initiated and sustained *independently* of a micro-organism.

That not merely a transfusion of the products from an externally putrefying source to the interior has occurred is evidenced by the fact that once intitiated the gas being released from pieces of meat continues to be released even after contact with the external fluid has been broken and the membrane has been coated with sealing wax.[5]

Helmholtz summarized the results of his experiments in three conclusions: First, that putrefaction is a chemical process involving the breakdown of proteinous and glutinous materials; Second, that

it can exist independently of life and in fact offers the most fruitful medium for the development and nourishment of living beings, through the presence of which it is modified in its appearance. Fermentation is a form of putrefaction bound to and modified by the presence of an organism.[6]

The position advocated by Helmholtz was similar in many respects to the vital materialist theory of an organic *Grundstoff* except that in contrast to Müller he regarded life as dependent not on a particular material basis but rather on a set of chemical processes which served as nourishment for growth and development. Organisms were viewed as chemical factories capable of modifying this substrate. Fermentation was regarded as a modification of the processes conditioning putrefaction. But unlike his mentor, Helmholtz did not regard the processes at the basis of life as dependent upon a vital force imparting a unique mode of organization. Helmholtz was arguing in the very strongest terms that the precondition for life is a substrate of chemical interactions indistinguishable from chemical reactions in an inorganic setting.

In his third conclusion Helmholtz underscored his conviction that while capable of existing independently of an organic body the processes connected with putrefaction are indistinguishable from the phenomena of life:

> It is notably similar to vital processes in the identity of the materials in which it has its seat, through the manner of its reproduction, and through the identity of conditions necessary for its continuation or destruction.[7]

Having succeeded to his satisfaction in eliminating through careful experimental technique the presence of micro-organisms from certain instances of organic decomposition, it followed of necessity that some of the essential functional features of living systems are replicated by chemical processes not under the control of a living body, and that the explanation of the phenomena of life does not require principles beyond those furnished by physics and chemistry.[8]

Although Helmholtz's paper of 1843 dealt ostensibly with a technical question regarding the distinction between putrefactions and fermentations, there was little disguising the fact that his main interest in this problem was to establish that certain organic processes involving living organisms are indistinguishable from processes in which an organized body is not present. While he claimed to have established that vital processes of growth and nutrition are dependent upon a set of chemical interactions free from the regulative control of an organized body, he had not been entirely able to exclude a role to purposive organization in the phenomena of life, for he had alllowed that fermentation requires the presence of a living organism. He had fallen short of his goal of demonstrating unequivocally that special vital directive agents and purposive organization have no role in physiology.

Helmholtz renewed his attack on this problem in a paper published in 1845 in Müller's *Archiv*. This time he chose a subject and constructed a set of experiments which attacked the issue of purposive organization directly leaving no room for an alternative interpretation. The subject he chose was the one raised by Liebig and Lotze of determining experimentally the conditions of material exchange in an organic system, and in order to study it Helmholtz investigated the consumption of mater in muscle action.

Helmholtz laid all his cards on the table in the opening paragraph of the paper:

> One of the most important questions of physiology, one immediately concerning the nature of the *Lebenskraft*, is whether the life of organic bodies is the effect of a special self-generating, purposive force or whether it is the result of forces active in inorganic nature which are specially modified through the manner of their interactions.[9]

In light of Liebig's recent success in accounting for vital phenomena in terms of known chemical–physical forces, this problem could be restated in a more precise manner; namely:

> ... Whether the mechanical force and heat generated in the organism can be completely deduced from the process of material exchange or not.[10]

Muscle action was ideally suited for the investigation of this question. Of course, many physiologists had long assumed that muscle action is dependent in some manner upon metabolic processes. Thus Lehmann and Simons had demonstrated through urinalysis that increased quantities of nitrogenous compounds as well as phosphates and sulphates appear in the urine after strenuous exercise.[11] But such results left a great deal of uncertainty. It was not at all clear, for instance, what the metabolic processes are and whether they take place in the muscle itself. Moreover, it was not at all clear from such analysis that material processes of exchange alone were sufficient to account entirely for muscle action, that some contribution of a special vital agent might not also be required. Helmholtz did not believe that each step in the generation of muscle action could be understood given the state or organic chemistry in 1845, but recent advances in experimental technique could reduce the uncertainty considerably. In fact, it could be established that *only* material processes of exchange are involved.

In order to establish this claim Helmholtz conducted a series of careful experiments on froglegs. He severed the legs from several frogs, and removed the feet and epidermis. The legs were then washed in distilled water to remove the excess blood, carefully dried and weighed. After weighing, one leg from each pair was placed on a glass plate, the legs being arranged in series touching end to end, while the others were stored in an empty container. Next the legs on the glass plate were connected to a Leyden jar. After 400-500 discharges of the Leyden jar, the muscles demonstrated no further signs of irritability. They were removed from the glass plate and placed in distilled water to remove any further signs of irritability. The same washing procedure was done to the control legs being held in storage. The muscles were next separated from the bones; the bones were dried and weighed; and these weights were subtracted from the initial weights, thus leaving the muscle weight as remainder.

In the next part of his procedure Helmholtz removed and collected the albumen from the muscles by multiple washings in distilled water. The differences by percentage of total weight (approximately 0.2%) between electrified and non-electrified muscles was so small that an effective role of albumen in muscle action was able to be ruled out.

He next turned to an examination of the muscle constituents themselves. Extracts were prepared by boiling the muscles in distilled water for a very brief period. (The water remaining from the previous washings was also included in order to catch any possible muscle extract released during removal of the albumen). Three different types of extract were prepared from these solutions. One type was obtained by drying the solution and then preparing an extract from the resulting powder with pure alcohol as solvent. Helmholtz named this, appropriately, the 'alcohol extract'. The other two extracts were obtained by allowing the frogleg broth to be dried to a syrupy consistency and mixing it with a 90 percent solution of alcohol. The precipitate formed from this solution was filtered out, and since it was soluble in water, Helmholtz called it the 'water extract'. The remaining broth and alcohol solution was dried and weighed (warm to prevent water absorption), leaving an extract soluble in ethyl alcohol (spirit of alcohol), which Helmholtz labeled the 'spiritous extract'.

After recording the weights of the different extracts Helmholtz observed that the quantity of 'water extract' was less in the froglegs subjected to electric current than in those not, and that the amounts of 'spiritous' alcohol extracts prepared from flesh exposed to the electric current was greater than those same extracts prepared from the non-electrified flesh. What was even more significant:

It should be noted that the difference in the water extract, the mean value of which is 0.3, corresponds pretty nearly to the mean difference of 0.24 of the spiritous extract.[12]

Clearly, by switching the current on and off, a process of material conversion had taken place in the muscle. Moreover the data seemed to indicate that the decrease in one set of constituents, the water extract, was compensated by the increase in another set, namely the ethanol extract.

Since the muscle fibres participate directly in the material conversions, it was conceivable that some small breakdown of muscle fibre had also occurred, but his experiments could not decide this issue directly:

I must leave unresolved one of the most important questions, namely, whether the muscle fibre participates in the reaction. This seems apriori highly probable, for we find that protein compounds are universally the bearers of the highest vital energies; and especially in our case, the appearance of an increased quantity of sulfates and phosphates in the urine after strenuous muscle activity speaks for a breakdown of phosphoric and sulphuric compounds. ... Against a breakdown of the muscle fibre, however, speaks the fact that for the most part the quantity of water extract lost corresponds to the increase in spiritous extract.[13]

Helmholtz had arrived at this conclusion after performing several tests on his extracts. Since Lehmann's experiments had indicated the presence of sulfates in the urine, Helmholtz ran barium chloride through the solution of the water extract. No traces of sulphur could be detected in the extract.

Helmholtz felt confident in having demonstrated two things from his experiments: First that muscle action is the result of a measurable chemical conversion of material substance; and second that the process is localized in the muscle. The muscles are chemical factories extracting force from matter through chemical conversion. This was merely a rough first attempt at understanding the process of muscle action, however. Helmholtz wished to go further by giving a more detailed chemical analysis of the components in his extracts and the exact chemical pathway of the transformation. From such an analysis, which he believed himself capable of offering in the not distant future, a deeper understanding of the process would be forthcoming. But that was not all. There was a very deep internal connection between all of these phenomena, which seemed to offer insight into operations of nature encompassing for more than muscle action:

The results of these experiments remain, however, isolated and without internal connection. I have restricted myself here merely to their presentation, for my further investigations on these points require a more exact analysis of the material extracts than I have presented here, and from there a much deeper understanding of the process may emerge which appears to permit an even more exact foundation and more precise execution of the analysis. I have therefore postponed their publication.[14]

The researches he referred to in this passage were not published immediately. They were first presented in two papers to his co-workers on the reductionist program in the Physikalischen Gesellschaft zu Berlin on July 23 and November 23, 1847. The first paper was titled 'Über die Wärmeentwickelung bei der Muskelaction', a paper which clearly indicated the path Helmholtz had taken in establishing the conservation of energy.

Before turning to the more general treatment which would establish the internal connections between the conversion of forces transpiring in muscle action, Helmholtz wrote a paper assessing the current state of knowledge concerning animal heat. Evaluating and synthesizing the latest results of physiological research on animal heat, Helmholtz made a direct assault on the theory of the *Lebenskraft* in an article for the *Encyklopädisches Handwörterbuch der medicinischen Wissenschaften* entitled 'Wärme, physiologisch'. This work was published by the medical faculty of the University of Berlin in 1845, the same year in which his paper on material conversion during muscle action was published.

The essence of Helmholtz's position was elaborated in a section of the article concerning the origin of animal heat. There he related the problem of animal heat to the general theory of heat then being debated among physicists. There were two alternative models. Heat was either the effect of a material substance, caloric, or it was the result of molecular motion. After summarizing the caloric theory and the manner in which it accounts for such phenomena as latent heat, Helmholtz rejected it. The caloric theory, he observed, is incapable of satisfactorily explaining the continuous source of heat produced by friction, because in the material theory of heat "the constancy of the quantity of caloric would be the most necessary requirement".[15]

But the relationships of free and latent heat set forth in the language of the materialistic theory remain the same if in place of the quantity of matter we put the constant quantity of motion in accordance with the laws of mechanics. The only difference enters where it concerns the generation of heat through other motive forces and where it concerns the equivalent of heat that can be produced by a particular quantity of a mechanical or electrical force.[16]

Having opted for the mechanical theory of heat, Helmholtz went on to state a condition that insures the possibility of determining in principle the relationship of heat generated to motive force expended:

It is clear in the first place that the forces present in an organized body can only produce a particular quantity of heat [Wärmebewegüng], and that it must always remain the same in all the complications of its action, because it is determined by the laws of mechanics that a particular quantity of a moving force [bewegende Kraft] can only produce the same definite quantity whatever the complication of its mechanism might be.[17]

The question to be decided, therefore, was whether the heat generated by metabolism of foodstuffs and respiration is equivalent to the local heat liberated by the animal body, or whether in addition to the heat generated by these means a special Lebenskraft capable of producing force indefinitely is not also required. "Precisely here it is clear how important the question of the origin of animal heat is for the entire theoretical consideration of vital phenomena, for in it (i.e. animal heat) by far the greatest part of the force equivalent contained in the chemical forces of the ingesta is manifested".[18]

Assuming that the rise in body temperature is equal to the difference between the total latent heat of ingested materials and the latent heat of the egesta, Helmholtz proceeded to catalogue the various types of substances taken in and excreted by the body. In order to limit his investigation Helmholtz eliminated substances from consideration that seemed to make a negligible contribution to the total outcome. He focused on three basic constituents. "If we determine how much heat is evolved when foodstuffs are broken down and oxidized into carbonic acid, carbon dioxide, water and ammonia, the quantity thus determined will correspond within a few percent to the heat generated in the organism by chemical processes".[19]

Quickly Helmholtz turned his sights on Liebig's theory of animal heat production. Liebig had maintained that the site of heat generation in the body is the lungs and that the quantity of heat generated by the oxidation of organic compounds containing carbon and hydrogen is equivalent to the heat generated by burning carbon and hydrogen in a free state. There were several difficulties with this position. Foremost among these were the older experiments by Dulong and Despretz which had indicated that such an hypothesis could not account for the total quantity of heat evolved by the animal. Unlike Liebig, who sought to dismiss these results, Helmholtz viewed them as crucial. There was a way of avoiding this difficulty, however, that had escaped Liebig's notice; namely that hydrogen and oxygen in the nutrient compounds themselves might combine through metabolism in the tissues contributing thereby to the generation of heat. While this assumption would save Liebig's goal of accounting for animal heat as a purely chemical process, it entailed jettisoning his principal assumption that respiration is its sole source. This was thoroughly consistent with the idea Helmholtz had been pursuing since his own inital attack on the Lebenskraft, however. Following up the line of reasoning he had begun a few months earlier on muscle action, it was essential to Helmholtz's plan of attack to demonstrate that heat is generated through chemical processes and that it does not take place in the lungs primarily but throughout the body in the tissues.

There seemed to be some very strong empirical support for the view that sources other than respiration account for the total quantity of animal heat. In particular, from the work of Favre and Silbermann it could be shown that starches and sugars could in many instances generate 10 percent more heat through fermentation than is liberated by respiration.[20] This figure agreed nicely with the results of Despretz and Dulong. After careful re-examination of their older data and comparison with the results Liebig had obtained, Helmholtz concluded that on the average the quantity of heat generated by an animal is about 11 percent greater than that liberated by respiration. It was conceivable, therefore, that metabolism of nutrients alone might account for almost the entire quantity (within one percent) of animal heat:

> ... a result which is completely in accord with the consequences of the chemical theory of organic heat, as we have deduced it above from the most probable assumptions. To carry the analysis further empirically and determine whether the agreement between the entire quantity of heat generated chemically and organic heat is as exact as it is possible to determine experimentally might founder due to the insufficiency of the physical preparations now available. Hence we must be temporarily satisfied with the fact that at least very nearly as much heat is generated through chemical processes in the animal as we find in it and its output, and that we must regard it as experimentally demonstrated that by far the greatest part of organic heat is a product of chemical processes.[21]

Helmholtz's previous discussion had established that very nearly the entire quantity of animal heat can be accounted for in terms of chemical processes. He had been able to say nothing, however, about the particular processes through which heat is generated and constancy of temperature maintained. Moreover, the sites of heat production were untouched by his analysis. Helmholtz was confident that two conclusions at least could be drawn concerning these matters from the work of physiologists over the previous two decades: the lungs are not the sole site of respiration but rather heat seems to be generated throughout the arterial system; and secondly, constancy of heat production seems to require the activity of the central nervous system. Evidence for this latter claim had been accumulating since Brodie had first established in papers published in 1811 and 1812 that the temperature falls in the region of the lungs of decapitated animals on artificial respiration. Similarly, it had been shown by various other experiments that a continual temperature loss accompanies a gradual severing of the medula oblongata or through the action of poisons and narcotics on the central nervous system. However, one could not infer, as Brodie had wished to conclude from such experiments, that the sole function of respiration is to cool the organism and that the source of animal heat is the nervous system. It had been shown by Davy, Becquerel and Breschet, for instance, that venous blood is on the average between 0.75 to 0.25°C lower than arterial blood, and that this finding holds equally for arterial blood in transit between the capillaries of the lung and the heart. Clearly the exchange in the lungs must have contributed to this temperature differential. But the findings of Gustav Magnus and Johannes Müller had demonstrated that the lungs are not the only site of heat production. Magnus' classic experiments had shown that oxygen and carbonic acid are present throughout the blood stream either in solution or in an easily separable chemical combination. Müller had established furthermore that frogs respiring in an oxygen-free environment exhale carbon dioxide, the carbon and oxygen necessarily being provided in this case from within the organism itself. From such evidence Helmholtz concluded that the most reasonable assumption was that the lungs are not the principal site of heat production but that it takes place continuously in the blood.[22]

> The question concerning the generation of animal heat from respiration would seem to be answerable in terms of present data in the following manner: Respiration and digestion are its primary causes insofar as they provide the material to the organism which, in undergoing several intermediate stages in arriving at its final configuration, releases the entire quantity of heat present in the animal body, and the former, namely the changes in the constituents of the blood taking place directly in the lungs, provides only one, and as it appears very small, part of the total quantity of heat.[23]

One aspect of Helmholtz's researches on animal heat continued to thwart his goal of eliminating the *Lebenskraft* from physiology. He had succeeded in establishing that chemical processes of exchange could account for *almost* the entire quantity of heat generated by the animal body. Helmholtz wished to leave no possible hiding place for the elusive *Lebenskraft*, however. More critical than this, however, was the possibility, not excluded by his previous work, that the nervous system provides some portion of the total quantity of heat. Here was a potential niche which the vitalists might turn into a veritable fortress. It could be maintained, for example, that the small portion of heat unaccounted for by the best experimental technique is generated by the nervous system; and this fact combined with experimental findings concerning the loss of thermal constancy resulting from depressing the activity of the central nervous system could conceivably be construed as evidence for the activity of a directive and constitutive vital force. Helmholtz had pursued this question with dogged persistence and he was not about to relinquish victory at this point. He took up the problem in the paper delivered on November 12, 1847, before his colleagues in the Berlin physikalische Gesellschaft entitled 'Über die Wärmeentwickelung bei der Muskleaction'. The paper was published in Müller's *Archiv* in 1848. This work was the logical extension and culmination of all his previous investigations on animal heat.

Throughout his earlier paper on animal heat Helmholtz had brought forth evidence supportive of the view that respiration is not the only source of heat but that a greater source of heat was in fact to be found in metabolism of starches, sugars, and other nutrients. Essential to his plan of attack on animal heat was to demonstrate that the heat present in the tissues is not transported to them somehow through respiration, through the influx of the blood, or through the activity of the nerves, but that heat is generated directly in the tissues themselves through the process of material exchange. To demonstrate this point he again took up the question of muscle action, this time focusing on the generation of heat.

The problem of demonstrating quantitatively that heat is produced *in* the muscle tissue was a difficult one necessitating a very sensitive experimental apparatus. Helmholtz devised an instrument combining a thermocouple with DuBois-Reymond's astatic galvanometer. The device was truly ingenious, and its very construction embodies the conservation of energy in such an obvious way that it can scarcely have escaped Helmholtz's notice.

The well-known principle of the thermocouple is that an electric current can be generated in a metal loop consisting of two conductors such as iron, nickle or copper, whenever there is a temperature differential at the ends of the loop. The strength of the current will depend on the temperature differential. Furthermore, since the current generates a magnetic field, Helmholtz reasoned that the deflection of a compass needle positioned perpendicularly to the direction of current flow would povide a measure of the strength of the current and hence the temperature differential across the ends of the loop. Now the temperature grandients he was dealing with were very small, and hence, with even a very good thermocouple, the current generated was very weak. Accordingly, Helmholtz, magnified the effect through the application of a multiplicator, or coil. By connecting the thermocouple to a multiplicator with fifty turns of copper wire, he was able to generate a sizeable effect on the deflection of the needle with a weak current. Helmholtz added one further component to his experimental apparatus. He was unable to obtain galvanized copper wire completely free of all traces of iron. As a result the arrangement of coil and needle in his apparatus produced a deflection of the needle by about 10°C before the current was even initiated. In order to counteract this effect and make the needle function in its optimal range (i.e., near the normal), Helmholtz added a compensator to the instrument, a horseshoe magnet positioned with its ends close enough together so that the magnetic field would deflect the needle back to zero. In addition to these details Helmholtz went on to consider a host of factors, such as proper geomagnetic and heat insulation, the effect of the cross-sectional area of the wire, and the sensitivity of heat change of the parts of his apparatus, all of which could affect the sensitivity and accuracy of the instrument, and he took special precautions wherever necessary in order to insure the reliability of his instrument.

Having devised an instrument extremely sensitive to small temperature differentials, Helmholtz turned his attention to the most important and interesting problem of all, namely that of calirating the instrument. This required the correlation of small increments of temperature with current intensities. It was required in turn to correlate these current intensities with deflections of the needle in the magnetic field. This latter aspect of the problem was complicated by the fact that while the torque on the needle is always proportional to the current intensity, the relationship between current intensity and change of indicator position can vary from one multiplicator to the next, depending upon the number of turns, quality and thickness of the wire used. The exact correlation and variation with position had to be determined experimentally for the particular instrument in question. Through careful attention to detail, Helmholtz was able to construct his multiplicator and compass needle indicator such that within deflection angles of 0° to 20°, the 'deflecting force' [ablenkenden Kräfte] of the current remained very nearly constant (within three percent) within the range of 0° to 20°. Thus the deflection of the needle produced by the current was directly proportional to the current intensity, so that for deflection of 10° the value 10 could be assigned to the current intensity, while a deflection of 5° corresponded to an intensity of 5, and so forth. The current intensities thus established were proportional to the temperature differentials at the ends of the thermocouple. In the final step of his preparations Helmholtz proceeded to determine the magnitudes of these temperatures. He did this by placing the heater strips of his thermocouple in mercury baths at different temperatures, which he could measure very accurately. These experiments resulted in value of 0.037°C for a current strength of 1. Since it was easily possible to distinguish deflections of the needle within one tenth of a degree, the instrument was able to distinguish temperature differentials of less than one thousandth of a degree centigrade.

Just as in his previous work on material exchanges associated with muscle action, Helmholtz once again experimented on froglegs. He removed one leg completely while leaving the other attached only by means of the spinal nerve. This was done in order to produce the strongest contractions of the thigh muscles, but also to evaluate the contribution of the nerves to the total quantity of heat. A current was passed through the nerve for several minutes generating a strong muscular contraction, the leads of the thermocouple-galvanometer apparatus having beforehand been attached to the contracting and non-contracting muscles of the two test legs. A contraction lasting 2–3 minutes produced a deflection of 7–8° of the needle corresponding to a temperature differential of about 0.14–0.18°C. In a second set of experiments Helmholtz severed both legs completely from the frog leaving no connection to the spinal nerve, and the contraction-generating current was passed directly into the muscle tissue. Once again a noteable temperature differential was recorded. In a final set of experiments Helmholtz examined the quantity of heat generated by excitement of the nerve alone. The result of this test was extremely significant for his theory. He found that the amount of heat generated in the nerve would deflect the galvanometer needle no more than about 0.5°. This meant that the temperatures generated in the nerve could only be a few thousandths of a degree, an infinitesimal contribution to the temperature generated by the muscle contraction.[26]

Helmholtz's experiments on heat generated by muscle action led to three important general conclusions. First he had shown unequivocally that heat is generated directly in the muscle tissue itself and that its origins are due to chemical processes in the muscles. Secondly he had established that animal heat can be generated independently of nerve stimulation and that the contribution of the nervous system to the total quantity of animal heat is negligible. With these two results Helmholtz had effectively destroyed the possibility of assigning any vestige of a constitutive role to the *Lebenskraft* in physiology. It could not make a material contribution to the functioning of the animal because the physical forces known to be involved in material exchange could account for the relation of input to output in organisms well within acceptable limits of experimental accuracy. Nor was there any use in attributing to the *Lebenskraft* an immaterial directive role in animal function. For it could only do so through the intermediary of material agents incapable of bearing witness to its hidden presence.

The third implication of Helmholtz's work on the heat generated by muscle action far transcends the problem of eliminating vital forces from physiology. Clearly as he began to design and work through his experiments on animal heat Helmholtz was confronted with the conservation of energy in a more immediate and general way than anyone before him, including Mayer and Liebig. For several years the working assumption of nearly everyone had been that the forces of nature are mutually interdependent. As we have seen, Liebig believed they were interconnected through the most general laws of motion and that the investigation of the forces of nature in terms of such a general framework would reveal a common measure establishing their interconnections. His vision limited to organic and physiological chemistry and lacking the necessary understanding of mechanics to carry his idea through, Liebig was incapable of establishing the conservation of energy in anything other than a very suggestive metaphoric sense. Helmholtz was fortunate in having a thorough understanding of several of the areas at issue in the question of the interrelationships and interconvertibility of natural forces. Not only did he understand physiological chemistry, but he was in command of mechanics, electrical theory and the theory of heat. Certainly Helmholtz's wider knowledge and deeper understanding of some of the related problems gave him a critical advantage in attacking the conservation of force, but this aspect of Helmholtz's background was not the only factor in his breakthrough. There were two other critical factors. Just as important as his theoretical ability was Helmholtz's skill as an experimentalist, his concern with the problems of laboratory measurement, his ability to devise an instrument in a manner capable of answering in a very precise way a specifically focused set of questions. Helmholtz had a very special sense of how a machine operates, what factors can affect its reliability, how to go about determining the limits of the experimental apparatus and improving upon it. Like Kepler in pursuit of those elusive eight minutes of arc, Helmholtz's persistence in tracking down and identifying the specific sources of animal heat through refinements in experimental technique and technical apparatus was an important source of his developing understanding of the conservation of energy.

The other factor, not to be discounted, was the problem of eliminating the *Lebenskraft* from physiology. In order to silence his vitalist opponents, Helmholtz felt it necessary to close off every avenue of escape, no matter how small. For he knew that unless every detail of the vitalist position could be accounted for in terms of material processes of exchange his opponents could still find refuge for their immaterial force, the ghost in a perpetual motion machine. Within the problem context of contemporary physiology, Helmholtz's experiments on the heat generated by muscle action and the attendant problem of discounting a role for the nerves in heat production were the logical culminating step in his four year pursuit of the *Lebenskraft*. But the most significant aspect of those experiments is not only that they involved a direct conversion of several types of force but that all the elements of the puzzle were assembled in a manner which clearly pointed to a means of transcending the impasse Liebig and others had encountered concerning the interconvertibility of forces. Others before Helmholtz had shown that chemical forces are related to heat and electricity. Still others had shown that electricity is connected to magnetism, while others had shown that mechanical forces are related to heat. Helmholtz's experiments were nearly unique in that conversions between all of these forces were directly involved in a single experiment: a set of chemical reactions produced heat, which in turn induced a current in a thermocouple, giving rise to the deflection of an indicator in the magnectic field produced by the current. By itself that piece of experimental 'gee-whizery' was not of such major import as the fact that in order to establish the claim he wished to defend against vitalistic physiology, a definite and precise *quantitative* measure of the heat generated in muscle and nerve was required. From the point of view of the conservation of energy the critical step in Helmholtz's argument was when he sought to show not just that the heat generated in the muscle is proportional in some sense to the deflection of the galvanometer, but that specific values of temperature are quantitatively correlated to specific current intensities and deflections of the indicator. This was the major breakthrough. Others concerned with the interconvertability of force had not been able to hit upon a convenient measure for quantitatively demonstrating it. As a result of his experiments Helmholtz had established a basis for carrying through an idea that he had first ennunciated in a theoretical context in his paper of 1845 on animal heat: he now joined Joule and Mayer in realizing that the mechanical equivalent of heat held the key to the elusive demonstration of the conservation of force.

The physiology of muscle action laid before Helmholtz all the elements of the conservation of energy. Having adopted the mechanical theory of heat in his article of 1845, Helmholtz was attuned to the importance of work and *vis viva* as the key elements of the general theory of motion required for conceptualizing the problem. His 1845 sketch of the theoretical requirements for the solution of the problem had also revealed the importance of the mechanical equivalent of heat as the key to establishing the interconvertibility of electrical, chemical and mechanical forces involved in muscular activity. With these requirements in mind Helmholtz had turned in January of 1846 to detailed experiments on heat generated in muscle action. By the end of 1846 he had, in close collaboration with DuBois-Reymond, succeeded in devising the sensitive thermometric instrument combining the thermocouple and galvanometer, and he had performed experiments "on the composition of nerves with respect to their possible alteration by muscular contraction".[27] Although these various researches were not completed in their final form until November of 1847, his work on heat generated by muscle action had presented Helmholtz with a sequence of energy conversions and had demonstrated the feasibility of using the mechanical equivalent of heat as a means for measuring them. Independently of these experimental researches Helmholtz was also engaged after 1845 in deepening his grasp of the different branches of theoretical mechanics, to such a degree in fact, that when he set forth his ideas on the conservation of energy, DuBois-Reymond remarked that Helmholtz revealed himself to the surprise of even his friends to have become a master of mathemetical physics in a single bound.[28] Indeed the period between early 1846 and mid 1847 could be justly termed the *annus mirabilus* of Hermann Helmholtz. The various discoveries of that marvelous year were brought together by Helmholtz in 1847 in a memoire titled 'Über die Erhaltung der Kraft'.[29]

The treatise began with a discussion of the relationship between matter and force based on Kant's *Metaphysische Anfangsgründe der Naturwissenschaften*. Two abstract concepts, 'matter' and 'force', underlie all our perceptions of the external world, Helmholtz wrote, and neither of them could be thought independently of the other. Experience of the external world depends upon differences in the effects exercised by matter on us and these effects are what we term 'force'. Matter and force cannot be conceived as separate therefore. Pure matter incapable of exercising an effect would be inconsequential, for it could never change the state of another body. Similarly force cannot be something real existing independently of matter; for existence requires 'Dasein', which is what the term 'Materie' expresses. The qualitative basis of all experience, therefore, must in the final analysis be reduced to elemental matters with different forces. The only changes possible among a system of material particles are changes in spatial relation, and the forces expressing the effects of these changes can only be forces of motion [Bewegungskräfte] dependent upon spacial relations:

The phenomena of nature are to be reduced to motions of material particles with invariable forces of motion which are dependent upon spacial relations.[30]

The result of Helmholtz' reflections on the meaning of the terms 'matter' and 'force' as the conceptual basis for a possible experience of nature was that all natural phenomena are to be treated in terms of material particles exercising forces of attraction and repulsion as a function of the distance separating them. From the point of view of mechanics this was a very restrictive assumption, but as we shall see, it was imposed by Helmholtz with the clear aim of demolishing the teleomechanical theories of the likes of Liebig, Lotze, and Bergmann.

The principal question, which formed the substance of Helmholtz' monograph, was how to generalize this limited conception of force, the measure of which was to be conceived only as a function of the distance separating two material particles, so that complex systems of interacting particles seemingly incapable of a reduction to an idealized two-particle system could in fact be arranged in terms of the conceptual framework within which the critical analysis of the opening section had revealed experience of nature to be possible. In addressing this problem Helmholtz started from the assumption that no combination of natural bodies can ever generate force continually. The problem of rendering the phenomena of nature intelligible in terms of the concepts 'matter' and 'force' would be resolved by showing how the impossibility of creating a *perpetuum mobile* was to be extended to all classes of natural phenomena:

> This principle can be expressed as follows: If we consider a system of natural bodies which have determinate spacial relations and which are set in motion under the influence of the forces exercised by these bodies with respect to one another until they are situated differently, then we can consider their acquired velocities as a certain quantity of mechanical work, and transform them into such. If we want to allow the same forces to be effective a second time in order to obtain the same quantity of work once again, we must somehow return the bodies to their original disposition through the application of other forces at our command; and in order to do that, we will have to expend a certain quantity of work of these latter forces. In this case our principle demands that the quantity of work which is obtained when the bodes go from the initial state into the second, and which is lost when they are returned from the second to the primitive condition, is always the same irrespective of the path or the speed of this transition. For if the quantity of work were greater through one path than another, we could use the former for obtaining work and the latter for restoring the system to its initial state through a diversion of a portion of the work obtained; and in this way we would obtain an indeterminate amount of mechanical force; we would have constructed a *perpetuum mobile*, which not only maintains itself in motion but also is able to create force externally.[31]

In Helmholtz's view the *Lebenskraft* was such a *perpetuum mobile*, for in its most careful formulation it was a force generated by the systematic interaction of the organic parts, which was independent of the mechanical processes generating it, and which could therefore be used to maintain those mechanical processes in their systematic state of integration. In order to see how Helmholtz eliminated this interpretation of Lebenskraft we must follow his derivation of the conservation of energy.

Helmholtz first implemented the principle that motion cannot be perpetually generated by noting that the measure of work required can be considered in terms of the well known law of the conservation of *vis visa* [lebendigen Kraft]. In the case of falling bodies, the work obtained by falling through a distance h is $1/2\, mv^2 = mgh$, through which the body obtains a velocity, $v = \sqrt{2gh}$, sufficient to return it to the same height if it could be restored to the body without dissipation. According to the principle enunciated above, whenever a particle m under the influence of a central force exercised by a system of particles A has the same position relative to A, the quantity of work generated by the motion of m toward A is the same as that required to restore it to its initial position; or equivalently, the quantity of *vis viva*, $1/2\ mv^2$, is the same whenever it has the same position relative to A. After relating the tangential components of velocity and momentum under this condition, Helmholtz derived the extremely significant result that:

> In systems obeying the conservation of *vis viva* the simple forces of the material particles must be central forces.[32]

Thus mechanical systems conserving *vis viva* could be brought into the framework of matter and force set out in the opening section of the monograph.

In the following section Helmholtz generalized this result. Where Q, q represent tangential velocities at distances R, r and ϕ is the intensity of the central force, Helmholtz derived the equation

$$1/2\ mQ^2 - 1/2\ mq^2 = \int_r^R \phi\, dr$$

The left member of the equation represents the difference between the *vis viva* at points r and R, whereas the integral on the right represents the sum of all the intensities of force [Kraftsintensitäten] that act between the points R and r. "If we call the forces that attempt to move the particle m tensional forces [Spannkräfte] as long as they do not actually effect the motion, in contrast to what is called *vis viva* in mechanics, then we can designate the quantity, $\int_r^R \phi\, dr$, as the sum of the tensional forces between r and R and the aforementioned law will become: the increase of the *vis viva* of a point-mass moving under the influence of a central force is equal to the sum of the changes in the tensional forces corresponding to the change in the distances".[33] Therefore, in all cases of particles moving under the influence of a central force, the intensity of which is a function of the distance, the loss in the quantity of tensional force is equivalent to the increase of *vis viva*, and the loss of *vis viva* is similarly linked to an increase in the tensional force:

> We could therefore express the law as follows: ... The sum of the *vis viva* and the tensional forces is always constant. In this most general form we may designate our law as the principle of the conservation of force.[34]

Having established the general form of the conservation of natural forces, Helmholtz turned to a consideration of the force equivalent of heat. Similarly to the treatment in his 1845 paper on animal heat, Helmholtz pointed out the inadequacies of the material or caloric theory. Then, in a careful discussion of the work of Joule, Clapeyron, and Holtzmann, he explained that in the kinetic theory that which had hitherto been called free heat corresponds to the *vis viva* of the thermal motion of molecules, while the quantity of 'tensional forces' in these molecules which can be converted to *vis viva* through a change of their arrangement is what had previously been called latent heat. The fact that the two components of the general theory of the conservation of force, namely kinetic energy or *vis viva* and tensional force or potential energy, have direct interpretations in terms of free and latent heat provided a clear strategy for demonstrating the interconnection between the various forces of nature. The thermal effects of those forces could be used as the common measure for discussing their interrelations. In the remainder of the memoire Helmholtz pursued this strategy by casting the laws of the mechanical effects of electricity, electromagnetism and electrochemical action in a form in which it was possible to deduce their effects in terms of the generation of heat. The specific manner in which Helmholtz carried out this analysis has been the subject of numerous investigations,[35] and it is not of direct concern to the relationship of Helmholtz's work to the vitalist-mechanist debate.

THE ANTI-VITALIST CAMPAIGN AND THE DEMISE OF TELEOMECHANISM

There were three forms of vitalism in early 19th century biology. One form was the unscientific vitalism of the Naturphilosophen which treated organization as the result of the immanent material presence of mind in nature. This notion was rejected by the early contributors to the teleomechanist tradition. As we have seen, they constructed a much more fruitful account which treats the Lebenskraft as an irreducible, but nonetheless emergent property of the systematic integration of the animal machine. It was this version of the vitalist argument that Kielmeyer, Weber, and Johannes Müller, had defended in terms of the best evidence then available coming from the side of organic chemistry. In his early career von Baer had been the able spokesman of this view as well, as is particularly evident in his discussion of the *Gestaltungskraft* in the mammalian ovum. Helmholtz's treatise showed that this vitalistic interpretation was no longer tenable. This implication of Helmholtz's short treatise was spelled out clearly by Emil DuBois-Reymond in the preface of his *Untersuchungen über tierische Elektrizität* (1848).

Like Helmholtz, DuBois-Reymond argued that the only sense the term 'Kraft' can have in both the inorganic and organic realms is that of the central forces described in mechanics. It is a mistake, he argued, to treat 'force' as the cause of motion. But even allowing this error to stand for the moment, "since motions are supposed to take place in the direction of the forces, so it is already tacitly assumed that in the organic as well as in the inorganic realm no forces can exist whose ultimate components are not either simple attractive or repulsive forces, so-called central forces".[36] 'Force', DuBois-Reymond emphasized is the measure, not the cause of motion; "It is the second derivative with respect to time of the path of a particle in variable motion".[37] Starting from these considerations Helmholtz had shown that the sum of the *vis viva* and tensional forces in all interconnected systems of material particles is *constant*. For physiology that implied there was no source from which the 'emergent' *Lebenskraft* could possibly emerge.

What the *Erhaltung der Kraft* demonstrated to DuBois-Reymond is that there cannot be a single, systemic organizing force, but rather "innumerably many forces in infinitely many directions acting in the most manifold ways, which emanate from material particles in order to act on other material particles".[38]

It can therefore no longer remain doubtful, what is to be made of the question, whether there exists a single recognizable difference between the processes of inorganic and organic nature. No difference exists. No new forces are attributable to organic forms, no forces which are not also active outside them. There are no forces therefore, which deserve the title *Lebenskräfte*. ... There are no *Lebenskräfte* in this sense because the effects ascribed to them are be reduced [zerlegen] to those which originate from the central forces of material particles. There are no such *Lebenskräfte*, for forces cannot exist independently ...[39]

Once it had been shown that central forces are the only forces possible among material particles, and that the interaction of these forces and the motions to which they give rise are conservative, the *Lebenskraft* conceived as an emergent yet directive agency had to be abandoned.

No argument could be advanced by the vital materialists and developmental morphologists to save the special status of the *Lebenskraft*. In making a constitutive use of a methodological principle, that is, by attributing a physical existence to the principle of biological organization rather than declaring as inexplicable the source of the order found in biological mechanisms, the vital materialists had in fact prepared the ground for DuBois-Reymond's attack. The vitalist position defended by Kielmeyer, Weber, von Baer and Johannes Müller would have to be abandoned. But this point had been clear for several years following the critical analysis published by Liebig and Lotze in 1842; and in response to this critical problem in the teleomechanist program these two men along with Bergmann and Leuckart attempted to fashion a more consistent position which employed only strict mechanical modes of causation but uniting them within a teleological framework of explanation. In 1848 the real scientific debate centered on this third form of vitalism. Both Helmholtz and DuBois-Reymond wanted to carry the critique of vitalism one step further by cutting off the possibility of retreating to a teleomechanical position of the sort constructed by the functional morphologists. For, they argued, although we may not yet know how life first originated and what principle must be used in ordering mechanical processes in order to produce a living organism, who is to say that we may not one day be in a position to reproduce experimentally the *conditions* under which life first emerged? As DuBois-Reymond noted, the reductionists were willing to include the problem of how consciousness first emerged among the seven *Welträtsel*, but they were not willing to include the problem of how organized matter first arose as one of them.[40]

In the years following the publication of Helmholtz's memoire the conservation of energy became the central weapon in their arsenal as the reductionists waged their anti-vitalist campaign, which quickly assumed the characteristics of an acrimonious debate.[41] The first public expression of the gathering storm was unleashed by an outsider who bore the scar of a quarrelsome youth proudly on his face. The man was Carl Ludwig.

Ludwig had openly attacked the claims of the morphologists in a review of Rudolph Leuckart's work, *Über die Morphologie und die Verwandschaftsverhältnisse der wirbellosen Thiere* (Braunschweig, 1848).[42] In this book Leuckart had attempted to set forth the principles of morphology. The science of morphology, he argued, was grounded in the union of von Baer's embryological methods with the functional approach of Cuvier, which he and Bergmann were in the process of expanding. Central to their view of biological sciences was the notion of plans of organization, the venerable teleomechanist concept of the morphotype.

Ludwig attacked this approach to organization as utter and consumate nonsense. It was not science at all, but rather an artificial exercise in metaphysics or perhaps something more sinisterly approximating a weird religion of nature. The best thing that could happen to zoology was that

Leuckart's book might find no readers. For zoology could only advance, in Ludwig's view, by completely throwing out the foolish approach of the morphologists and being recast on the solid foundations of the physiological point of view. Zoology should take its starting point from the analysis of the elementary building blocks and functions of the animal machine. Only then could sense be made of the question of how these parts and functions are ordered into a harmonious whole. The morphologists had turned the proper order of scientific investigation on its head; the problem of the whole comes last in zoology, not first.

Leuckart responded to this review like the stunned recipient of a bolt from the blue. In an anguished reply he attempted to brush off the criticism by asking whether morphology does not after all possess *bona fide* scientific credentials.[43] Zoology cannot do without the standpoint of physiology. That much Leuckart was more than prepared to admit. He was, after all, heavily engaged with Bergmann in just the sort of physiological research held in high esteem by Ludwig. But that view must be supplemented by the morphological perspective, he argued, "Can he [Ludwig] understand the structure of the vertebrate, the differences which are manifested therein in the lowest forms of fish, without the morphological perspective? And yet morphology is supposed to have no scientific justification?"[44]

Leuckart was not alone in seeing foreboding clouds on the horizon. Even among the old guard of developmental morphologists no time was lost in joining the defense. Literally outraged by Ludwig's attack, Heinrich Rathke fired off a letter of support to Leuckart which was published as part of the reply to Ludwig's review. Ludwig notwithstanding, Rathke saw Leuckart's work as the cutting edge of scientific zoology. "For you have attempted, like Cuvier, to investigate and systematically order the general plan of structure, or, what is the same thing, the architecture or type according to which animals are modeled".[45]

The debate, of course, was not so quickly ended by an appeal to the older authorities. Ludwig went on the attack once again in the introduction to the first volume of his *Lehrbuch der Physiologie des Menschen*, where he set forth the essential elements of the reductionist program. The cornerstone of the program was the conservation of energy and the law of causality which it supposedly expressed, the one scientific truth that made it at least plausible that even though the reductionist approach must proceed with hypothetical foundations, they would only be temporary:

... all appearances produced by the animal body are a consequence of the simple attractions and repulsion that are observed by conjunction of those elementary being [atoms]. This conclusion will be irrefutable when it becomes possible to demonstrate with mathematical precision that these elementary conditions are arranged in the animal body according to mass, time and direction so that the effects of both living and dead organisms will follow from their interaction. ... This point of view demands in accordance with the law of causality, to which we must adhere if we want to think about anything at all, that an object contains the cause of its effects within itself.[46]

Ludwig's critics felt that far from eliminating metaphysics from physiology he had rather attempted to explain processes that were not well understood by others even more mysterious.

One might argue about the merits of different foundations for physiology interminably, but the last part of the statement quoted above concerning the law of causality clashed violently with the notion of cause the teleomechanists had defended as necessary for understanding the functioning of biological organisms. If the discussion were limited to the analysis of particular material transformations, such as the transformation of starches and sugars to fat, Ludwig's perspective might be defended as perfectly adequate. But as soon as the discussion was focussed on even the most primitive functional unit of organization, things began to look quite differently. Rather than haggle about metaphysical issues and the hardcores of their respective research programs, a strategy followed by some teleomechanists was to strike the reductionists at what seemed to be their weakest point of defense, the cell theory.

Ludwig had applied his program of atomistic reductionism to the cell theory in volume two of his *Lehrbuch*.[47] A critique of Ludwig's treatment of cellular organization provided Carl Reichert with an opportunity to enter the fray. In his 'Berichte über die Fortschritte der mikroskopischen Anatomie' for 1854, which was published in Müller's *Archiv für Anatomie, Physiologie und Medicin* in 1855. Reichert, who had been a student of Müller and longtime admirer of von Baer, sought to delimit the different methodological perspective of the two competing schools of physiology. In the same year Virchow's epoch–making paper on cellular pathology was published in *Virchow's Archiv*. This work provided Reichert with all the ammunition he needed to attack Ludwig once again in the 'Jahresbericht' for 1855; for Virchow had set forth a new version of the cell theory in complete accord with the teleomechanist view of biological organization. In expanding the defense of morphology attempted earlier by Leuckart and Rathke, Reichert's discussion provides one of the most penetrating analyses of the differences in approach to biological organization advocated by teleomechanists and reductionists.

The fundamental point of reference throughout Reichert's discussion was his emphasis on the *systematic unity* of biological organisms. "In the organization of an individual, wrote Reichert, "we have a systematic product before us, ... in which the intimate interconnections of the constituent parts have reached their highest degree. When we think about a system, we normally picture to ourselves precisely this form of systematic product. Concerning such systems Kant said that the parts only exist with reference to the whole and the whole, on the other hand, only appears to exist for the sake of the parts'.[48]

In order to investigate the 'systematic ground character' of biological organisms Reichert reminded his readers that it was necessary to have a method appropriate to the subject. It could not be hoped that in time through persistent application of a mechanistic-reductionistic experimental framework the systematic character of the organism would at last be revealed as the final link in the causal chain. On the contrary, "the systematic perspective and method has its own pure logic which is valid at all times. ..."[49] This by no means implied that the systematist paid no attention to chemistry, physics, or the mathematical treatment of his material. But these were to be subjected to the guiding principle of their interrelation in the 'systematic whole' which is the constant source of insight into vital phenomena. With this perspective in mind the researcher would be in a better position to employ the principles of inductive logic which are the heart and soul of all sound investigation of nature:

namely that according to the universally valid method of inductive logic, natural objects must in all cases by collected and further investigated in the bonds and interconnections in which they are given.[50]

The mechanistic-reductionist approach, by contrast, attempts in Reichert's view to impose upon the organism an artificial framework which is only suited for understanding the production of "ships, clocks and pots – through aggregation in accordance with an arbitrary schema, only that the master craftsman is assumed to be sitting in the cell".[51] Whereas the systematist knows that at all times when investigating any part of an organism he is dealing with a component element related to a fundamental regulative unity, the atomist views the central nervous system, for instance, as "an aggregate of cells and their derivatives which have grouped together in a particular manner, originating perhaps in the lower end, in the spinal chord or even in the brain".[52] The mechanistic approach, in short, seeks to isolate organic subsystems from the whole, the *innige Verband* of interconnections in which they are given; and having thus sundered the relation to the regulative unity of the whole, reason is left to fabricate its own picture of organ function.

Reichert could envision only one method suited to the investigation of the living organism which avoids disrupting the intimate interconnections of its parts:

The systematist is aware both that he proceeds genetically and that he must proceed genetically. He is aware that the structure of an organism consists in the systematic division or dissection of the germ, which receives a particular systematic unity through inheritance, makes it explicit through development and transmits it further through procreation.[53]

As we have seen, this statement expresses one of the core methodological doctrines of the morphologists: To understand the systematic unity of the organism, one must examine its development. A further aspect of this position, which we have seen expressed by Müller but which was central to von Baer's embryological work as well, is that development always proceeds from a homogeneous undifferentiated mass to the particular, fully developed individual. Development as von Baer viewed it, proceeds from a center to a periphery. In the process of development the more general characters precede and guide the development of the more specialized structures, which are viewed not as aggregates of cells but rather as *divisions* of primordial generalized structural elements, von Baer's primitive organs.

Since the strict mechanist, such as Ludwig, views organization completely as a result of inter-atomic processes unconditioned by any relation to the regulative unity of the systematic whole, the most important organic processes must remain unintelligible to him:

The concepts of a germ, generation, development, organization, function, and irritability in the broader sense of the term, all of which lie rooted in the ground of the systematic being in organic nature, are in part simply incapable of definition within the atomistic framework and in part only in a fashion which does not at all correspond to the actual relations of organic nature.[54]

When the atomist refers to 'systematic unity', in Reichert's view, he can only mean an artificially contrived collection of mechanisms.

Rather than conceiving it as the basic atomic building block, the starting point of structure as Ludwig recommended, the cell, according to Reichert, is viewed by the systematist as the terminus of development.

If the cell is viewed as the servant in the organic creation, then we stand with it on the systematic standpoint. ... The cell is then not the starting point of each further consideration, rather it is the organized endpoint in the systematic division of organic being generally and of its individuals in particular.[55]

For the atomists the cell is an easily handled atom. For all who would rather construct organic beings artificially than dissect them, the cell is a suitable universally distributed building material. For the systematist the cell is that unitary, indifferent although organized foundation upon which and by the mediation of which the complex organic generative system makes itself explicit through development, propagates itself and presents itself as fully developed.[56]

Morphologists such as Müller and Reichert did not reject the cell theory as incompatible with their understanding of biological science, but they viewed its significance in terms of the systematic unity of biological organization in service of which it functioned.

Reichert was not interested in simply describing the differences between the systematic and atomistic viewpoints as exemplified by their interpretation of the cell. Rather he had two goals in mind: First he intended to show that whether consciously or not the systematic perspective is really smuggled into Ludwig's discussion of the cell without his ever calling attention to it. The material itself dictates this point of view. The cell simply is organized, and the mechanistic framework has neither language adequate to describe nor means appropriate to explain the fact of organization. The second point Reichert wanted to make was that while Ludwig and the reductionists were loudly proclaiming the cell theory as a victory for the atomistic approach to organization, it is more compatible with the morphological perspective and in fact filled an important gap in earlier theories advanced by the developmental morphologists.

To make his first point Reichert noted that while Ludwig and the reductionists attempted to compare the formation of cellular structures to crystal formation and other complex processes of precipitation, this language was at best metaphorical. Nothing of the sort happens in cell formation and Ludwig well knew it. Ludwig wanted to account for these complex processes in terms of the geometrical properties of the surfaces of the chemical constituents, the intermolecular forces between particles, and external conditioning factors in the surrounding milieu. But at the crucial moment in his discussion Ludwig made a leap to a completely different level of interpretation:

The particular shapes in the animal body, he adds, must necessarily be the consequences of form-giving structures – a template [Prägung]. ... The organism is, accordingly, the product of molds [Pregeanstalten], similarly to the galvano-plastic templates in factories, in the mint, or even the molds in the confectioner's shop.[57]

While the most important feature of the cell – namely its organizational unity – escaped the framework of a mechanical explanation, it seemed to fit naturally into the teleomechanist view of organization. One of the long-standing assumptions of the tradition we have been discussing, dating to the work of Kant and Blumenbach in the 1790s, was that while the germ was considered a homogeneous mass of material, it was nonetheless regarded as *organized*, containing its full developmental potential in the order and arrangement of its constituent elements. A major advance was made by von Baer in the investigation of this organic system. He demonstrated that the first structural manifestation of that potential was the 'primitive organs' of the germ, in which the type and rudimentary organ systems were generated. Missing from von Baer's embryological theories, however, was an organic mechanism for actually completing the developmental process of differentiation and individuation of the generalized primitive organs. A wide gulf separated the first rudiments of the animal in which its most general characteristics were present and the actual, particular inividual with its full complement of subordinate characters. The cell filled this gap:

Before the discovery of the cell ... a hiatus existed between the formless organic material and the complex generative system, which was represented in all its components to the last member by formed, organized elements – a hiatus which was filled by a sort of *generation originaria*, but which is now filled by the cell.[58]

A consequence of the view emphasized by the teleomechanists was that an understanding of the organism could not be obtained from a rational synthesis of component organic parts or structural units such as the cell. In order to preserve the continuity of developmental form and function as well as the *unity* of organization, emphasis had to be placed on *division* of a primitive, highly organized but structurally undifferentiated unity. Viewed in this light the cell was not an atomic building block with an independent life of its own. Rather it was a servant of the whole, an organized element in a systematic unity, receiving its instructions from the regulative unity of the whole, an agent carrying out the task of differentiation and individuation in which the life of the whole consists.

From the developmental history of each and every organism we conclude that a real system develops through division, or (according to a different scientific terminology), differentiation of a unitary indifferent foundation, the cell, and that this foundation traverses a series of developmental stages which are characterized by a given sequential increase in internal division and differentiation.[59]

In light of these considerations it was not Ludwig's artificial mechanical interpretation but Virchow's version of the cell theory that Reichert praised. Virchow had treated the cell not as having an independent life but as a component part of a systematic unity. Virchow "conceives the complex living organism as composed of multitudes of living individual cells, which remain united because they are interrelated to one another just as members of the same state".[60] In language heavily laden with then-current political imagry the mechanists stood convicted by advocating a liberal bourgeois conception of biological organization in which the march of technology breaks a truly organic society into factions and ultimately atomic individuals with no connection to one another or the organic whole which is the only source of meaningful existence.

Together with the principle of natural selection clarification of the mechanism of cellular development figured to be the two most critical issues in the debate over biological organization. For just as natural selection concerned the role and types of causal explanation to be employed in interpreting organization at the macro-level, the mechanism of cellular development concerned the interpretation of causal mechanisms at the micro-level of organization. If the reductionists were correct in viewing cellular development as completely accountable in terms of – albeit complex – linear modes of mechanical interactions between atomic elements, then these basic units of organization would simply be the results of concatenating independent series of physico-chemical processes. There would be no need to invoke an overarching set of organizational principles defining the conditions for linking mechanical processes in a manner appropriate to functional requirements. The organization of the cell and its particular function would be completely determined by its atomic constituents. A soup of the appropriate chemicals warmed at the proper temperature would produce a sequence of transformations resulting in a cell ready to serve a specified function. If, on the other hand, cellular development could not plausibly be interpreted in terms of standard modes of mechanical causality, however one might attempt to qualify the crystallization metaphor in order to render it applicable to the physiological context, it would become necessary to employ a different framework of causal relations as the only reasonable alternative for mediating the physico-chemical processes upon which the life of the organism is dependent. Virchow made what Reichert viewed as the crucial contribution to this entire debate. For his observations of cellular development removed the issue from the abstract domain of philosophical interpretation by introducing some new facts that seemed incompatible with the atomistic approach. But Virchow contributed much more than crucial observations. His analysis of the phenomena brought cellular development into line with the conceptual foundations of functional morphology as set forth by Liebig and Lotze; and in the course of doing so he resolved some problematic aspects in the views of his mentor, Johannes Müller.

Virchow's approach to the problem of biological organization ranks alongside the work of Bergmann and Leuckart as one of the best examples of careful teleological reasoning within a strict mechanical framework. Like Schwann and Vogt he sought to eliminate the notion of a teleological agent as a directive force and to identify every vital phenomenon with an integrated harmony of observable physical structures. Schwann, however, had overstepped the bounds of mechanistic explanation concerning biological organization. Virchow argued that earlier treatments which assumed that biological organization required a special organizing set of potencies residing in an organic fluid, such as Schwann's cytoblastema, could not serve as the basis for a truly biological theory, for ironically, they did not exclude the possibility of spontaneous generation from an accidental concurrence of events, and biological organization was not to be explained in terms of random associations of physico-chemical processes. Virchow's own observation of cellular development led him to the conclusion that the intracellular substance within which new cells form is the product of pre-existent cells.[61] Hence the substance from which cell structure is constituted must itself be the product of a pre-existent cell:

Absolutely no development originates *de novo*. ... We must ... reject the assumption therefore of spontaneous generation. ... Where one cell arises, there must have been a previously existing cell (omnis cellula e cellula).[62]

Schwann and those who followed him in assuming the construction of cells in a cytoblastema had erred in their interpretation of the relationship of the nucleus and nucleolus as well as in the process of nucleation. The nucleolus does not appear first and then serve as the basis for the construction of the nucleus, which forms around the nucleolus like an hourglass; rather a new nucleus is always formed from the division of a pre-existent nucleus according to Virchow. The process of division is indicated first by the enlarged appearance of the nucleolus, then by a gastrulation of the nucleolus and finally the appearance of two separate nucleoli. While this process is going on, the cell nucleus itself is passing through an exactly parallel set of stages, so that *simultaneous* with the appearance of two separate nucleoli, the cell nucleus has itself divided. Shortly after the division of the nucleus, the cell divides.[63] This was no trivial observation. For it seemed incompatible with a strictly deterministic form of mechanical causality of the sort required by Ludwig. The problem was that the different processes linked with cell division did not follow one upon the other in a sequential temporal series, but rather each process seemed to be reciprocally interconnected with all the others so that each one was both end and means of the other. One set of processes did not result in the emergence of a structure, which in turn initiated another series of transformations, and so forth. The structure emerged and divided *in toto*, in a developmental process characterized by auto-regulation. The whole and its parts were inseparable. There are no independent subunits of structure out of which the cell is assembled. It is a complex structural unity by itself incapable of undergoing dissolution without losing its essential vitality.

Müller too had defended a holistic interpretation of the cell, but he had, as we have seen, required an emergent non-material *Lebenskraft* as the agent orchestrating intra-cellular chemistry. Virchow replaced this teleologic agent with purposively organized irreducible structures. In Virchow's view the cellular material itself rather than the extra-cellular fluid organized by an emergent force serves as the source of the new cell. The investigation of the particular mechanisms within the cell which initiate these processes are a proper object for future chemical and physical research, he wrote, but one thing at least was certain:

We as biologists have at least established that the vital activity of fully formed structural elements produces other new elements, and that this activity is inseparable from the elements themselves even if external stimuli are required to set it in motion. These formative stimuli can be of various sorts, physical, mechanical, or chemical. Just as the spermatozoa stimulate the egg cells to their plastic activity, so are there other catalytic materials which can excite cells.[64]

Virchow served notice that his conception of biological organization was identical with that set forth by Liebig and Lotze. The viewpoint stressed throughout Virchow's work is that vital function is inseparable from a structural basis and that the fundamental elements must be an integrated structural unit rather than 'potencies' of an organizational principle. Thus where Müller, von Baer, and Döllinger had taken recourse to a directive agency localized in matter beneath the germinal disc as the source of cellular differentiation within the germ layers, Virchow demanded that unobservable material differences pre-exist in the apparently homogeneous germ cells:

I hold it for probable that in the germ cells fine internal differences exist which pre-determine the subsequent transformation to a determinate substance; not differences which are mere potencies present in the germ cells, but rather actual material differences which are so fine that we have as yet been unable to demonstrate them.[65]

Although he sought to eliminate any role for an "organizational force or an organizational idea as viewed from a higher standpoint"[66] and strictly limit biological organization and the source of vital activity to observable physical structures, Virchow recognized that causal-mechanism as an explanatory tool in biology must stand under the guidance of a regulative principle; namely the concept of integrative unity as the fundamental characteristic of life:

I demand above all else the recognition that the cell is the final form-element of all living phenomena in healthy as well as in diseased conditions, and that all vital activity issues from it. To some it may not appear justified that life is treated in this way as something completely special; it may even appear to many as a sort of biological mysticism if life is separated from the rest of natural phenomena and not immediately dissolved into chemistry and physics. In the course of my investigations everyone will be able to convince himself that it is scarcely possible to think more mechanistically than I do, whenever the illumination of the processes within the final elements of structure are at issue. But however much the exchange of matter within the cell is connected to particular structural parts, the cell is always the seat of vital activity, the sub-domain [Elementargebiet] upon which the manner of activity is dependent, and it retains its significance as a living element, only as long as it presents an undisturbed whole.[67]

It is important to emphasize – as indeed Reichert indicated – the extent to which this statement is consistent with the original aims of teleomechanism as formulated by Kant. What Virchow offered was a positivistic interpretation of the core doctrine of that program, vowing complete agnosticism with respect to the causes of organization. We have seen that Kant argued for the necessary assumption of certain *Keime* and *Anlagen* at the basis of biological oranization which are the source of the determinate structure of the organism. So long as observation was incapable of establishing an integrated set of primitive structures that remained constant throughout all the changes in the development of the organism, the vital materialists and developmental morphologists had been forced to treat these *Keime* and *Anlagen* as potencies residing in the organized generative fluid; and in order to explain the source of the organization of these potencies within a materialistic framework, they had required the notion of a *Lebenskraft* as a special formative force. Having established that all the vital processes are traceable to a set of structures that are never dissolved into an organic 'cytoblastema', Virchow had no need of assuming that the *Keime* and *Anlagen* were 'potencies' for structure. For him they were simply the material elements of cell structure. This did not mean, of course, that physical and chemical investigation could go no further. Indeed, as we have seen, he asserted that the mechanism employed in nucleation and cell division as well as in the differential physiological functions of different cell types was an important field for chemical and physical investigation. But those investigations were not conceived by him as providing an explanation for organization; rather they were understood as explaining how a body given as organized functioned.

Life begins with the cell and all its structures fully constituted – all of a sudden and at a single bound, not by a gradual sequence of physico-chemical tranformation. It was the special importance of an integrated set of structures as an organized whole at the basis of this conception that led Virchow to emphasize the extraordinary role of the nucleus as the locus of organization:

> The nucleus plays an extraordinarily important role in the cell, a role which ... is related less to the specific performance of the elements as much more to the preservation and multiplication of the elements as vital parts. The specific – in a narrower sense, animal – function is manifested most clearly in the muscle, nerve or glandular cells, but the special activities of contraction, sensation, or secretion appear not to have any direct relationship to the nucleus. On the other hand, that in the midst of its functioning an element remains an element that is not destroyed through continuous activity, this appears to be bound essentially to the existence of the nucleus. All cellular structures which lose their nuclei are frail, they are destroyed, they disappear, die and are dissolved.[68]

Johannes Müller had argued, as we have seen, that two processes are necessary to an organized being; the constant regeneration of the germinal material which maintains the organization of the whole as such, and the specialized agents of physiological function. Müller had never resolved on a satisfactory mechanism for these two problems. As an answer Virchow proposed the organizational unity of the cell nucleus as the constant source of integration in the absence of which specialized function of the cell was unthinkable.

THE SCIENCE OF LIFE AND THE NEWTONIAN WORLD PICTURE

At first glance the tenacity with which the teleomechanists continued to defend their position may seem surprising. For as we have seen, the teleomechanists were the ones to first point out the insufficiency of the older notion of the Lebenskraft; and it was they who had defined the critical theoretical problem confronting the advance of physiology in the 1840s as the need for a general analysis of the conditions of material exchange. The solution of this poblem, they had argued, would eliminate the role of a special vital force and place physiology on the sound theoretical and empirical foundations of physics and chemistry required for its future advance. Helmholtz had solved that problem by establishing the conservation of energy, but now the teleomechanists were not prepared to support all the conclusions DuBois-Reymond wanted to draw from it. Without doubt their reasons transcended the specific issues of the scientific debate and included the intersection of a variety of political and religious concerns, but also at stake in their refusal was a set of issues deeply internal to the entire conceptual fabric of mid-nineteenth century science. At stake was the Newtonian character of natural science, which although apparently invincible in its advance on all fronts, was about to be severely challenged, its core concepts about to suffer radical revision and transformation. Central to this conceptual revolution were the two areas of science directly at issue in the teleomechanist-reductionist debate: namely, theories of organic form and transformation, particularly the Darwinian theory of evolution by natural selection and the developing field of energetics. Neither reductionists nor teleomechanists were aware of the transformation about to take place in the foundations of classical Newtonian science. Both sides believed that the principles of Newtonian science could unite in a consistent manner the analysis of form and function in biology with the investigation of the atomic-molecular processes of material exchange. But in reviewing the positions defended by the two sides, we cannot help sensing that we are in one of the antechambers of great historical change.

Since the early decades of the nineteenth century German biologists had been seeking a Newtonian theory of animal form and function. Both teleomechanists and reductionists believed they had hit upon the correct set of principles and assumptions required to succeed in this aim. In essence their debate concerned how best to construct a Newtonian biology and where its limits might reside. While both sides agreed that a scientific, Newtonian biology must employ only natural forces and the principles of mechanics, many prominent biologists, including Liebig, Lotze, Bergmann, and Virchow believed the wiser strategy to be the one offered by teleomechanism. Their position was stated by T.L.W. Bischoff in 1858 in his eulogy of Johannes Müller. No paraphrase of the position could represent it more clearly than Bischoff's own original statement, which I quote here in full:

The material body once fashioned, the organ once constructed in such and such a manner with such and such a composition, the fluid once constituted in such and such a fashion, all are subject to the general laws of matter with which chemistry and physics have made us familiar to date or will yet reveal to us in the future. All changes within them, all phenomena produced by them, all so-called functions are the products of material changes and interactions. They can be investigated, therefore, in accordance with their determining conditions, and they can be explored in terms of their generation. They are determined by laws of necessity given with the material conditions of their existence, and the determination of these laws must be the sole object of our pursuit if we wish to be the masters of those phenomena. Let me give an example: I think that mankind will never succeed in completely researching the forces that generate a muscle, and the forces that determine and maintain it. However, once matter has assumed the form and composition which we find given in the muscle, which we can study with the aid of anatomy and according to the rules of physics and chemistry, then when we understand those things completely, it will be possible for us to say which phenomena, which so-called functions will necessarily be made manifest in the muscle if this or that change is effected in it. The so-called functions of an organ, the motions or other activities which we observe in it or in an entire organism are therefore not direct effects of specific vital or organic forces, rather they are the necessary consequences of material conditions. But that the organ is what it is and that the organism possesses this form and composition and no other, this is the effect of special forces which control matter with its characteristics up to a certain degree and are capable of bringing it in this or that form and relation which it is incapable of assuming in inorganic nature merely in accordance with the conditions lying within it. The latter forces are accessible to our research and study; the former are not and probably never will be.[69]

The version of teleomechanism set forth by Bischoff differed from the older versions of the program in one crucial respect. Functional morphology did not require the assumption of special vital forces. Mechanical, chemical and electromagnetical forces were the only tools at the disposal of the functional morphologists. While he did not postulate the existence of special organic forces, the functional morphologists did, however, demand the existence of special modes of organizing and arranging the forces of the inorganic realm into a functional organ. In Bischoff's view muscle, for example, is constituted from and functions only by means of physico-chemical forces, so long as one examines it within the organic context of a living animal. But when abstracted from this context, some agency other than those physico-chemical forces is required to generate that organ, to arrange those forces into a functional organ. In effect he was asserting the existence of a special set of *organic laws* for relating physico-chemical forces, but he strictly denied the possibility of ever attaining a causal explanation for them. This position was clearly a modification of the view held earlier by Kielmeyer, Weber, Müller and von Baer. As we have seen, von Baer had assumed in his early work the existence of a special developmental force, the *Gestaltungskraft*, whose effects could be investigated directly. Von Baer had even gone so far as to characterize this force as a central force in terms analogous to the gravitational force. Ontological status was now to be denied that developmental force. It could never be the object of empirical research.

Instead of assuming the existence of a special set of biological entities in addition to the forces of the inorganic realm, Bischoff was arguing that the problem of biological organization could be solved by the assumption of certain biological laws for relating the forces of inorganic nature and that these are manifest in certain ground states of organization which are the given starting points of physiological research.

This assumption of special laws was the real source of contention between Helmholtz and the functional morphologists. In fact, some of the peculiar features of Helmholtz's formulation of the conservation of energy can be interpreted as a conscious attempt to reject both functional morphology and developmental morphology by showing that they could be equated to the same position. Helmholtz's statement of energy conservation, as we have seen, rested upon an extremely idiosyncratic assumption. He argued that the *only* forces in nature are constant, central, action-at-a-distance forces.[70] Developmental morphologists, von Baer in particular, had universally treated the vital force as a central force. Helmholtz's memoire demonstrated the impossibility of this assumption because the effects it was assumed to account for could all be explained in terms of mechanical action in accordance with the conservation and interconvertability of all central forces. If Helmholtz's initial premises were accepted, then the conclusion was inevitable: Vital materialism must certainly go. But Helmholtz would also not permit the functional morphologists' assumption of organic laws to stand. Whenever he discussed the philosophical foundations of the physical sciences, he went to great lengths to impress upon his readers and listeners that *any regularity in nature which we designate as a law, must be the effect of a force*:

... A law of nature, however, is not merely a logical concept that we have adopted as a kind of *memoria technica* to enable us more readily to remember facts. ...

We experience the laws of nature as objective forces.[71]

To remain consistent with this requirement energy cannot be considered to designate a specific entity such as a reservoire of power or force, of which other forces in nature are only special manifestations. Helmholtz was determined to eliminate the possibility that through the proper arrangement of physical forces an 'emergent force' might be generated as a sort of epiphenomenon of the original set of forces. This strategy had been employed, for instance, by Müller and von Baer. Consequently, 'energy' in Helmholtz's view does not designate a force but a *constancy of relation* between motive forces; hence his persistence in designating it the 'conservation of force'. More accurately stated, the conservation of energy in Helmholz's view is much more akin to a *principle* than to a law as he described it in the passage quoted above. Like the principle of least action, it expressed a constancy of relation rather than the effect of a physical entity.

This manner of viewing energy had a number of important consequences. One of these consequences was that Helmholtz was never able to accept the notion of a field, for it entailed the existence of energy diffused in space, which was not the effect of a relation between two or more bodies. The notion of a physically existent field surrounding a single isolated point mass was in Helmholz's view simply not possible.[72] But this relationship between laws and forces also had significant consequence for biology, and indeed Helmholtz had these consequences explicitly in mind when he first insisted that laws must be the expression of forces. For if that claim is admitted, it effectively annihilates the functionalist position. There cannot be special organic laws, because any 'organic law', such as a special law regulating muscle action, or a law of development would entail the existence of a 'vital' force producing it, and the argument establishing the conservation of force had excluded any possibility of such forces. Teleomechanism could not assume the existence of special organic laws under the notion of 'Zweckmässigkeit' without implicitly postulating vital forces. Consequently Helmholtz felt secure in concluding that functional morphology was in essence another (rather uncritical) version of the old vital materialist approach, and as such, it too must be rejected.

In reconstructing a debate over a theory that has long since become scientific dogma, forming as it were the cornerstone of every elementary science textbook, we are apt to view the historical debate through the eyes of the eventual victors. In doing so we run the risk of neglecting the merits of the positions held by their opponents. It is a tired cliché that we must view any debate in its historical context by taking into account the constellation of factors forming its total problem domain as well as the historical actors involved. Applying this well-worn caveat to the current case, we must keep two facts clearly in mind. First, Helmholtz and DuBois-Reymond were not defending a position against physiologists. Had their opponents been physiologists pure and simple there would have been no debate. In fact one of the major strengths of the campaign waged by the reductionists was that they managed to contain the issues discussed within the confines of physiology. But, as we have seen, after 1842 no German physiologist worth his salt really believed that *on physiological grounds* it was necessary to invoke a special vital force to explain the functioning of the animal machine. Thus even

Leuckart and Bergmann, two of the most significant functional morphologists, argued that every detail of ontogenesis could in principle be incorporated within a mechanical model of explanation. But the audience Helmholtz had to contend with was not strictly composed of physiologists, they were *zoologists* and *morphologists*. These were a unique breed of polymath individuals, fostored no doubt by the peculiar institutional requirements of the German bio-medical sciences of the mid-nineteenth century. For them physiology was a central element of an entire theory of animal form. Consequently any major innovation in physiology was immediately evaluated in terms of its consequences for other related domains, especially its implications for the general theory of organic form so dear to German zoologists.

A second fact is also to be born in mind when reviewing the split between teleomechanists and reductionists. Helmholtz published his memoire in 1847 and DuBois-Reymond was quick to seize upon it in the debate over vitalism as early as 1848. But once the scope of the debate widened to include the implications of energy conservation for the theory of animal organization a host of serious problems arose that could only be resolved by the theory of natural selection. Darwin's *Origin of Species* was published in 1859 and the full force of its argument was not instantly appreciated in Germany. The argument Helmholtz launched against the assumption of purposive organization in biology was on safe ground so long as he confined it to physiology, but in order to extend it to the general theory of animal organization and the interconnection of forms it was necessary to conjoin the implications of energy conservation with the principle of natural selection. In the interim between 1848 and 1859, therefore, there were serious unresolved questions. In this context the objections of the teleomechanists were not the ravings of the – perhaps conservative and politically motivated – lunatic fringe. They were in fact rational objections of zoologists concerned to place their science on its most consistent theoretical foundations.

Apart from the brief discussion of the conservation of force in organic contexts in the famous memoire of 1847, Helmholtz did not treat the problem in any detail in his scientific papers. There were, however, a number of crucial questions concerning the full application of the principle in biology. The only place in his writings Helmholtz devoted to these subjects was in his lectures for popular consumption. The popular lectures reveal a master of the internal politics of science at work, keenly aware of the critical problems confronting his research program, deftly slipping past troublesome areas, and conflating the objections of intelligent opponents with views held by strawmen and whippingboys. In his popular lectures Helmholtz managed to dismiss the approach of the teleomechanists by classifying their position along with that of Müller as identical in all essential to the views of Schelling and Stahl, views that were considered anathema to the progress of science by Helmholz's contemporaries.

It is clear that Helmholtz was aware of the problem confronting the full acceptance of energy conservation in biology. But one of the critical problems carefully avoided in the popular lectures was the problem of entropy, although he was certainly aware of it. In all his discussions of the application of the principle of conservation of energy in organic contexts, Helmholtz compared the animal body to a heat engine. As early as 1854 Helmholtz, however, was underscoring the directedness of thermodynamic processes:

> The heat of a body which we cannot cool further cannot be changed into another form of force – into electric or chemical force for example. Thus in our steam engines we convert a portion of the heat of the glowing coal into work by permitting it to pass into the less warm water of the boiler. If, however, all the bodies in nature had the same temperature, it would be impossible to convert any portion of their heat into mechanical work. According to this we can divide the total store of force in the universe into two parts, one of which is heat and must continue to be such; the other, to which a portion of the heat of the warmer bodies and the total supply of chemical, mechanical, electrical and magnetical forces belong, is capable of the most varied changes of form and constitutes the whole wealth of change which takes place in nature.[73]

Helmholtz then continued with the consequences drawn by William Thomson from the necessary distinction between these two parts of the total energy:

> But the heat of warmer bodies strives perpetually to pass to bodies less warm by radiation and conduction, and thus to establish an equilibrium of temperature. At each motion of a terresrial body a portion of mechanical force passes by friction or collision into heat, of which only a part can be converted back again into mechanical force. This is also generally the case in every electrical and chemical process. From this is follows that the first portion of the store of force, the unchangeable heat, is augmented by every natural process, while the second portion, mechanical, electrical and chemical force, must be diminished; so that if the universe be delivered over to the undisturbed action of its physical processes, all force will eventually pass into the form of heat, and all heat come into a state of equilibrium. ... In short the universe from that time forward would be condemned to a state of eternal rest.[74]

These sobering thoughts were expressed in a lecture on the interaction of forces in nature in 1854 and they have obvious implications for biological phenomena. But when Helmholtz later lectured on the application of the law of conservation of energy to organic nature at the Royal Institute in 1861, he failed to mention matters relating to entropy. The energy principle was clearly stated in that paper: if mechanical power is produced by heat, a certain amount of heat is lost and this is proportional to the quantity of work produced. Having discussed the importance of the first law of thermodynamics, Helmholtz failed to examine the relationship of entropy to biophysical systems. This was a curious, perhaps strategic omission. A legitimate concern to most German zoologists and physiologists, however, was a demonstration that progressive evolution was not negated by the existence and necessary increase in entropy. Viewing the animal machine as a heat engine, Bergmann, for instance, had argued that the homiothermal plan of organization is more complex than polikilothermy and that it therefore represents a higher initial potential energy state, i.e. one capable of generating more mechanical work. In general, since the potential energy required to sustain more complex states of organization requires an increase of information, it would seem that entropy considerations must rule out progressive evolution on the grand scale in which it appears to have occurred in nature.[75] Bergmann and Leuckart's strategy of assuming that variations on the same organizational plan and regression of forms from higher to lower states of organization was consistent with the physical principles Helmholtz had espoused. The idea that the vertebrates might have a polikilotherm as an ancestor seemed to stand in sharp contradiction to those physical principles.

Although he did not discuss entropy in the passage quoted earlier, Bischoff's formulation of the position advocated by the teleomechanists was sensitive precisely to the issues of relating conservation of force, biological organization, and complexity; and in 1860 this position was a rational alternative to physical reductionism. The teleomechanist position was that no matter how many times the conservation of energy is confirmed by biophysical systems – a proposition they all held as indisputable – it says nothing about how biophysical systems get organized in the first place. Muscle action can be explained completely in terms of processes of material exchange, but, as Bischoff indicated, the laws of physics do not explain how such intricately organized systems are generated as organized and capable of functioning in the manner they do, nor are the laws of chemistry and physics able to account for the obvious 'Zweckmässigkeit', the teleonomy built into biophysical systems. Bischoff and the teleomechanists argued that in the current state of science, it seemed better to assume the existence of certain ground states of organization and then investigate the manner in which they vary. Since the principles of mechanics could not account for the assemblage of complex physical systems into an organic machine, it necessarily followed that a progression of increasingly more highly organized and complex organic machines could also not be explained.

By 1869 Helmholtz had assimilated the Darwinian theory of evolution into the service of his reductionist perspective and he made his first public assault on the teleomechanist position in its broader zoological dimensions. In his lecture, 'Über das Ziel und die Fortschritte der Naturwissenschaft', Helmholtz argued that even though a number of German physiologists in recent decades (our functional morphologists) had restricted the active agents in physiological explanations strictly to chemical and physical forces alone, they had nonetheless allowed the animal body to develop according to a plan. However much they sought to avoid the appelation, as Helmholtz saw it, they were vitalists:[76]

If, thus, the law of the conservation of force holds good for living beings too, it follows that the physical and chemical forces of the material employed in building up the body are in continuous action without interruption and without choice and that their strict conformity to law never suffers a moment's interruption.[77]

The cutting edge of this remark was in the phrase 'without choice', for functional morphologists could easily defend themselves against the other objections in this passage. They too held that physiological processes occur in uninterrupted conformity to natural law. But they required a specific order and arrangement of these forces to be given. Purposive organization for them was a necessary assumption in physiological explanations, which could not be further explicated in terms of known natural laws. Although they did not postulate an organizing agent in nature, functionalists, dependent as they were upon the notion of 'plans of organization', had not succeeded in eliminating something analogous to 'choice' from their theories. For, in modern terms, their theories rested ultimately on the input of a great deal of 'information' in order to establish the potential required to manifest the sequence of motions identified with the phenomena of development and physiological function. In effect they seemed to be saying that physiology cannot operate in actual practice without the postulation of an intelligent universe, but that all mention of this postulate must be carefully excluded from the explanatory framework itself. Just as Kant had assumed an irreducible categorical framework for the operation of the understanding, so the teleomechanist ultimately assumed a framework of irreducible bionomic rules prescribing the interrelations of natural laws.

Helmholtz was keenly aware of the dilemma the functionalists had created. There was only one solution: complete physical reductionism. "The wonderful – and, through the growth of science, the more and more evident – purposiveness in the structure and function of living beings was indeed the main reason behind the comparison of the vital processes with the behavior of something functioning under the control of an entity like a soul. In the whole world we know of but one series of phenomena possessing similar characteristics, namely the acts and activities of intelligent human beings; and we must admit that in innumerable instances organic purposiveness appears to be so extraordinarily superior to the capacities of human intelligence that sometimes we are inclined to ascribe a higher rather than a lower level to it".[78] But organic purposiveness could be dismissed from physiology altogether in Helmholtz's view, for:

> ... Darwin's theory of the evolution of organic forms, ... provides the possibility of an entirely new interpretation of organic purposiveness.[79]

An explanation completely in accord with the conservation of energy and with physical reductionism.

Helmholtz championed the view that Darwin had explained biological adaptation as the result of a blind law of nature rather than as the manipulation of a pre-programmed and untestable original purposive organization. The 'law of nature' he attributed to Darwin here, however, was not natural selection, rather Helmholtz referred to "the law of transmission of individual characteristics from parents to offspring".[80]

> If both parents have characteristics in common, the majority of their offspring will also possess these characteristics; and even though some among the offspring may display them to a less marked degree, there will always be others (given a large enough number) in which they have been intensified. If the latter offspring are selected to propagate, a gradual intensification of the characteristics may be obtained and transmitted.[81]

Having established that animal breeding provides experimental evidence for the truth of his 'law', Darwin had gone on to demonstrate its applicability to animals and plants in the wild state.

Helmholtz's formulation of Darwin's theory reflects his outlook as a Newtonian physicist and physiologist. As a strict Newtonian he emphasized that evolution depends on a law governing material interactions: it is a law of inheritance describing the behavior of the genetic material. This 'law' explains natural selection. In Helmholtz's view natural selection is not a principle or a law of nature, rather it is an effect of the laws of inheritance. The key to evolution is, in his view, variation in the 'intensity' of traits among individuals. These variations are inheritable, and through proper selection of mates, they can be further intensified. In the wild state Helmholtz argued, these differing intensities correspond to organisms with a slight advantage in the struggle for existence, and "the individuals distinguished by some advantageous qualities are most likely to produce offspring and transmit their advantageous quality".[82] Thus in Helmholtz's view, given variability among individuals and given the inheritance and transmission of that variability, the result will be a differential reproductive advantage in the struggle for existence, adaptation and finally the transmutation of species.

Living in the wake of the great victories of Darwinian evolution and the successes of molecular biology we are apt to see profound wisdom in Helmholtz's discussion of Darwin's theory. But there was something amiss about his formulation that sounded a loud alarm in the mind of every strict Newtonian mechanist. Helmholtz knew exactly what that was:

An animated controversy still continues, of course, concerning the truth or the probability of Darwin's theory, though the controversy now centers mainly on the scope that should be assigned to variations of species.[83]

Variation. That was the heart of the problem. While Helmholtz spoke as though the law of inheritance were a good deterministic natural law, he could not hide the fact that the whole weight of Darwin's theory rested ultimately upon the innate, unexplained, chance variation of biological individuals. Moreover, Helmholtz was certainly aware of the seriousness of this problem in overcoming the resistance to Darwinian evolution even by well informed Newtonian scientists. As he indicated, by the late 1860s the debate had focussed on the apparent limits on variability of organisms, these arguments being developed most forcefully by Fleeming Jenkin. Helmholtz was well aware of Jenkin's arguments, just as he was aware that Jenkin's associate and his own good friend William Thomson (Lord Kelvin) strongly rejected Darwin's theory on physical grounds. Here were good Newtonians who fully understood the conservation of energy but who flatly rejected Darwin's mechanism for explaining the diversity of organic forms.[84] Could Darwin's world and Newton's world really be made consistent with one another?

In defense of Darwin Helmholtz went on to cite the growing body of paleontological evidence demonstrating the transition of some forms into others. He pointed furthermore to the harmonious manner in which Darwin's theory synthesizes the results of embryology, comparative anatomy, and geographical distribution. This, of course, was not the issue at all, certainly not in Germany, but equally for a large body of British and French biologists.[85] The German teleomechanists were evolutionists. They did not need to be convinced that species had been transformed. Indeed it was a central doctrine of all their biological theories. But they were not Darwinian evolutionists. Limiting the transmutation of species to organisms constructed on the same organizational plan, they rejected the Darwinian theory of the community of descent.

Helmholtz was defending a position fraught with paradox. In order to eliminate vital forces and active teleological agents from physiology he had insisted on a strict mechanistic determinism as the only mode of analyzing the interactions of material particles. In such a universe chance cannot assume the status of a cause. It can only indicate the failure of some cause to act, and this 'failure' must ultimately be explained in terms of a further, unobserved mechanism which has suppressed the expected course of events. By improving our understanding of these microprocesses, the possibility of providing a strict deterministic prediction lies open; predictability in terms of necessary mechanistic causation increases, while the domain of uncertainty, accident, and chance is progressively eliminated. There may be statistical regularities in random processes, but for a strict Newtonian determinist those regularities can never be raised to the level of a natural law. In drawing upon the Darwinian theory of evolution in order to eliminate the last vestige of teleological thinking from biology and thereby setting the life sciences on the same Newtonian conceptual foundations as the inorganic sciences, therefore, Helmholtz was forced to violate the very conceptual foundation of the whole enterprise. For in arguing that the whole of biological science could be reduced to the conservation of energy and the laws of inheritance, he was attributing a direct causal role to chance.

In order to see the irony of Helmholtz's position as viewed from the perspective of the Newtonian methodological principles he espoused, one need only ask the same question of his 'law of inheritance' that he had demanded the teleomechanists answer: To wit, since every law of nature is the effect of some force, what 'force' accounts for the random features of individual variation and inheritance?

In attempting to refashion biology as a truly Newtonian science, Helmholtz had discovered the very limits of the Newtonian causal-mechanical approach to nature. As necessary background assumptions to any scientific biological theory Helmholtz had assumed a mechanistically determinate universe in which energy is conserved and entropy is increasing. In such a universe the principles of Darwin can only be introduced at the cost of inconsistency at the heart of the conceptual foundations of science. Only in a universe operating according to probabilistic laws, a universe grounded in non-deterministic causal processes, is it possible to harmonize the evolution of sequences of more highly organized beings with the principles of mechanics.

Two paths lay open for providing a consistent and rigorous solution of this dilemma. One alternative is that of twentieth century science. It is simply to abandon the classical notion of cause in favor of a non-deterministic conception of causality. In the late nineteenth century this was not an acceptable strategy. To be sure statistical methods were being introduced into physics with great success, but prior to the quantum revolution in mechanics no one was prepared to assert the probabilistic nature of physical causes. To Helmholtz as well as such contemporary pioneers in statistical methods as Boltzmann, Clausius and Maxwell, the idea that the interactions of individual material particles might obey probabilistic laws was simply inconceivable.[86]

A second solution to this dilemma is that proposed by the teleomechanists. According to this interpretation rigidly deterministic causality can be retained, but then limits must be placed on the analysis of the ultimate origins of biological organization, and certain ground states of purposive or *zweckmässig* organization must be introduced. It was Kant who had first realized that a truly Newtonian biology must operate under these limiting assumptions. Kant, who was perhaps the greatest spokesman of the Newtonian mechanical universe, had penetrated deeply into the inner recesses of the classical, Newtonian notion of causality. In doing so, he sensed, as we have seen, that the mechanically deterministic notion of cause must be severely restricted in the biological realm. This did not mean, however, that Hume's probabalistic approach to all questions of causality should be explored as a means of resolving the dilemma. To Kant, of course, the proposal of jettisoning the classical notion of cause would not have been an acceptable alternative. For in opposition to Hume he held that the notion of cause as rigidly deterministic is a category of the human mind, and it cannot be dismissed at certain points as no longer relevant. There is no choice in this matter. It is the inalterable given starting point of all experience of nature, the ultimate foundation of natural science itself. As the true heirs of the Kantian biological tradition the teleomechanists could sense the deep lack of fit between their own view of biological science and that emerging from the work of Helmholtz and Darwin. They were bound to resist any attempt to introduce chance as a causal principle of biological organization.

The position defended by the teleomechanists was ultimately untenable. It could not bear the weight of the empirical evidence accumulating against it. As intermediate forms became known such as *Amphioxus* linking the vertebrates and invertebrates classes as well as forms such as *Archaopteryx* and *Compsognathus* linking the reptiles and birds within the vertebrate class, the theory that a few plans of organization underlie the animal kingdom became increasingly difficult to defend. At another level, however, there was a stalemate between the advocates of teleomechanism and the reductionists. For both sides were insistent on the Newtonian foundations of their respective approaches to biology. In the final analysis the only resolution of their impass was the construction of an entirely new set of conceptual foundations for both the biological and physical sciences which could cut the Gordian knot of chance and necessity. No one in Helmholtz's generation could foresee that statistical phenomena such as natural selection and the second law of thermodynamics held part of the key to those new conceptual foundations.

In breaking the impasse over the Newtonian foundations of biology, Helmholtz adopted the strategy of eliminating teleomechanism from consideration as a viable research program by arguing that it was in essence indistinguishable from a form of vitalistic physiology with which no scientifically minded physiologist wished to be associated. Schleiden adopted this strategy more openly and in much more polemical fashion. He selected statements from Virchow's works and held them up to ridicule.

In his *Vier Reden über Leben und Kranksein* (1862), Virchow had claimed:

It is useless to attempt to find a contradiction between life and mechanics.[87]

But on the same page of the lecture he remarked:

The cell is a self subsistent port in which the known chemical materials with their usual properties are arranged in a special manner and enter into activity in accordance with this arrangement and its properties.[89]

Having affirmed the teleomechanical view that physiological phenomena must be considered in terms of functional arrangements operating according to strict mechanical necessity, Virchow went on to underscore the further point that the sources of the apparently purposive organization could not be reduced to mechanical principles alone:

The mechanical conception of life is not materialism. For what can be meant by this term other than to claim that the direction of all existing things and occurences can be explained in terms of known material substances.[89]

Schleiden could barely find language harsh enough to express his contempt for this point of view. He could think of no one in the entire history of natural philosophy since Plato and Aristotle so unskilled in logic as to even consider endowing matter acting according to mechanical necessity with a built-in organizational plan. "I exclude Hegel, whom I imagine capable of every sort of nonsense".[90]

> Does Virchow not know anything of Kant, his *Metaphysische Anfangsgründe der Naturwissenschaft*, Kant the man who comprehended the entirety of mathematics and natural science of his own day? Impossible! Otherwise he would not have made such poor judgement.[91]

Here we see one of the key elements in the strategical polemic waged by the reductionists. They all claimed Kant as the intellectual father of their program. Theirs was a very selective reading of Kant, however, which characteristically emphasized those positivistic features of Kant's epistemological writings consistent with a stringent reductionism. Helmholtz, for instance, reflected long and deeply on the sections on space and time as well as the deduction of the categories in Kant's *Kritik der reinen Vernunft*. He also devoted considerable attention to Kant's *Metaphysische Anfangsgründe*. These works were frequent objects of discussion in Helmholtz's writings, and his own careful study of Kant led him to claim that he was an even better Kantian than Kant himself, for he had worked out the implications of Kant's work for the natural sciences in a manner more consistent with the 'spirit' of the system than Kant had succeeded in doing.[92] Nowhere in the writings of Schleiden, Helmholtz or DuBois-Reymond, however, is mention made of the *Kritik der Urteilskraft* and of the need for supplementing the mechanical with a teleological framework in biology. As we have seen the framework of causality that had been capable of grounding mechanics in the *Kritik der reinen Vernunft* and *Metaphysische Anfangsgründe* was regarded by Kant as an insufficient basis for the life sciences, and he had carefully examined the difficulty in the *Kritik der Urteilskraft*. Even assuming that the second critique was no longer included among the standard repertoire of Kant scholars in the 1850s and 60s, Helmholtz, DuBois-Reymond, and Schleiden certainly knew the problems raised by Kant and the arguments he set forth concerning teleomechanical frameworks in biology, for as we have also seen, Johannes Müller's discussion of vital forces and teleology in his *Handbuch* was explicitly based on those sections of the *Kritik der Urteilskraft*, and Müller made direct use of the definitions of vital function set forth by Kant in that work. This selective omission from the writings of the reductionists was indeed significant.

Helmholtz, DuBois-Reymond, and Schleiden won their campaign against the teleomechanists. Several factors played a role in their victory. Not the least of these was an unshakeable conviction in their own competing research program. Armed with the conservation of energy and the principle of natural selection, they argued that the principles of mechanism were not only sufficient for treating the problem of material exchange in organic systems but that they were also sufficient for solving the question of *Zweckmässigkeit*, of purposive organization as well. Here implacable conviction succeeded where genuine arguments could not. Certainly another factor in their victory was the overwhelming success of their own science. Helmholtz's work on muscle action and the conservation of energy and DuBois-Reymond's work on animal electricity were indisputable as major scientific achievements even if all the consequences they wished to draw from them were not. Moreover, Helmholtz and DuBois-Reymond became the reigning deans of German science. Their views were not just the considered opinion of extremely competent researchers, rather they carried all the weight of the highest authority within the scientific establishment.

Teleomechanism was dead but not forgotten. In the last decade of the nineteenth century it resurfaced once again, being resurrected by the 'neo-vitalists', Driesch and Hertwig. In the works of these men DuBois-Reymond recognized his old enemy, and it is instructive to consider his response to them, for it not only illuminates the position he had defended along with Helmholtz, Brücke, Ludwig and Schleiden in the 1850s and 60s, but also the brand of physiology they considered to be the true enemy of science.

Dubois-Reymond expressed his views in a lecture delivered at the Berlin Academy in 1894, the year in which Helmholtz died. It was entitled, 'Über Neo-Vitalismus', and it bore a motto from the works of Schleiden:

The savage who calls a locomotive a live animal is no less wise than the scientist who speaks of the vital force in the organism.[93]

After describing the turbulent days of the 1840s and 50s when Helmholtz and Darwin had succeeded in curing physiology of its vitalistic disorder and placing the science on the royal road to health, DuBois-Reymond turned his stern and critical glance toward the new generation of upstart vitalists. Recently, a comprehensive analysis of data from physiology, histology, cellular biology and embryology had led Hans Driesch to the conclusion that biology must be conceived as a self-subsistent, independent science based on bionomic laws not subservient to the laws of chemistry and physics.[94] "He emphasizes", wrote DuBois-Reymond, "how limited the perspective is that sees in 'life' a problem in principle solvable not only mechanistically but even physico-chemically. He sees, furthermore, no objection to the inclusion of teleological considerations as part of natural science, whereby he does not simply mean their employment as occasional heuristic guides, but rather as the purposive ground of organic structure and function".[95] In agreement with both DuBois-Reymond and Helmholtz, Driesch held that where physical causality ceases natural science must also come to an end. But citing Kant's discussion in the *Kritik der Urteilskraft*, Driesch argued that in the case of causality in the biological realm, "it should not be forgotten that at this juncture the connection to the teleological form of judgement must begin".[96] DuBois-Reymond was unmoved:

However great the authority of Kant is otherwise to be estimated, ... he is not to be considered infallible in the domain of natural science.[97]

After all, DuBois-Reymond argued, Kant had no understanding of the conservation of energy. Had he understood that principle, he would never have underwritten a teleomechanical approach to form and function. DuBois-Reymond was in fundamental error in this assertion, for it stood in stark contradiction with the core of Kant's understanding of causality, the heart and soul of his entire critical philosophy. Kant, as we have seen, held a rigidly deterministic notion of causality. But in analyzing the classical Newtonian conception of cause, which he took to be the fundamental category of human understanding, he had discovered its limitations. He had discovered that it cannot account for biological organization.

Perhaps no one better epitomized the view of the teleomechanists concerning the limitation of causal explanations in biology than Karl Ernst von Baer, who may have been the most sensitive philosophical mind among them. Von Baer, writing one year after the appearance of the *Origin of Species*, expressed the central difficulty with the reductionist program in a paper entitled, 'Welche Auffassung der lebenden Natur ist die Richtige?' (1860):

The body of even the highest animal forms such as ours arises from the simplest materials on the earth.... We are unable, therefore, to form a picture of living bodies on other planets so long as we are ignorant of the materials from which those planets are constructed. But even if we were somehow in possession of that knowledge, we would still only be able to form judgements concerning the chemical constituents of their inhabitants; in no way, however, would we be able to judge concerning the vital processes or forms of transformation on those planets.[98]

At issue in von Baer's view was not that organic processes are somehow incommensurate with chemical and mechanical processes. But even if, as DuBois-Reymond had mused, it were somehow possible to be in possession of the original set of conditions prevailing when life emerged, no prediction could be made about the *forms* living organisms would take and the manner of their functioning. At its most fundamental level the causal principles of mechanical determinancy do not apply in biology so that in essence, biology must be a descriptive rather than a predictive science. We are, however capable of penetrating to the most fundamental levels of biological organization on our planet: For firstly, we are ourselves living organisms and we have a primitive understanding of what it means to be alive which no mechanical explanation can replace and which guides our investigation of organized being. Secondly, these organisms are given to us fully constituted in experience. Beginning from this experience we can proceed to analyze the mechanisms that are ordered into an organic whole. But for an ultimate mechanical explanation of the origin of life everything depends upon knowing the principle in terms of which the mechanical processes forming the material basis of life are ordered. The example in the passage above illustrates the view that on another planet in spite of a knowledge of the physico-mechanical basis of life, the *experience that* these mechanisms are ordered in a particular way to form organized beings would fail us, and accordingly we would not be in a position to predict the form organization would take. Where, in the absence of an original experience of organization as given, are we to find the principles of order without which life is unthinkable? Is it possible that the order we recognize as life can be reduced ultimately to conditions coming together accidentally, by chance: that order could somehow emerge from chaos? Having posed these questions in their strongest scientific terms, the teleomechanists and reductionists, the opposed 'schools of 1848' as they have been called by some historians, stood before one of the greatest watersheds in the history of Western thought. For the answers to these questions would only be provided by a fundamental revolution in our ways of thinking about nature.

CHAPTER 6

TELEOMECHANISM AND DARWIN'S THEORY

Karl Ernst von Baer was a proud Estonian nobleman. He was not easily provoked into controversy. In 1827, for instance, he had made one of the most important discoveries in the history of embryology. But when the Deutsche Naturforscher und Ärtzte assembled a few months later in September of 1828 not a single one of the older biologists mentioned his work on the mammalian ovum. Von Baer sent a copy of his monograph to the Prussian minister, von Altenstein, who sent a reply congratulating him on having *rediscovered* the ovum. As he later reported in his autobiography von Baer was mortally offended by these displays of ignorance and professional jealousy.[1] But he did not take his revenge, as he might have, by demeaning the work of others. That was not his style. Rather, without ever once publicly venting his anger and frustration, he simply shut the door of his laboratory in Königsberg and departed for St. Petersburg, never again to consider an appointment in a Prussian university. That wound festered for six years – until 1834 – before von Baer finally decided to leave.

A similar pattern was repeated in the affair over Darwin's theory. Von Baer first learned of Darwin's theory of transmutation in 1859 from T.H. Huxley and Richard Owen when he visited London to report on some interesting skulls in the St. Petersburg collection. Darwin's book had not yet been published, but rumors of its contents were widely circulating in scientific circles. Von Baer read the full account of the evolutionary hypothesis the following year when the book arrived in St. Petersburg. From what he had learned in London, however, von Baer did not expect to be favorably disposed to Darwin's hypothesis, and his expectations were not disappointed. But he was nonetheless impressed by the theory. This was not one of the lighweight evolutionary schemes he had seen come and go in his long career. Darwin presented a wealth of empirical evidence to support his view and he attempted to anchor the hypothesis firmly in the new materialism.

Von Baer did not react impulsively to the new doctrine. He studied it carefully – for several months. He discussed Darwin's views among friends and in a public sitting of the St. Petersburg Academy in 1861. Without directly challenging Darwin's evolutionary model von Baer began to attack its reductionist underpinnings in 1866 in a lengthy discussion of the importance of teleology for understanding the organic world. But these considerations remained unpublished.

The situation changed decisively in 1867, however. In that year the Von Baer Prize, which had been endowed at von Baer's golden jubilee two years earlier, was awarded for the first time. Von Baer sat on the prize committee, and with his blessing the award was made to the young Russian embryologist Alexander Kowalewsky for his study of ascidian larvae. At the end of that piece Kowalewsky boldly announced that the development of the ascidians follows the pattern of the vertebrate type and that, accordingly, the ascidians could be viewed as the evolutionary link between the vertebrate and invertebrate classes. While voting to award the prize to Kowalewsky, von Baer dissented from his conclusion. Darwinism had now struck very close to home; but still von Baer did not publish.

The final straw came in 1871 when Darwin published the *Descent of Man* in which Kowalewsky's work appeared prominently as a principal support of the evolutionary hypothesis. In private, having been encouraged for some time to do so by his friends, von Baer now resolved to present the fullest scientific refutation of Darwin's views he could muster. Though an old man, almost completely blind, who preferred to avoid controversy, von Baer decided to enter the lists in order to set the record straight. In 1873 he published a paper which re-examined Kowalewsky's claims in a careful discussion of the embryology of the ascidians. And in the last year of his life von Baer added a second part to his defense of teleology. He dictated a book-length monograph attacking Darwin's theory of evolution entitled 'Über Darwins Lehre'.

Von Baer's examination of the problems of mechanistic reductionism and Darwin's theory of evolution forms a fitting conclusion to the present study. For these papers permit us to see the principal differences between the two major approaches to biological organization in the nineteenth century. As the most important architect of the teleomechanist program, no one was better prepared to defend it than Karl Ernst von Bear; and as a tireless observer in virtually every major research area relevant to Darwin's thesis, including comparative anatomy, embryology, anthropology, the geographical distribution of species, and even geomorphology, no one was better qualified to assess the merits of Darwin's theory than this old and trusted confidant of Nature. The reflections made by von Baer at the end of an active scientific career spanning sixty years of the most important developments in the emergence of biology as a scientific discipline thus carry special weight. For these were not the vituperative outbursts of a man entering senescence. They represented a fair and sober assessment of a competing program which was based on ten years of careful consideration of its core doctrines and the evidence supporting them. In the course of assessing Darwin's theory von Baer was forced to re-examine his own views to see if and in what respect they had changed as a result of the weight of new evidence accumulated over the years. What he discovered was that during his last years in Königsberg, between 1828 and 1834, he had sketched a broad framework for investigating biological organization with which he found himself still in agreement. Central to that earlier program was a theory of limited biological evolution. His own extensive research as well as the latest findings of paleontologists and work on the geographical distribution of species confirmed that theory which von Baer now wished to elaborate more fully. It was a theory of evolution, but one strongly anti-Darwinian in character. Von Baer wished to appear as neither completely for nor completely opposed to the theory of evolution. For that reason he chose to write not *gegen* Darwin but *Über* Darwins Lehre.[2]

VON BAER'S EXAMINATION OF DARWINIAN EVOLUTION

The object of von Baer's criticism throughout was the principle of natural selection. That a succession of life forms had emerged in the long course of the ages von Baer did not dispute. That was a fact no zoologist could deny. Moreover, that later forms had evolved from the earlier ones could also not be denied. Von Baer found absurd the notion that a special act of generation was required for each different species of organism. Every competent zoologist was aware that a theory of transmutation was the only sensible and economic solution to a scientific treatent of biological organization. The problem was to offer a satisfactory mechanism for bringing about this transmutation. Such a mechanism had to satisfy two different sets of demands. On the one hand, it had to be generally consistent; the mechanism itself had to be plausible. But, on the other hand, it also had to agree with the empirical data.

Darwin's great contribution had been to offer the first complete theoretical account for effecting the transmutation of species. Darwin, however, had allowed himself to be led too strongly in his search for an acceptable mechanism by the metaphysical heuristic principle that one should always look for the simplest explanation. This is a logical demand which is valid only so long as it agrees with the facts. Once transmutation is accepted, the simplest hypothesis is that all life forms have evolved by descent from an original form. In von Baer's view, however, Darwin's mechanism for bringing such a transmutation about was not a plausible one. And secondly, the empirical evidence concerning both the history of life as well as the relationships between related species currently existing did not support Darwin's hypothesis.[3]

Von Baer found the theory of evolution by natural selection to be a truly ingenious method for generating descent by modification. But it rested on some very ambiguous explanatory principles. The mainspring of the argument, according to von Baer, was the variability of organisms. Slight, imperceptible variations accumulate to produce varieties and ultimately new species. These variations have an internal source independent of any predetermined fit to the environment. Due to the natural tendency of organisms to multiply, the process of speciation must go on even within a relatively stable environment. In order to counterbalance the total chaos of forms that would result from this mechanism, Darwin, von Baer explained, had invoked the selective pressure of the environment, the struggle for existence. The 'successful' variants are preserved in the 'Kampf des Daseins' and passed on to future generations as adaptive modifications.

Von Baer had several problems with the mechanism of natural selection and with such Darwinian notions as 'selective advantage', 'successful variants', and the like. As he saw it, Darwin's theory implied that the longer any particular constellation of characters remained intact from generation to generation, the more determinate and fixed those forms must become; a kind of momentum must build up locking them into relatively fixed and invariant paths, increasingly less subject to modification. Furthermore, even assuming that these forms are only relatively fixed and are in reality constantly changing due to gradual shifts in the circumstances of life as a result of geological or ecological change, there must be considerable evidence of transitionary forms. If the most successful variants get preserved over the long run in the struggle for existence, then, thought von Baer, each divergent form of an original stock must be at least as successful as its predecessor; for according to the theory, it can only survive and reproduce if it is successful. Therefore, von Baer concluded, the evidence of transmutation should be everywhere immediately to hand. In establishing an evolutionary series from some animal form, A, to another form F, for instance, since each form has its own particular selective advantage, the intermediate links B, C, D, E should be relatively easy to find. This ought to be true especially for the end of the series. The difficulty should be to find the earlier, – and hence less successful – members of an evolutionary series.

In making his case, von Baer drew upon an example constructed by Heinrich Bronn, the German translator of the *Origin*. In an appendix to his translation Bronn had discussed the advantage of harbor rats over their weaker competitor, the house rat. Whenever the harbor rat is introduced into territory occupied by the house rat, the latter is driven out within a few generations. This is, von Baer noted, a good example illustrating the role of adaptation in the struggle for existence. Presumably both species had originated from the same stem and had gradually diverged through generations of accumulated variation. But like Bronn, von Baer wondered:

How is it that precisely the intermediate links are the ones that are missing? According to Darwin the transitionary forms should have been made in very small, almost imperceptible steps and later have perished. But even the transitionary forms had to have some advantage for existence. How is it possible that these have completely disappeared and the weakest form survived, particularly in the interior of the continent?[4]

Furthermore, these two species of rat differ from one another in characteristics, such as ear and tail length, which offer no apparent advantage in the struggle for existence.

The same argument applied to the geological record. The theory of natural selection seemed to imply that intermediate forms, having their own special advantage in the struggle for existence, ought to be at least as plentiful as the animals they displace. But in order to bring his account into line with the evidence of paleontology, Darwin had to claim the exact reverse to be the case. In von Baer's view the defenders of Darwinian evolution were being unscientific. In claiming that the intermediate links ought to be the most difficult to find, they were defending a position which seemed to violate one of the consequences of their own theory.

In von Baer's view a more serious problem confronting the theory concerned the assumption that minute, chance variations provided all the material necessary for generating a continuous source of evolutionary change. Von Baer found particularly troublesome the notion that variation was to be generated spontaneously with neither a direct nor indirect causal link to the external environment. "How is the summation of these small variations possible in the absence of a continuously acting ground of variation"? he asked. "Otherwise variations in different directions must in succeeding generations cancel the effect of the first."[5] At the heart of the problem was a host of seemingly insurmountable obstacles to be overcome by chance. Variations produced by chance in one generation, even if advantageous, were not on Darwin's theory immune from being eliminated by other chance variations in the next. In order to insure a sufficient number of adaptively significant variations it had further to be assumed that numerous individuals spontaneously varied in the same direction.

But even that was not sufficient; for some mechanism was required that maintained these variations within the population. With the aid of an able mathematician, Johannes Huber had shown that if four out of one hundred individuals possess the same variant trait, by random pairing of all individuals in the population, the number of individuals possessing the trait would increase in the next generation, "but in the third only a few would possess the trait in its full strength, and by the fourth generation it would disappear altogether".[6] This 'demonstration' rested, to be sure, on a blending model of inheritance, but Darwin did not altogether disallow blending. In von Baer's opinion Huber's mathematical model incorporated the features significant to Darwinian selection relating to chance variation and the possibility of their accumulation through equally random processes. The argument seemed to demonstrate that the mechanism of natural selection could not generate continuous evolutionary change.

One possible solution to these difficulties was that offered by Moritz Wagner, the renegade brother of Rudolph Wagner. A convert to Darwinian evolution, Moritz Wagner had explained the preservation of favorable variations through geographical isolation. He supposed that the variant individuals were segregated off together with only a few members of the original stock.[7] To von Baer, however, this approach further taxed the credulity of even the most indulgent critic. Only through extraordinarily improbable circumstances would the variants in a given population be isolated together. Geographical isolation, therefore, could hardly be relied upon as a standard mechanism for overcoming the necessary results of blending inheritance. Rather than multiplying auxilliary hypotheses in order to correct defects in the practical application of Darwin's theory, von Baer recommended attending to its principal defect: Darwin's model had no consistent mechanism for preserving and channeling the random sources of variability upon which it fed. Somehow variability had to be linked to the conditions of existence defined by the external environment rather than to an isolated internal source.[8]

While he found Darwin's mechanism of natural selection inadequate, von Baer did not dispute the reality of transmutation. He devoted a lengthy discussion to various sorts of evidence which can only be explained in terms of a model of descent by modification. The strongest evidence, von Baer argued, is supplied by the geographical distribution of closely related forms and the comparison of living and extinct animals.

Of particular interest are genera with numerous species but which exist only on one continent. Present day sloths, von Baer noted, whether they have three or only two toes on their forelimbs live only in South America. Similarly the anteaters which differ considerably from one another in their size and structure are only present in South America. Antilopes with non-branching antlers, rhinocerouses, giraffes, and many other genera on the other hand are present only in the old world. Do not such patterns of distribution lead to the conclusion that:

> the now distinct species of a genus originally emerged out of a common stem [Stammform] which has been broken up into different forms; but where the two landmasses are widely separated forms on one could not go over to the other?[9]

This notion is further strengthened, von Baer noted, by the distribution of closely related subgenera within a genus. Thus antilopes with forked antlers appear only in America, and in this region of the world there are no other types of antilopes. Antilopes with twisted horns only live in Africa, and distinct subgenera are also peculiar to other parts of Africa and Asia. The equine mammals exhibit a similar pattern. Striped horses such as the zebra, quagga (now extinct), and mountain zebra live only in Africa, while horses without stripes are of Asiatic origin. The distribution of monkeys also supports this evolutionary scheme. In spite of the close structural relationship between old and new world monkeys, all old world species have five molars on each mandible while American species of monkeys all have six. Differences also exist between tail types and other structures. The cats are distributed throughout all continents of the globe except Australia. In von Baer's view they too are probably all descendants of a common form, the branches of which extend across land masses previously united.[10] Such phenomena are only explicable von Baer concluded through the "variation of certain ground forms into specialized forms; that is through transmutation".[11]

A particularly suggestive example of circumstances surrounding the emergence and radiation of these ground forms into potentially considerable numbers of varieties and species was provided by a genus of land snails, the Achantilla, peculiar to the Hawaiian Islands. Ludwig Pfeiffer had found no less than 207 species of Achantilla on the island chain. Drawing a parallel to varieties of a species of snail in Ceylon which are localized to particular gardens, von Baer reasoned that each particular valley in the mountainous Hawaiian Island group might have its own distinct forms. What especially attracted von Baer in this example was the relationship between the organism and its environment. The great distances separating the islands from the nearest mainland seemed to preclude colonization of the sort Darwin had described on the Galapagos. On the other hand the Achantilla is found nowhere else on earth; varieties of species seemed to be potentially correlated to extremely localized environments. "How were these forms transported to the islands?" he asked. "Or did they come into being here?"[12] As we shall see, von Baer preferred the latter alternative, and he found parallel instances from paleontology to support it.

The doctrine of the transmutation of species also received ample support from the fossil record. Von Baer lamented the fact the discovery of links between presently existing and extinct forms was largely a matter of chance, but there was nonetheless some important evidence in favor of evolution. The most beautiful example was the evolution of the horse established by Vladamir Kowalewsky and O.C. Marsh. Von Baer regarded Kowalewsky's careful reconstruction of the evolutionary sequence of animals originating with the paleotherium medium as paradigmatic of the procedures that should generally be followed in establishing phylogenies. For no step in the reconstruction was left to be filled in by conjecture. Not what *must* be in the fossil record viewed through Darwinian spectacles, but rather what actually *is* there should be the point of reference for accepting or rejecting a scientific theory. Kowalewsky and Marsh had not only established a plausible transition of forms based on the functional correlation of teeth, skull, neck and vertebral column, and foreshortening of toes, which established a progressive transformation through several genera of animals, including the paleotherium medium, orohippus, miohippus, anchitherium, hipparion, up to the modern day horse. But they had also demonstrated that these fossil forms gradually follow one another sequentially in the geological record.

Von Baer was convinced that similar phylogenies could be established for other present day organisms. But the fossil record revealed unmistakeable limits to these possible phylogenies.

That a transformation of related forms out of a ground form has occurred seems to me from the above considerations very probable. However this probability seems to me to have been demonstrated only for the evolution of separate species of a genus out of a ground form, and at most for closely related genera out of a common ground form. On the other hand all evidence of a transformation of animals in the higher classes is still lacking. ...[13]

Von Baer drew the boundaries of interspecific evolution sharply around classes of organisms. Fish, amphibians and reptiles, birds and ultimately mammals were not genetically related to one another by descent from a common set of ancestors.

The Darwinians were all guilty of the same violation of sound scientific method in von Baer's view; "they assume that the necessity of certain events is demonstrated and in accordance with this assumption explain what must have happened".[14] This was particularly true of the doctrine of the unity of descent from an original life form. As he had emphasized earlier, one can be so overcommitted to the simplicity of one point of view that other alternatives appear illogical and even impossible. At this point a fine line separates scientific logic from fantasy. Consider, von Baer asked, a total and irrational commitment to 'usefulness' as the criterion of evolutionary change. There have been no doubt many circumstances in the long history of life in which the survival of some phylogenetic line of vertebrates would have been aided by the evolution of wings in addition to two pairs of extremities. No such variations have ever been produced, however; and the reason is that certain internal factors other than variability condition the production of animal forms.[15] In biology what is possible and what is necessary cannot be decided by untestable metaphysical principles but rather by nature and the facts of existence alone. As a young man von Baer had seen another school of zealots make this same fatal mistake; they were the *Naturphilosophen*.[16] Although their theory was not be confused with the doctrines of the Naturphilosphen, the Darwinians were nonetheless committing the same error of deciding the most difficult and critical issues in terms of a single all-embracing logical principle for which no direct empirical support could be provided. And in order to immunize that principle they had surrounded it with a host of *ad hoc* auxilliary hypotheses such as those concerning the improbabilities surrounding the preservation of important links in the evolutionary sequence.

Throughout his discussion of Darwin's theory, von Baer carefully selected examples designed to establish the point that transmutation had occurred but not in the manner supposed by Darwin. Relying upon embryology, the geographical distribution of species and the evidence of the fossil record, von Baer intended to emphasize three points which would form the basis for his own limited theory of evolution: First, he wanted to establish evolution within the confines of large genera and closely related groups of genera; but secondly, he wanted to exclude the possibility of the evolution of different classes from a single ancestral form; and third, he attempted to present evidence from both paleontology and animal geography that supported the conclusion that the different classes of organisms such as fish, birds, reptiles, mammals, etc., had independent and possibly multiple origins. He wished to establish that the principal forms of life depend upon internal laws of biological organization closely correlated to definite conditions of existence.

Von Baer applied this criticism especially to the presumed links between the various types and the classes of organisms within them such as fish, reptiles, birds and mammals of the vertebrate type. There were, he conceded, transitionary forms strongly suggestsive of a phylogenetic linkage between the different classes. The fossils in the Solenhofen gypsum quarries were particularly revealing. One of them, Archaeopteryx, reconstructed by Richard Owen, could be, he was willing to admit, the *Stammvater* of the birds. This bipedal creature had feet structured similar to those of the birds. It not only had wings but the clear impression of feather shafts on its lizard-like tail clearly points in the direction of the avian type. Other genera strengthen the possible link between the different classes. Fossil specimens of Ondontornis discovered in America and England suggest a connection to reptiles and possibly mammals. Some species of this bird had teeth which were firmly attached to the lower mandible like the teeth of fish and most reptiles; other species, however, had teeth sitting in hollows like crocodiles and mammals. Another genus of these toothed birds, the Icthyornis had depressions in its vertebrae like those of fish. Other possible links between reptiles and mammals had been discovered such as the Pterodactyls, Icthyosaurae and Plesiosaurae, which seem to point in the direction of crocodiles on the one hand and to bear a close resemblance, von Baer thought, to the whales, porpoises and dophins on the other.

The extraordinary importance of these creatures for the theory of evolution was not to be denied. But they did not without further adieu support the Darwinian theory of evolution. The source of the difficulty was illustrated in von Baer's view by the dipnoi, the lungfish living in the warm rivers of South America, Africa and Australia. One genus, the African Protoptera, are similar to the larvae of salamanders in that they have externally protruding gills. But even more remarkable is the presence in this organism of either a single lung or a pair of lungs in place of the swim bladder found in fish. Moreover, venous blood is pumped to the lungs through the aortal arch and is circulated back as arterial blood to the heart chamber which is partially divided into two parts. This organism is suited, in short, for survival on land and in water; and indeed during dry seasons when the streams in which they are living dry up these remarkable animals dig themselves into the mud and become completely air-breathing.

While such organisms indicate probable routes taken in bridging the gap between water and land animals, von Baer cautioned against automatically assuming that such transformations must have taken place in the absence of empirical evidence of the sort provided by Kowalewsky and Marsh demonstrating the actual evolutionary sequence. For as von Baer saw it, numerous problems stood in the way of even theoreticaly conjecturing this sort of major evolutionary change. In order to render the whole thing plausible the organs of locomotion must undergo considerable structural reorganization. This transformation was effected, to be sure, but as in the ontogeny of present-day amphibians the transition seems to have taken place very rapidly.[17] In von Baer's mind this was the critical observation, and it would later appear as central to his own theory of limited evolution. Moreover, salamanders and frogs "have already received from their ancestors the *Anlage* for developing four extremities with fingers and toes. These developed in the larval stage of frogs at a time when they swim with the aid of a long tail"[18] It was not at all clear, von Baer continued, how anything similar could evolve in organisms similar to lungfish. The fins of these organisms would have to become much broader and somehow be restructured into segmented extremities without ever having received the Anlage for such structures. Even the transformation of a relatively small sea creature into a land animal capable of supporting its weight and moving requires an enormous feat of engineering.[19] Of course the Darwinians would throw up their usual defense at this point by calling upon limitless expanses of time to carry out this reorganization. Von Baer concurred on the view of many zoologists that the lungfish forms an intermediate link between fish, and amphibians and reptiles, particularly the Ceratodus found in New Zealand, which agrees in numerous respects with a genus of ganoid fish of the mezozoic era.[20] But then at least some record of the numerous genera of intermediate forms required to effect this gradual sequential change ought to be available, although none were known to von Baer.[21] If such changes were made, they were effected very rapidly; a point which did not appear compatible with natural selection as the mechanism of evolutionary change.

Similar problems confronted those who imagined the common ancestor of the various placental mammals to be some sort of marsupial. In von Baer's view such a transition seemed the most probable on theoretical grounds:

But how the transformation from other classes of vertebrates to the first marsupials was effected, and how furthermore the transformation of the marsupials into true placental mammals is supposed to have come about, has not in the least been demonstrated. I freely admit that I wish such a demonstration would be given, for then I believe we would be able to see deeper into the processes of nature. But I know of no such explanation. ...[22]

The doctrine of the unity of descent depended upon establishing transitionary forms linking the most generalized taxonomic groupings. In particular it depended upon establising evolutionary links between the invertebrate and vertebrate types. Since Darwin had drawn upon embryological criteria for establishing this crucial aspect of his theory, von Baer felt confident in staging his attack in what surely had to be regarded as his own special domain. Von Baer's critique of Darwin's use of embryology is briefly summarized in 'Über Darwins Lehre', but it is to the separate treatise published in 1873, entitled 'Entwickelt sich die Larve der einfachen Ascidien in der ersten Zeit Nach dem Typus der Wirbelthiere', that we must turn to appreciate the full force of this, the most crucial aspect of von Baer's argument.

As mentioned above, von Baer's discussion had been prompted by the appearance of the very important paper by A. Kowalewsky, 'Enwickelung der einfachen Ascidien', in the *Memoires* of the St. Petersburg Academy for 1866. In this paper Kowalewsky had argued, on the basis of the embryology of the ascidians, that a direct analogy could be drawn between the formation of the nervous system of these organisms and that of the vertebrates. Darwin was, understandably, quick to seize upon this result. In the *Descent of Man* (1871), Darwin had drawn upon Kowalewsky's findings, coming to the conclusion that with the discovery of the primitive notochord in the ascidians the source from which the vertebrates had emerged had been found. Kowalewsky's embryological work had justified the assumption, Darwin claimed, that at some very distant point in the past a group of animals had existed having many similarities to the larvae of currently existing ascidians and that two branches had diverged evolutionary from these animals. One branch had regressed, bringing forth the present ascidians, while the other progressively evolved into the present vertebrates.

Von Baer regarded the careful repetition of Kowalewsky's observations to be a critical issue in biology, for it concerned the tenability of his entire conception of biological organization together with the assumptions and methods that must be employed in investigating it:

A great amount of interest has been generated recently by the claim of very experienced observers that the ascidians, which differ markedly from the vertebrates in their fully developed stage, develop in the earliest stage of their existence according to the norm of the vertebrates and that as embryos and in the larval stage, they are built on the plan of the vertebrates. If this claim could be established, then the current interest in this problem would be completely justified, for then the bold hypothesis of Darwin that that higher forms of animal organization have evolved in the course of time from the most completely divergent and lowest forms of animal life would have received very considerable support.[23]

After examining the claim made by Darwin, von Baer argued that as an hypothesis it was indeed flexible. One could just as easily argue the converse from exactly the same evidence. "According to the customary *raisonnement*, that which is exhibited very early in embryological development is assumed to be inherited from the earliest ancestors. Accordingly, the ascidians would have to have evolved from the vertebrates and not vice versa. But the burden of the argument is to show that the vertebrates have evolved from lower forms".[24] Von Baer feared that those who supported Darwin in this case had in fact already accepted the Darwinian evolutionary hypothesis as true before they set to the task of observing embryos.[25]

Although von Baer was willing to acknowledge a certain apparent similarly between the neuraxis of the ascidians and the notochord of the vertebrates, there were differences in the symmetry of the parts evident from the very earliest stage of development which in his view argued against an agreement with the vertebrate plan of organization. "That the supposed *chorda* does not run throughout the entire body, that the two pigment patches i.e. the rudimentary eye and statocyst are neither symmetric nor are situated behind one another; that the arrangement of the neural tube is completely foreign to the arrangement of the vertebrates".[26] The main lines of von Baer's attack are clearly visible here; the arrangement of the ascidians in their embryonic stages does not point toward the bilateral symmetry of the vertebrates.

Following this plan of attack, von Baer examined the morphological basis for maintaining a relation between the ascidians and the vertebrates. If there is such a connection, it must be capable of being established through a direct series of materially connected homologies. Since the point at issue is the presumed homologous connection between the ascidian notochord and the neuraxis of the vertebrates, and since "in the vertebrates everything above the notochord, namely the central nervous system, the brain and spinal chord, belongs to the dorsal side of the animal",[27] the hypothesis advanced by Kowalewsky and Darwin required as a minimal condition of its correctness the demonstration "that the neural fold of the ascidian embryo lies in the dorsal region". After a careful comparative study of various species of ascidian embryos von Baer concluded that even this minimal condition cannot be fulfilled.

In order to establish his position von Baer laid down a general principle for distinguishing the dorsal and ventral sides of an animal. No animal is ever found in nature with its ventral side permanently attached. The intake of food would then be too limited. "I believe the reason for this lies much deeper in

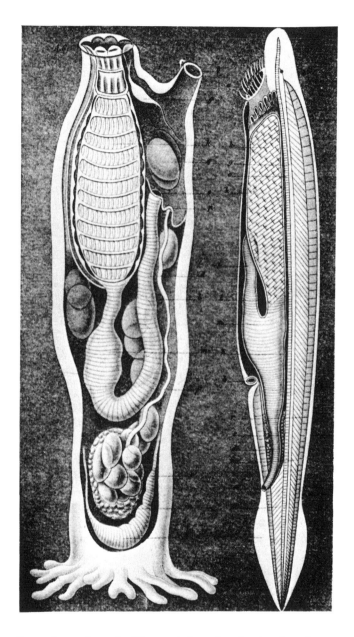

Comparison of an Ascidian and Lancelet showing homologous structures. Taken from Ernst Haeckel, *Schöpfungsgeschichte*.

the formative powers of nature. To be sure one could imagine a flexible shaft originating from the belly of the animal, which would permit the mouth a large area of access with its arms [Fangarmen]; but no such structure is to be found in nature; rather the shaft is always found on the side opposite to that of the mouth. Should we not suspect here a general law of formative nature?"[28] According to this princple, the intake and outlet openings of the animal must always lie in the ventral portion of the body. This turns out to be an absolutely crucial point, which many anatomists such as Bronn had overlooked,[29] for in some tunicates such as *Salpa pinata*, for example, the intake and outlet openings are on the surface *turned away* from the ocean floor. Thus (from our point of view), the ventral portion of the body is really turned upwards, in these organisms, he dorsal portion by contrast being turned downwards. In contrast to fish and other vertebrates, therefore, the dorsal side is not the 'top' of this organism.

The ascidians, which are large sac-like creatures living on the ocean floor, have a body covered by a tunic in which there are two openings, a terminal mouth and an atriopore, both carried upon siphons. According to the principle stated above that the intake and outlet openings of an organism are always in the ventral region, the siphons of the ascidians must be in the ventral side of the body. As the neural tube develops in ascidian larvae, the brain vesicle is transformed into a ganglion. As von Baer notes, this ganglion is to be regarded as functionally equivalent to a brain, for in the adult organism, nerves emerge from it leading to the visceral organs, the gills, and the buccal siphon.[30] In all ascidians, however this ganglion lies *between* the two siphons. Von Baer seeks to establish this point with considerable care through a comparative study of various tunicates, and he illustrates this point in graphic detail in the plates accompanying the monograph. But if this is indeed the case, then the central ganglion must belong to the ventral region of the ascidians. Therefore, von Baer concluded, even though the brain and nervous system of the ascidians is functionally equivalent to the brain and nervous system of the vertebrates, they cannot be morphologically homologous, because in the ascidian they emerge from and belong to the ventral portion of the body whereas in the vertebrates they belong to the dorsal region.[31]

Thus on grounds of both the larval arrangement of the different organs belonging to the nervous system as well as on their morphology in the adult organisms, von Baer argued that no basis existed for claiming a transformation of the organizational plan of the ascidians into one of the three other fundamental plans, and in particular for claiming the ascidians as ancestral forms of the vertebrates. But von Baer did not regard this result as a refutation of the transmutation of species. It was evidence against the *Darwinian* theory of evolution:

Not being in principle disposed to deny the doctrine of the transmutation of animal forms, but rather being disposed favorably toward it, I demand nonetheless a complete demonstration before I can accept a transformation of the vertebrate type into that of the molluscs.[32]

Von Baer added a short post script to this remarkable paper. In bold italics he said that while he had written the piece mainly for zoologists and anatomists, he realized that he had often gone into greater detail than the patience of good biologists, who were mostly familiar with this material, could certainly endure:

I had in this instance the many dilettantes in mind who believe in a complete transmutation of species, and who hold it for idle ignorance if one does not recognize in the ascidians the ancestors of man. I beg forgiveness that, in consideration of these dilettantes, some repetitious material has been included.[33]

Having praised Kowalewsky throughout the memoire as a careful and expert observer, there could be no mistaking that the intended recipient of this remark was Charles Darwin.

Although he rejected Darwin's theory of the community of descent, von Baer presented evidence from embryology and the geographical distribution of species which he regarded as consistent with the view that each class of organisms required an independent origin. The strongest evidence from the fossil record in support of the view that each class of organism arose independently in close connection with the conditions of itss existence and very swiftly radiated into numerous genera and species was provided by the trilobites. Von Baer placed great confidence in the work in Joachim Barrande, who had argued that the trilobites appear for the first time in the Silurian system. Significant in von Baer's view is that, according to Barrande, the trilobites seem to emerge suddenly with no previous ancestors, and they leap onto the geo-temporal map in the earliest Silurian stratum already with 7 genera consisting of 27 distinct species.[34] These expanded to 127 species in 27 genera by the middle of the Silurian system, although several of the earlier genera had become completely extinct by then. The trilobite had become completely extinct by the end of the Paleozoic era. Von Baer reported similar findings for brachiopods, cephalopods, ganoids and sharks in the upper Silurian and Cambrian systems. They all appeared to emerge suddenly in several genera and a variety of species with no apparent transitionary forms linking them to other types of organisms.[35]

Most significant in von Baer's view was the apparently enormous gap between the animal forms in the upper cretaceous strata and those in the cenozoic era. Suddenly in the gypsum quaries, which belong to the early cenozoic, numerous mammals emerge with no obvious link to pre-existing land forms. The most probable form of the first mammals, according to von Baer, is the marsupials, duckbills and echidna being the earliest forms. But marsupials are very rare in the remains of the cretaceous and lower tertiary strata von Baer notes. Moreover, these animals were extremely small. Considering the collosal reptiles of the jurasic and lower cretaceous period, the earliest mammals would have to be extremely large if the transition were to transpire in a gradual evolutionary sequence. While some large mammals do appear, the first mammals were, on the average, about the size of present day rats, and the precious few marsupials were also very small. An exception to this pattern is met in New Zealand, where the remains of enormous marsupials had been discovered. Based on the close similarity of plant forms, von Baer did not think it altogether improbable that New Zealand had been connected at one point to Asia, hence providing a possible source of origin for the first mammals. Against that assumption in von Baer's view however, is the fact that no large, true placental mammals with a fully developed uterus currently inhabit the island group. Even assuming an earlier land bridge, it seemed unlikely, that all of the descendants of these transitionary forms would have migrated, and if they had, it is surprising that none re-colonized the region.[36]

Summarizing the evidence in support of evolution gathered from the fossil record and animal geography, von Baer concluded:

> ... it appears that many animal groups have emerged completely without traceable transitionary forms. Thus the trilobites and somewhat later the cephalopods of the Silurian era. ... In the same manner the fish emerge later with no demonstrable transformation. ... From the fish to the reptiles numerous probable transitions have been established, in part from the ganoids, but also from fishes with scales. Intermediate steps between the reptiles and birds have been demonstrated in a few but nonetheless rare instances. The transitions to the mammals, however, are completely absent. ... Almost suddenly we find a great multiplicity of mammals in the Eocene, particularly ungulates. Where did these come from?[37]

They did not migrate from Asia in von Baer's view. Rather, as we shall see, they had a completely unique origin.

GENERATION, EMBRYOGENESIS AND THE THEORY OF LIMITED EVOLUTION

For all his criticism of the untestable hypothetical elements in Darwin's theory, von Baer's alternative evolutionary model was certainly not immune from attack on exactly the same ground. To biologists increasingly committted to reducing the number of events unaccountable in terms of known physico-chemical laws, von Baer's proposal must have seemed – as no doubt it does today – to be taking a giant step backward. For von Baer proposed to explain the emergence of each class of organism in terms of an original, *primitive generation*. His rationale for adopting this extraordinary position was that the facts simply seemed to require it. Fully aware that the idea would be repulsive to late nineteenth century minds enamoured with mechanisms, he reminded his readers of the fact that Darwin had required a primitive generation:

> It is possible that the great gaps that now exist in our knowledge of transitionary forms will be filled in with time. In part at least one ought to hope for such a completion. Until then, however, persons who support their convictions with reasoned argument cannot say where the boundaries between transformation and original generation lie. It appears more correct and scientific to acknowledge our ignorance in this matter. ...
>
> Should I express my own opinion concerning both forms of generation, I must admit at the outset that I believe in both sorts. An original emergence of organisms without descent must have taken place. That is demonstrated by the success, even if one cannot determine more precisely the nature of the process. But absolutely no reason whatsoever can be given for why this new formation should not have been repeated.[38]

As he explicitly stated, von Baer's position had nothing whatever to do with the suspension of natural laws through the intervention of a creator.[39] Rather his view was identical with what we have seen defended repeatedly by teleomechanists. The origins of life are not to be dissociated from the organization of matter obeying physico-chemical laws; but the principles of organization cannot be themselves reduced to physicochemical laws. These bionomic laws are, in von Baer's view, inseparable from the conditions of existence. Rather than seeing the emergence and transformation of life forms completely in terms of a strictly mechanistic reductionism, von Baer's perspective as a teleologist and developmental morphologist provided him with a completely different set of categories and models for explaining the diversity and interrelation of organic forms. He summarized his own views on biological evolution in the following terms:

Above all we must combat Darwin's view that the entire history of organisms is only the success of material influences rather than a development. To us it appears unmistakeable that the gradual evolution of organisms to higher forms and finally to man was a development, a progress toward a goal, which one may conceive as more relative than absolute.

If I consider the gradual appearance of the different animal forms as a development, that is as a process which leads to a determinate goal, it then appears understandable, even necessary, that the present era differs from the past and that in earlier periods a greater productive force [Productionskraft] held sway than now. The lesser productive force of the present age is manifested by the fact that no new animal forms emerge and that the variations do not depart from the major stems in order to form independent developmental series, whereas it appears indisputable to me that the emergence of numerous species of a higher animal genus has come about through the variation of a ground form. A stronger variability appears, therefore, to have existed earlier. And even primitive generation had without doubt more powerful success in the earliest times than later.[40]

Von Baer viewed biological evolution in terms of development, and embryogenesis served as the model for interpreting the history of life. Although he did not think that ontogeny recapitulates phylogeny, a point he emphasized repeatedly throughout his long career, von Baer did nonetheless see a direct parallel between the successive appearance of related animal forms on the earth and the successive development stages in the life of an individual organism. The beginnings of life are without visible structure, but are contained in the organization of the maternal and paternal fluids. In a process both incapable of duplication and beyond the limits of physico-chemical models of explanation, the union of these primitive conditions results in the emergence of a productive force [Productionskraft]. In the early stages of its development the organism passes rapidly through a series of different forms. In the earliest period these forms are very numerous and subject to a great deal of variation. In this period structures emerge which are functionally necessary for the life of the foetus in its particular embryonic environment. Some of these structures eventually disappear but they serve as the foundation for the development of others to follow. Still other structures remain throughout the entire foetal life of the organism and disappear only when it emerges from this stage. Very early in its development, however, the essential parts of the organism are already present, and as development proceeds, these parts, the principle determinants of its form, become stronger and more complex. After this early period of growth the organism becomes stabilized, is less subject to major variations, and in this state it remains for the most considerable period of its life.

Von Baer intended the evidence presented throughout his consideration of Darwin's theory to suggest a direct parallel between the general pattern of ontogenesis and organic evolution. Thus, he emphasized the sudden appearance of the earliest forms in each particular class of animals. Like the process of fertilization, resulting in a 'productive force', these organisms appeared by a form of primitive generation governed by the conditions of existence and laws of organization. In the earliest period this productive force is strongest, being expressed in numerous variations of the original ground form, the various genera in a class of genetically related organisms. As in the case of the foetus, these different forms of organization were all closely correlated to the conditions of existence in the external milieu, and as these changed, so too did the organisms. Some in fact became extinct, while at the same time serving as the basis for the further structural development of other organisms within the same class. Gradually, however, as the fossil record shows according to von Baer, these variations all stabilized around a single *Hauptform*. This explains why varieties show reversion to type and never, in our own day, proceed beyond certain definite boundaries.

This model is, of course, familiar to us. It was the epochal theory sketched metaphorically by Kant, expounded upon and brought directly into connection with embryogenesis by Kielmeyer, and it was the model described by von Baer himself in the lecture delivered in Königsberg in 1834 entitled 'Das allgemeinste Gesetz der Natur in aller Entwickelung'. Rather than rejecting his earlier view, von Baer believed that subsequent developments in paleontology, studies of animal geography and embryology had only served to confirm it, and in the last days of his life, though convinced that his words would fall on deaf ears, von Baer wished to reassert his earlier view more strongly than ever.[41]

To point to the broad similarities between ontogeny and biological evolution, however, was only to underscore the use of embryogenesis as a metaphor for interpreting the history of life. In order to demonstrate its importance as a superior alternative to Darwin's theory as a model for understanding evolution, von Baer had to settle several important questions. First he had to offer some explanation for the transmutation of species. Darwin had proposed the mechanism of natural selection modeled after the principles of successful animal breeders. Could embryology offer a mechanism of its own? Secondly, central to the embryological perspective was the position, emphasized repeatedly by von Baer, that transmutation does not occur at the level of classes of organisms. Yet he had himself noted that intermediate forms existed between birds and reptiles and between reptiles and fish. What plausible explanation could be given for these forms if they were not to be viewed as genetically related steps in an evolutionary series?

A fundamental law governs the production and potential variation of animal forms, according to von Baer; that is Cuvier's law of the correlation of parts:

> In point of fact the development of organisms is based on something which may be compared to a mathematical formula. That is the relationship that has been called the correlation of parts.[42]

This law implied that any slight modification of an organism necessarily entails a total restructuring of its interrelated parts and a corresponding change in its development path. "If we knew the inner conditions determining the path of development", von Baer wrote, "then we would know the 'mathematical formula' for each form of plant and animal".[43] The correlation of parts implied furthermore that evolutionary change should be discontinuous. Von Baer attempted to illustrate his point of view by drawing a parallel to the formula of organic compounds: Chemical combinations occur only in mathematically determinate relations of the component materials, and as long as those relationships remain the same, he noted, the same compound will result whether or not the quantities of the constituents are altered. But change the *relationship* between the component elements and a completely different type of matter will be generated.

> In a similar fashion the seeds of plants and animals could become discontinuously different from one another when different chemical influences act upon them. These different influences, however, might have been provided by the different geological periods of the earth. I will not go further into the different periods of the earth, for I do not know how to properly illuminate either them or chemical influences in general. But it appears to me indisputable that in the germs of animals certain fundamental relationships [Grundverhältnisse] must exist out of which the 'formulae' develop. The correlation of parts points to that fact, although the different fundamental formulas will be much more complicated than those of simple chemical constituents.[44]

As we have seen, both the chemical and mathematical metaphors had been used often by teleomechanists from Kielmeyer to Bergmann and Leuckart as a means of representing transmutation. According to this view an organizational form has the built-in capacity to respond to changes in its environment, not through a continuous and gradual series of different forms, but rather by generating a discontinuous series of forms. In fact Kielmeyer's mathematical metaphor of the 'integral of life' aptly represented the notion von Baer had in mind. Each partial fraction in the expanded integral is itself a species of organism adapted to its environment, and the integral itself is the 'mathematical formula' containing the general rule for generating each discrete member of the series. When viewed in this light, organic evolution thus takes the image of a caleidoscope of discontinuous forms, all variations on the same organizational theme.

These considerations implied a theory of limited evolution. According to this theory, the basic structural organization of the animal body, the type, would be determined by the requirements of functional organization and the conditions of existence. No sort of modification could ever succeed in transforming animals of one organizational type into another. Accordingly each type must have separate origins produced by an independent set of generative conditions. This argument applied equally to the different classes within the type. As von Baer attempted to illustrate through his consideration of the fossil record, the different classes of vertebrates are not descendants of a single ancestral form, a hypothetical Urchordate for instance, having characteristics of provertebrates such as Amphioxus on the one hand and tunicates on the other. On the contrary, the correlation of parts demands that each vertebrate class had to have separate origins. There can be no true intermediate evolutionary stages between fish and amphibians, or between birds, reptiles, and mammals. To be sure there are both living and fossil forms that seem to *suggest* such transitions, but they are, in von Baer's view, really the results of convergent or, as he called it, 'vicarious evolution'. Due to the conditions of existence in which they are forced to live and seek their nourishment, animals of diverse origins can assume similar structural characteristics in order to adapt to their environment.[45]

The greater groups of animals are quite evidently distinct; the transitions from one group to another are in part seldom; and they never fill in the gap continuously... Recently some examples have been advanced in support of a transition between the reptiles and birds, which have been discussed above. But these examples are extremely rare, and if we were in complete possession of them, we would certainly not be in doubt that the forms in question were either reptiles or birds, not really standing somewhere in the middle. Between reptiles and fish intermediate steps are much more numerous and they are not completely absent in our own era, but the zoologists are scarcely ever in doubt concerning the class to which these organisms are to be assigned. A similar situation holds for the invertebrates as well.[46]

Transmutation by modification has only taken place, in von Baer's view, within large groups of closely related genera. The problem to be faced, therefore, was to propose a plausible model to account for the emergence of discontinuous groups of organisms built on the same general organizational plan.

Von Baer had first proposed the theory of limited evolution in his lecture of 1834. In the intervening years several related embryological phenomena had been discovered which seemed to offer the means of turning the metaphors used in that older version of the theory into concrete models. The phenomena in question related to so-called 'heterogeneous generation', paedogenesis, the alternation of generations and metamorphosis. Von Baer was especially attracted to the phenomena of alternating generations as a potential model for generating discontinuous evolution. A paradigm example of the phenomena of alternating generations is provided by the jellyfish, medusae, whose larval stages were investigated by Albert Kölliker, Ernst Haeckel, and Fritz Müller. After emerging from the egg these organisms have the form of polyps, that is, organisms of a lower class, the tubularians. This larval stage produces a second generation, which when fully developed once again assume the form of the medusa. What von Baer found further impressive about this second generation was that they frequently have forms different from the original parent stock. Thus a six-tentacled medusa may produce offspring having eight tentacles, and a twelve tentacled form may be the offspring of a medusa with eight. These offspring are, however, fully independent forms capable of reproducing themselves through sexual generation. "Here", von Baer concluded, "one particular life process is generated out of another".[47] This was a perfect example of 'heterogeneous generation', the discontinuous emergence of several related forms from a single ground form. Von Baer wanted to generalize it into a model of evolutionary change for higher classes of animals.

A different sort of heterogeneous generation, possibly capable of serving as a model for the emergence of new animal forms, was provided in a study by Rudolphy Leuckart. Leuckart had shown that when the small white earthworm, Rhabditis, is swallowed by a frog, it settles in the lungs of the frog and is there metamorphosed into a much larger worm, the fat portion of which is nearly black in color. This worm, commonly called Ascaris migrovenosa, is capable of sexually reproducing young similar in all respects to the first generation of dark worms. Still another form of Rhabditis is transformed into the intestinal worm, Leptodera, found in many animals. For von Baer's purposes these examples demonstrated the capability possessed by lower organisms of undergoing a sudden and discontinuous reorganization in response to altered environmental circumstances:

Here we have two different lifeforms of the same original species of animal which both reproduce themselves. Both life forms are only conditioned by the abode and the different nutrients.[48]

Von Baer realized the difficulties in basing a mechanism for the transmutation of species upon patterns of development and generation that are found only in lower organisms, particularly invertebrates. In particular nothing analogous to the production of new forms through either asexual generation or the alternation of generations is observed in higher classes of organisms. To bypass this problem von Baer offered a conjecture originally proposed by Blumenbach in his *Beiträge zur Naturgeschichte*; namely that through external influences the developmental path of the organism might be altered. But he could suggest no mechanism for such an alternation, nor did he provide evidence that such alternations actually occur. Nevertheless he thought that some form of heterogeneous generation was the only means possible for explaining evolutionary change. The fundamental law of biological organization, the correlation of parts, seemed to require it:

> It is important, however, not to overlook the fact that all of these approaches or actual successful instances of heterogeneous generation are now only observed in lower organisms. But since it cannot be denied that in an earlier remote period on the earth a stronger formation and transformation of organisms must have taken place than occurs presently, is it not natural to assume a much larger extent for heterogeneous generation? To be sure, it seems less likely that the original formation of higher animals was effected through budding than through an altered developmental path in the egg. But even now we observe that the summer generation of Chermes abietis has wings, while the winter generation is wingless. Moreover, the emergence of the early animals supports the notion that immediately after their first appearance their offspring demonstrated great variability, and that afterward, when the act of fertilization had been effective for numerous generations, the special form of development and with it the form of the body became fixed. The numerous forms which the trilobites and later the cephalopods demonstrate in the otherwise doubtlessly uniform silurian seas lead to this conclusion.[49]

With our advantage of hindsight, which reveals how successful Darwin's model of natural selection working on minute individual variations has been in accounting for evolution, we are apt to judge von Baer's approach to the problem as a futile grasping at straws, an *ad hoc* attempt to save a dying program. But there is subtle logic to von Baer's position which can be easily overlooked if it is considered out of the context of the problems faced by Darwin's theory during the 1860s and 1870s. As von Baer pointed out on several occasions in his discussions, biological evolution was a fact which no reasonable zoologist could deny. The problem was to determine its limits and to propose a plausible mechanism for it. Darwin's great achievement, in von Baer's opinion, was to have developed a full-blown theory of organic evolution by focusing on the immense variability of biological individuals and constructing a model for turning that variability into transpecific change by generalizing upon certain features of artificial selection. This was an extremely clever proposal, von Baer conceded, "and this cleverness will have to be recognized in the future, just as it is undeniable that his work has already powerfully advanced zoology, particularly in Germany. The German scientists had badly neglected varieties, which point to different species, and they had completely overlooked the importance of variability".[50] But variability was not the only biological phenomenon one could focus upon for the purposes of constructing a model of evolutionary change, and indeed von Baer did not think it was even the most natural candidate. Rather, first consideration seemed more appropriately to belong to embryogenesis and development processes in general. While extolling the strengths of this approach, von Baer readily admitted some of its central weaknesses. But there were defects, by no means trivial, in Darwin's mechanism of natural selection. Most damaging among these was the lack of any satisfactory explanation of how variations might be accumulated to produce gradual transitions of forms. More troubling to von Baer, the morphologist par excellence, was the fact that a mechanism based on minute changes seemed directly to contradict the demands of the law of correlation of parts:

In the construction of an organized body nature is directed by the goal or the necessity of producing a viable organism. That is the deeper meaning of the correlation of parts.[51]

It is much easier, von Baer argued in consideration of this law, to produce completely new forms than to produce a series of intermediate forms. This fact and the groupwise divisions of natural species lead, in von Baer's view, inexorably to the assumption of a model of evolution based on heterogeneous generation.[52]

CHANCE, NECESSITY AND TELEOLOGY

Von Baer's disagreement with Darwin concerned a set of issues much more fundamental than either the limits or mechanisms of biological evolution. At the heart of the matter was the rejection by Darwin and his followers of teleological thinking in biology and the replacement of a teleological framework of explanation by the reign of chance and necessity. "In Darwin's hypothesis", wrote von Baer, "all goal-directedness is avoided as much as possible".[53] Throughout his examination of Darwin's theory von Baer objected to the notion that chance variations having an internal source independently of any relation to the external environment could ever lead to functionally adapted organisms.

Against the charge that Darwin had raised chance to the status of a causal principle in interpreting the order of nature some proponents of the theory had responded by claiming just the opposite: there was neither chance nor purpose in nature they said; everything is the result of blind necessity. This position had been advanced most forcefully by Ernst Haeckel:

> According to our view chance collapses together with teleology into nothing. For 'chance' no more exists than does purpose [Zweck] in nature or a so-called 'free will'. On the contrary each effect is necessarily conditioned by previously existing causes, and each cause has necessary effects in its consequences. In our view 'chance' in nature, as well as purpose and free will, is replaced by absolute necessity, $\alpha\nu\alpha\gamma\kappa\eta$ (necessity).[54]

Von Baer discussed this point of view along with the scientific status of teleology in a work begun in 1866 but finally completed in 1874, titled 'Über den Zweck in den Vorgängen der Nature', which formed part of his extended critique of Darwinian evolution. The main line of von Baer's attack on the position set forth by Haeckel above was that if the only principles admissible in science are those deriving from mechanistic necessity, then the most fundamental questions of zoology, namely those concerned with organization, generation, development, function, and adaptation, must remain ultimately unintelligible; for they must reduce to an accidental concatenation of mechanical processes without a common ground for their necessary interconnection. Only a teleological framework could serve as a corrective to this problem. In order to make his point von Baer reviewed the basic principles of teleomechanism that had guided so much important nineteenth century biology, and he explained why it must continue to guide the life sciences in the future.

At the conclusion of his discussion of teleology von Baer pointed out that such considerations would have scarcely been necessary if modern biologists had not lost sight of fundamentals of the problem as set forth long ago by Immanuel Kant:

> Nearly a century ago Kant taught that in an organism all the parts must be viewed as both ends and means [Zweck und mittel] at the same time. We would rather say: goals and means [Ziele und Mittel]. Now it is announced loudly and confidently: Ends do not exist in nature, there are in it only necessities; and it is not even recognized that precisely these necessities are the means for reaching certain goals. Becoming [ein Werden] without a goal is simply unintelligible.[55]

Von Baer wanted to correct Kant's choice of terms. The problem with Kant's formulation was that in German the term 'Zweck' carries the connotation of an intentional act initiated by a rational agent.[56] 'Ziel' on the other hand implied a "prescribed result which can be achieved by necessity".[57]

Considered in this light, Nature has no purposes [Zwecke], von Baer argued; but goals it certainly does have. "Every organism in the process of coming into being has a goal. Without goals how could anything subject to *regulation* come about?"[58]

In part the failure to appreciate the potential role of teleological thinking in science was due to the poverty of most Western languages. They are simply incapable of distinghuishing ends rationally selected for some human purpose from ends which are the results of ordered necessity. In part, however, the problem had to do with the unfortunate confusion of teleology with natural theology. "Manifestly the attacks on teleology are based on the rejection of one of its particular forms, whereby an anthropmorphic creator is envisioned who effectively arranges each individual event for the benefit of man completely independently of the laws of nature."[59] Von Baer concurred in rejecting this form of teleological argument, which conflated 'Ziele' and 'Zwecke'. Such uses of teleology had no place in natural science; but that did not mean that the notions of purposive organization and goal-directedness had no roles to play in biology.

In von Baer's view teleology and mechanism are not in the least opposed to one another. They simply refer to different aspects of a biologically organized being. Every process in the animal body takes place in accordance with the laws of physics and chemistry. The greatest advance in biology during the nineteenth century, von Baer stated, had been improvements made in understanding the processes of material exchange upon which organ function is dependent. Where deficiencies in the understanding of physiological chemistry continued to exist, biology could only advance, in his view, by attempting to overcome them through improved knowledge of physics and chemistry.[60] Reflecting on the state of the art at the beginning of his career von Baer recalled the considerable doubt that had prevailed concerning the ability of the forces of physics and chemistry to account for most physiological processes. Resource had been taken to a *Lebenskraft*, which was really only another name for the problem whose solution was being sought.[61] Since that time it had been demonstrated that no natural forces are required to understand individual physiological processes other than those of chemistry and physics. But the underlying problem that the *Lebenskraft* was originally intended to designate had not been removed. The problem of accounting for the order and direction of physico-chemical processes remained:

Without doubt the organism is a mechanical apparatus, a machine, which builds itself. The life process runs under uninterrupted chemical operations. Therefore the organism might also be called a chemical laboratory. But it is also the chemist in that it assembles the materials necessary for the continuation of the chemical operations from the external world. If it cannot have them, life ceases. However great the progress has been in recent times in understanding the individual operations in the life process, something has always remained behind which guides them and which controls the physico-chemical processes; life itself.[62]

Each individual vital process may be subject – indeed, according to von Baer, it is subject – to the laws of physics and chemistry, and yet without the notion of a common ground uniting them all into, a unified functional whole their biological significance remains unexplained. That common ground in terms of which those processes are ordered and arranged is the *life* of the individual organism and its preservation through the production of offspring.[63]

In order to understand biological organization something in addition to the necessary interactions described by physics and chemistry must be added. This 'something in addition' is a set of necessities of a different sort connected with the life of the organism. Accidental occurrences, von Baer said, are events that come together with no causal interconnection. Suppose now, he suggested, we follow Haeckel and interpret every event in the vital functioning of an organism in terms of necessities determined by physics and chemistry alone. Unless we add some further set of principles that determine a common ground for their combined and orderly interaction, the resultant is at best a product of chance:

If the individual 'absolute necessities' (or forces of nature for short) do not proceed from a common ground, they stand related to one another only by chance and they can only act upon one another destructively or at best have a limiting effect on one another.[64]

Cyclones, or volcanic eruptions von Baer observed in illustration of this point, are generated by some of the same physico-chemical principles that come into play in organic development, but because these natural forces are not regulated by a common goal, they are destructive merely, incapable of creating anything permanent:

Nothing happens without sufficient reason; that is certain. But natural forces which are not directed to an end cannot produce order, never a mathematically determinant form much less a complex organism.[65]

Von Baer conceded that one might be tempted to reject his position as an outdated superstition. But there was at least one biological phenomenon that would forever resist interpretation within any framework other than a teleological one, namely embryogenesis. His favorite example was the metamorphosis of the butterfly. In each stage of its metamorphosis the organism develops structures that are only of use in the stage to follow. Among the several examples singled out by von Baer throughout his discussion of teleology, the presence of the fat-bodies in the caterpillar is representative. Whereas most animals assimilate what they need and excrete the rest, the caterpillar stores a portion of the nutrient converted to animal matter within its digestive tract in this magazine. At about the same time that its external parts are completely developed the fat-bodies become full. The caterpillar loses its ravenous appetite, seeks out a place where it will spin its cocoon and begin the next developmental stage. During chrysalization a complete restructuring takes place, including a transformation of the nervous system, the further development of wings, antennae, and the long feet of the future butterfly, the rudiments of which were present before chrysalization began. During the pupal stage the intestinal tract is reshaped and development of the sex organs is completed along with the construction of the proboscis. During the entire time that these and other remarkable structures are being formed and transformed the fat-bodies gradually diminish in size. They have, in short, provided the material for all of these various structures.

Consideration of such phenomena led von Baer to ask how it is possible not to recognize that the entire pattern of physico-chemical events harmoniously interconnected in each developmental stage are not directed toward a result to be achieved in the future; namely the production of a butterfly. In each stage organs and the rudiments of structures are laid down that have significance only for the stage that is to follow:

> How is it possible to mistake that all of these operations are ordered with respect to a future need? They are directed to that which is to come into being. Such a relationship was designated by the Latin philosophers a *causa finalis*, a cause 'which lies in an end or result.[66]

Von Baer was not asserting the need for any special vital agent which overpowered the normal course of physical-chemical processes. Rather he was pointing to the fact that biological phenomena, particularly the phenomena of embryogenesis, required some further set of explanatory principles in addition to the laws of physics and chemistry in order to incorporate the most essential phenomena of life; namely the patently obvious and real existence of auto-regulation and goal–directedness within biological organisms. For the requirements of constructing a science of life, in short, the application of the laws of chemistry and physics must stand under the guidance of a set of archetechtonic principles incapable of further reduction. This was the teleomechanical framework outlined by Kant at the beginning of the nineteenth century that had been neglected by the new generation of materialists.

How should these principles be described if one wanted to escape the difficulties encountered by earlier vital materialists, such as Kielmeyer, with their conception of the *Lebenskraft*? The best suggestion von Baer could provide was a musical metaphor:

> What we have discussed here: the reciprocal interconnections of organisms with one another and their relationship to the universal materials that offer them the means for sustaining life, is what has been called the harmony of nature, that is a relationship of mutual regulation. Just as tones only give rise to a harmony when they are bound together in accordance with certain rules, so can the individual processes in the wholeness of nature only exist and endure if they stand in certain relationships to one another. Chance is unable to create anything enduring, rather it is only capable of destruction.[67]

As he had explained earlier in the Scholia to his famous *Entwickelungsgeschichte*, in the organic realm the principles of these 'harmonies' are the conditions of existence, the correlation of parts and the laws of development. These rules give rise to certain harmonies, namely to morphotypes; and their interconnection in turn constitutes the principal organizational themata of the animal kingdom. The task of zoology is to investigate and learn to recognize the variations on those themes.

Conjuring up images of a Bach fuge or Beethoven sonata, it was a grand, personal and somehow inspiring nature that von Baer depicted. When we contrast his metaphor with those of the steam engine or the electronic computer later to be suggestive of successful models for mastering the secrets of biological organization, we may be led to agree with von Baer that with the passing of his generation the sense and perhaps even the deeper meaning of life were being lost.

EPILOGUE

With the passing of Karl Ernst von Baer the unique vision of the science of life set forth by German biologists at the turn of the nineteenth century lost its most able spokesman. The debate over the relevance of teleological frameworks to biological explanations, of course, has continued up to the present day. But with von Baer, Bergmann, Bischoff, Virchow, and Leuckart the research tradition I have described ended. A research tradition is much more than a set of ideas and methods for investigating some domain of nature. It depends in part upon a shared commitment to a view about the order of things cemented by personal bonds. Throughout this study I have attempted to emphasize the close relationships between the individuals who developed the teleomechanist research program. In my view, the fact that Rathke, Rudolph Wagner, Leuckart, Bergmann, and von Baer, for instance, not only admired one another's work but were also linked by ties of friendship played an important role in preserving and transmitting the perspective of the program. The same could be said of the relationship between Müller, Rathke, Bischoff, Virchow, Reichert and others. While I do think that communalities of interest and systems of shared values serve to sustain the continued development of an approach to nature by keeping the memory of its less tangible core elements alive, it is not my view that social relations get mapped into scientific theories. Accordingly, I have focused the present study on the very real problems of zoology, physiology, and the issues related to mechanistic explanations in the life sciences during the first half of the nineteenth century, for they structured the development of the biological tradition I have treated here, and it was around these problems that personal alliances formed.

In contrast to one trend of thought that sees ever increased specialization and institutionalization of research traditions as the hallmark of scientific development in the early nineteenth century, I have argued that the mainstream of German biology was guided by a comprehensive program for constructing a unified science of life. The formulators ot this program distinguished three different levels of biological inquiry as essential to their enterprise: cause-effect relationships, historical relationships, and functional relationships. It is important to emphasize that these were diffferent *levels* of inquiry, for within the hierarchy cause-effect and historical relations stood under the guidance of functional or teleologic considerations. The logic for this ranking lay in the object itself; namely the wholeness of the biological individual. Von Baer actually raised this conception to the status of a definition of biology; for him biology is the investigation of the processes leading to the development of individuality in every respect. The workers in the teleomechanist research tradition never lost sight of this their principal focus; and whenever their colleagues or students faltered, the central figures of the research tradition admonished them and attempted to remind them of their proper goal.

Defined in terms of these different levels of inquiry the biologists in the teleomechanist tradition were committed to advancing the aims of their program in several specific research areas, all of which were to be pursued simultaneously and in concert. In the mature formulation of the program achieved in the 1840s, biological inquiry came to be divided as follows: Cause-effect relations were to be the province of physiology, organic and physiological chemistry. Insight into historical relations was to be provided by embryology, paleontology and animal geography. These areas were in effect the data bases for the master level of inquiry concerned with functional or teleological relations. These latter relations were to be examined by the science of morphology, and its object was to explore the principles of the ordered spatio-temporal patterns revealed by the investigations at the other levels. Examples of this morphologic science in practice were von Baer's work on developmental morphotypes, the law of correlation of parts revealed by comparative anatomy, and Bergmann's work with homiothermy and the principle of 'least waste of energy.' Since teleological relations were viewed as based upon, but irreducible to, the lower levels of inquiry, morphology was described by Ludwig and DuBois-Reymond as metaphysical, and hence dispensable.

By attributing the above typology to the mature formulation of the program I intend to call attention to another feature of the present study. The program evolved over time, achieving slightly altered formulations as a result of advances made in one or another of its subsidiary research components. The clearest example of this is the shifting position of embryology within the different versions of the program. For Meckel and Johannes Müller, for instance, embryology was a key to cause-effect relationships and was accordingly included by them as part of physiology. The reason behind this, as we have seen, was their commitment to the need for preserving a holistic perspective of the organized individual throughout and the inability of both organic chemistry and surgical technique to provide the tools necessary for the specific physiological questions of interest to them. After the breakthroughs in physiological and organic chemistry of the 1840s, however, embryology gradually came to be excluded from physiology. Cause-effect relations in the work of Bergmann and Leuckart, for example, became the exclusive preserve of physics and chemistry.

Recognition of the essential unity of the program through each of its successive improved versions and the unwavering commitment of its practitioners to a holistic, teleological perspective helps us to appreciate certain important features of their scientific careers. The teleomechanists were, almost to a man, prodigious contributors to a variety of different lines of research. Kielmeyer was not alone in working in physiology, comparative anatomy and organic chemistry. The same could also be said of Bergmann and Leuckart, who also made staggering contributions to embryology and systematic zoology. Von Baer's work in embryology was certainly dwarfed by his anthropology, and the originality of his contributions in this field continue to remain relatively unexplored. The paradigm example of the polymath tendencies of teleomechanists is provided, of course, by Johannes Müller. Müller not only made outstanding contributions to comparative anatomy, embryology and systematic zoology, but as his work on digestion with Schwann indicates, he was eager to pursue research in physiological chemistry whenever it was relevant to his interests as a teleologist. Müller's work in sensory physiology laid much of the groundwork for subsequent development. As a result of the incredible breadth of his contributions, Müller has been correctly portrayed as the Alexander of nineteenth century German biology. Like the empire of Alexander, Müller's famous chair at Berlin was divided among five successors. This commitment to pursuing a variety of different types of research was not exclusively a result of the fact that during its infancy biological research was less specialized, allowing one to acquire the skills necessary for advanced research in most areas in a short time. Rather it was much more an expression of commitment to the holistic conception of life at the heart of the teleomechanist research tradition.

There was a certain irony in the situation of the teleomechanists. Von Baer, Müller, and Leuckart, among others, were fond of pointing out that every type of organism has its own peculiar life cycle; that in achieving the fullest expression of its powers, it is at the same time sowing the seeds of its own destruction and preparing the ground for other organisms to follow. The point was no less applicable to their own research program. For teleomechanism almost certainly contained the seeds of its own demise deep within its guiding principles. In order to realize their vision of the science of life in practice, the teleomechanists required an ever expanding battery of experimental methods and mastery of a variety of technical and theoretical knowledge. This necessarily implied specialization.

The traces of this effect were already evident in Müller's lab. In order to generate a work of the magnitude of his *Handbuch der Physiologie des Menschen* and not relinguish the other lines of research important to the program, it was necessary to divide and conquer. Assistants such as Schwann and DuBois-Reymond were encouraged to pursue highly specialized lines of research. The production of each new improved edition of a work like the *Handbuch* could only emerge from an intensification of this drive toward specialized research. It was just a matter of time before students like Helmholtz would turn up; young men who understood the deepest workings of the program but rejected it, who vigorously attacked its central ideas and resolved to see them silenced. What would happen when these men became independent of their former teachers and began to head up research institutes and train students of their own?

The answer is obvious. Gradually the research areas in which the teleomechanist approach found sympathetic ears began to retract until the only followers it could claim were those working in embryology and zoology, the areas in which one is forced to consider the organism as a whole. As von Baer pointed out, however, even in zoology support would soon erode as a result of the triumphant march of Darwin. For Darwin had turned attention to the external problems of form, to those aspects most capable of submitting to the framework of a mechanistic explanation. Everywhere, in Sigfried Giedeon's phrase, mechanization was taking command.

These were factors internal to the working of the teleomechanist program that led to its demise. But there are also significant external factors that seem to have played a role in its decline. Concentrating as I have throughout on the problem context of the emergence and development of the teleomechanist program, these have not been relevant to my concerns. In discussing the reasons for the decline of the tradition, however, it is important to call attention to the potential importance of these external factors.

In recent years a considerable body of knowledge has accumulated on the professionalization of research in nineteenth century Germany and on the institutionalization of research specialities.[1] In a very convincing treatment of patterns of professionalization in German universities which draws together much of this research, Charles McClelland has pointed to two important factors. First, he argues, the ideal of *Wissenschaft*, of pure science, and the institutional structure through which it was realized, the *Seminar*, led to increased specialization of research. The path toward success and recognition in the new research–oriented educational structure became increasingly dependent upon advancing the frontiers of knowledge. Under the guild-like system of the seminar, 'apprentices' and 'journeymen' were challenged not only to acquire the skills of the master but to go beyond him through the development of specialized techniques capable of generating new kwowledge.[2] It was a wonderful heady environment, but it could lead to tense relations between students and teachers. For the students were quickly becoming superior to their mentors in the mastery of precisely those techniques upon which the advancement of Wissenschaft was becoming increasingly dependent. As the examples of Ludwig, Helmholtz, and DuBois-Reymond illustrate, the apprentices were often in command of powerful techniques which the masters could not even hope to duplicate. The situation contained all the elements of a classic conflict between generations in which the older generation defended its position in terms of a grander philosophical vision while the younger generation staked its claims on technical competency.

These sources of tension were aggravated by the structure of professional advancement within the German universities. The period covered by my study was marked by an enormous increase in the budget devoted to the universities. But while the enrollments doubled in the period from 1800 to 1835, for instance, the ratio of full professors to students sharply declined. To meet the needs of the larger number of students, an expanded use was made of the *Privatdozent* and *ausserordentliche* professorship.[3] This had both advantages and disadvantages. On the one hand, it provided employment, however meagre, for the increased numbers of highly trained and technically competent research being generated by the seminar system. But on the other hand, the virtual freeze on all positions of higher rank meant that one might have to occupy a holding pattern for years in wait for promotion to one of the few full professorships.[4] The competition between the *Privatdozenten* and assistants was naturally extremely intense, and expectation of eventual reward could only be reasonably hoped for by those who demonstrated the very highest level of technical expertise. The situation served as a potentially volatile wedge to be driven between generations; and the generation of young men most affected by these developments was precisely that of Ludwig, Helmholtz, and Du-Bois-Reymond. The anti-vitalist campaign waged by these men may have been strongly conditioned by such external factors. It is a hypothesis, at any rate, that deserves serious future consideration. If the results of my study are valid, the problem context generated by teleomechanism supplied the framework of conceptual issues within which these hostilities found their expression.

NOTES:

INTRODUCTION

[1] Joseph Schiller, *La notion d'organisation dans l'histoire de la biologie* (Maloine, Paris, 1978), pp. 1–6.
[2] Gotthelf Reinhold Treviranus, *Biologie. Oder Philosophie der lebenden Natur (Göttingen, 1802–22)*, Vol. 1, p. 4.
[3] William Coleman, *Biology in the Nineteenth Century. Problems of Form, Function and Transformation* (Cambridge University Press, Cambridge, 1977), p. 13.
[4] E.S. Russell, *Form and Function. A Contribution to the History of Animal Morphology* (John Murray, London, 1916).
[5] Hartwig Kuhlenbeck, *The Central Nervous System of Vertebrates* (Springer, New York, 1967), 5 Vols. See especially volume 1 for an interesting history of morphology.
[6] Michael T. Ghiselin, *The Triumph of the Darwinian Method* (The University of California Press, Berkeley, 1969), pp. 131–159.
[7] Michael Ruse, *The Darwinian Revolution. Science Red in Tooth and Claw* (University of Chicago Press, Chicago, 1979), p. 91.
[8] See, for example, Ghiselin, p. 137.
[9] This is discussed by Ruse, *The Darwinian Revolution*, pp. 94–131.
[10] Dov Ospovat, 'The influence of Karl Ernst von Baer's Embrology, 1828–1859: A Reappraisal in Light of Richard Owen's and William B. Carpenter's "Palaeontological Application of von Baer's Law",' *Journal of the History of Biology*, 9 (1976), pp. 1–28.

Dov Ospovat, 'Perfect Adaptation and Teleological Explanation: Approaches to the Problem of Life in the Mid-Nineteenth Century', *Studies in History of Biology*, 2 (1978), pp. 33–56.

Arleen Tuchman, 'Epigenetic Development and the Problem of Heredity: An Analysis of Martin Barry's Solution to the Problem', Unpublished Master's Thesis, University of Wisconsin, 1980.

[11] Thomas S. Kuhn, 'Energy Conservation as an Example of Simultaneous Discovery', in *Critical Problems in the History of Science*, ed. by M. Clagett (University of Wisconsin Press, Madison, Wisconsin, 1959), pp. 321–356.
[12] R.C. Stauffer, 'Speculation and Experiment in the Background of Oersted's Discovery of Electromagnetism', *Isis*, 48 (1957), pp. 33–50. D.M. Knight, 'The Physical Sciences and the Romantic Movement', *History of Science*, 9 (1970), pp. 54–75. H.A.M. Snelders, 'Romanticism and Naturphilosophie and the Inorganic Natural Sciences, 1798–1840. An introductory Survey', *Studies in Romanticism*, 9 (1970), pp. 193–215. Dietrich von Engelhardt, *Hegel und die Chemie* (Guido Pressler Verlag, Wiesbaden, 1976). Reinhard Löw, *Die Pflanzenchemie zwischen Lavoisier und Liebig* (Donau Verlag, Munich, 1977). Charles A. Culotta, 'German Biophysics, Objective Knowledge and Romanticism', *Historical Studies in the Physical Sciences*, Vol. 4 (1975), pp. 3–38.
[13] Timothy Lenoir, 'The Göttingen School and the Development of Transcendental Naturphilosophie in the Romantic Era', *Studies in History of Biology*, 5 (1981), pp. 111–205.
[14] Marjorie Grene, *A Portrait of Aristotle* (University of Chicago Press, Chicago, 1963), pp. 133–155, pp. 232–234.
[15] Paul A. Weiss, 'From Cell to Molecule', in *Dynamics of Development: Experiments and Inferences* (Academic Press, New York, 1968), pp. 24–95.

[16] Michael Polanyi, 'Life's Irreducible Structure', *Science*, 113 (1968), pp. 1308–12. Also see Polanyi's *The Tacit Dimension* (Doubleday, New York, 1966).

[17] Jacques Monod, *Chance and Necessity* (Alfred A. Knopf, New York, 1971), pp. 28–29.

[18] E.S. Rusell, *Form and Function*, p. 345.

[19] D'Arcy Thompson, *On Growth and Form* (Cambridge University Press, Cambridge, 1961), p. 9.

[20] Lila L. Gatlin, *Information Theory and the Living System* (Columbia University Press, New York, 1972).

[21] Such a scheme was wonderfully elaborated, for instance, by Georg August Goldfuss in his work, *Über die Entwicklungsstufen des Thieres* (Nürnberg, 1817).

[22] Eugene P. Wigner, 'Are We Machines?', *Proceedings of the American Philosophical Society*, 113 (1969), p. 100. A similar view was expressed, of course, by Erwin Schrödinger in his essay 'What is Life'.

[23] See Lila Gatlin, *Information Theory and the Living System*, pp. 14–23.

[24] Rupert Riedel, *Order in Living Organisms* (John Wiley & Sons, New York, 1978), pp. 272–274.

[25] Imre Lakatos, 'Falsification and the Methodology of Scientific Research Programmes', in *Philosophical Papers* (Cambridge University Press' Cambridge, 1978), Vol. 1, pp. 8–101. The notion of a 'research tradition' has been developed by Larry Laudan. See Larry Laudan, *Progress and its Problems*. (University of California Press, Berkeley, 1977), pp. 78–120.

[26] I have explored the applicability of Lakatos' model to the materials of this study in detail in my paper, 'Teleology without Regrets. The Transformation of Physiology in Germany: 1790–1847', *Studies in the History and Philosophy of Science*, 12 (1981).

[27] Lakatos, 'Falsification and the Methodology of Scientific Research Programmes', *Philosophical Papers*, Vol. 1, pp. 47–52

[28] *Ibid.*, pp. 32–33; and especially pp. 65–67.

[29] Oswei Temkin, 'Materialism in French and German Physiology of the Early Nineteenth Century', in *The Double Face of Janus and other Essays in the History of Medicine* (The Johns Hopkins University Press, Baltimore, 1977), pp. 340–344.

CHAPTER 1

[1] See H. Bräuning-Oktavio, 'Vom Zwischenkieferknochen zur Idee des Typus: Goethe als Naturforscher in den Jahren 1780–86', *Nova Acta Leopoldina*, New Series, Nr. 126, Vol. 18, especially pp. 79–86.

[2] Friedrich Wilhelm Joseph von Schelling, *Von der Weltseele, eine Hypothese der höheren Physik zur Erklärung des allgemeinen Organizmus* (1798), in *Schellings Werke* (Beck, Munich, 1927), ed. by Manfred Schröter, Vol. 1, p. 590.

[3] For a full discussion of the background to Blumenbach's work see Timothy Lenoir, 'The Göttingen School and the Development of Transcendental Naturphilosophie in the Romantic Era', *Studies in History of Biology*, Vol. 5 (1981), pp. 111–205.

[4] Blumenbach, *De generis humani varietati nativa* (Göttingen, 1775) in *The Anthropological Treatises of Johann Friedrich Blumenbach* (Longman, Green, Longman and Roberts, London, 1865), translated by Thomas Bendysche, p. 70.

[5] *Ibid.*, p. 70.

[6] *Ibid.*, p. 70.

[7] *Ibid.*, p. 71.

CHAPTER 1

[8] Blumenbach, *Über den Bildungstrieb und das Zeugungsgeschäfte* (Dietrich; Göttingen, 1781) reprinted in 1971 by Gustav Fisher Verlag, Stuttgart, p. 61. Kielmeyer also viewed the importance of Kölreuter's experiments as having refuted the pre-formationist theory. His account of the debate follows that given by Blumenbach exactly. See Kielmeyer, 'Die Württembergischen Reformatoren der Botanik', in *Natur und Kraft. C.F. Kielmeyers gesammelte Schriften* (W. Keiper, Berlin, 1938), ed. by F.H. Holler, pp. 268–272.

[9] For a discussion of the discovery of the polyp by Trembley and its importance for eighteenth century discussions on generation, see: Shirley A. Roe, *Matter, Life and Generation. 18th-Century Embryology and the Haller-Wolff Debate*, (Cambridge University press, Cambridge, 1981), pp. 9–10.

[10] Blumenbach, *Bildungstrieb*, p. 10.

[11] *Ibid.*, pp. 12–13.

[12] *Ibid.*, pp. 25–26.

[13] *Ibid.*, p. 14.

[14] Blumenbach, *Handbuch der Naturgeschichte* (Dietrich, Göttingen, 1797, fifth edition), p. 18.

[15] Blumenbach, *Handbuch der Naturgeschichte* (Dietrich, Göttingen, 1791), p. 10.

[16] *De generis varietati humani nativa*, in op. cit., pp. 75–76.

[17] Blumenbach, *Geschichte und Beschreibung der Knochen des menschlichen Körpers* (Dietrich, Göttingen, 1786), pp. 84–86n. This is discussed extensively in Lenoir, 'The Göttingen School', loc. cit., note 3 above.

[18] *Bildungstrieb*: pp. 62–63ff.

[19] Kant's views on regulative principles and teleology both with respect to system building (formal or subjective teleology) and insofar as it relates to biology (objective teleology) have been discussed by several authors: see Gerd Buchdahl, *Metaphysics and the Philosophy of Science* (M.I.T. Press, Cambridge, Mass., 1969), pp. 485–532. J.D. McFarland, *Kant's Concept of Teleology* (University of Edinburgh Press, Edinburgh, 1970), especially pp. 69–139.
The best treatment of Kant's teleology in its relation to biology is provided by: Reinhard Löw, *Philosophie des Lebendigen. Der Begriff des Organischen bei Kant. Sein Grund und Seine Aktualität* (Suhrkamp, Frankfurt am Main, 1980). Emil Ungerer, *Die Teleologie Kants und ihrer Bedeutung für die Logik der Biologie* (Karlsruhe, 1911).

[20] Timothy Lenoir, 'Kant, Blumenbach, and Vital Materialism in German Biology', *Isis*, Vol. 71 (1980), pp. 77–108.

[21] Kant, *Kants gesammelte Schriften* (Georg Reimer, Berlin, 1902–1923), ed., Königlichen Preussichen Akademie der Wissenschaften, Vol. XI, p. 176.

[22] Kant, *Kritik der Urteilskraft*, in *Kants gesammelte Schriften*, Vol. V (1908), p. 373. The translation is taken from J.H. Bernard, tr., *Immanuel Kant. Critique of Judgement* (Hafner, New York, 1951), p. 219.

[23] *Ibid.*, 371. Bernard, pp. 217–218.

[24] *Ibid.*, 360. Bernard, p. 206.

[25] *Ibid.*, 179. Berhard, p. 15.

[26] Kant, *Prolegomena*, in *Kants gesammelte Schriften*, Vol. 4, p. 306. That is to say, the principles of the categories establish the framework of possible objects in general. They are not, however, determinate with respect to particular objects of experience. Natural laws, such as Kepler's laws, Galileo's law of falling bodies, or Newton's law of universal gravitation are empirical laws arrived at by application of these general principles to the concrete data of experience. Kant states this clearly in the deduction of the categories:

> Pure understanding is not, however, in a position to prescribe to appearances any *a priori* laws other than those which are involved in a *nature in general*, that is, in conformity to laws of all appearence in space and time. Special laws, as concerning those appearances which are empirically determined, cannot, in their specific character be derived from the categories, although they are one and all subject to them.

Quoted from Kant, *Kritik der reinen Vernunft*, in *Kants gesammelte Schriften*, Vol. 3, p. 127 (B165), Norman Kemp Smith translation, *Critique of Pure Reason* (St. Martin's Press, New York, 1965), p. 173. On the relationship of the categories to empirical laws, see Gerd Buchdahl, *Metaphysics and the Philosophy of Science. The Classical Origins from Descartes to Kant* (M.I.T. Press, Cambridge, Mass., 1969), pp. 651–665.

[27] *Ibid.*, 182. Bernard, p. 19. My translation here differs from Bernard.

[28] *Ibid.*, 375. Bernard, pp. 221–222.

[29] *Ibid.*, 424 Bernard, p. 274.

[30] Kant, 'Über den Gebrauch teleologischer Prinzipien in der Philosophie', in *Kants gesammelte Schriften*, Vol. 8, p. 179.

[31] For a fuller discussion, see Lenoir, 'Kant, Blumenbach and Vital Materialism in German Biology', loc. cit., note 20 above.

[32] Kant, *Kritik der Urteilskraft*, p. 419 note. Bernard, pp. 268–269.

[33] *Ibid.*, 418. Bernard, pp. 267–268.

[34] *Ibid.*, 419. Bernard, p. 268. On the difference between the functional theory proposed by Kant and the descent theory of Darwin see: William Coleman, 'Morphology between Type Concept and Descent Theory', *Journal of the History of Medicine*, 31 (1976), pp. 149–175. For a detailed discussion of this passage and Kant's theory of systematics, see Reinhard Löw, *Philosophie des Lebendigen*, pp. 185–190.

[35] For Blumenbach's ultimate reconciliation of his views with this aspect of Kant's position, see Lenoir, 'Kant, Blumenbach, and Vital Materialism in German Biology', loc. cit., note 20 above.

[36] Shirley Roe has argued for the relevance of a similar distinction made by Caspar Friedrich Wolff. See Shirley Roe, 'Rationalism and Embryology: Caspar Friedrich Wolff's Theory of Epigenesis', *Journal of the History of Biology*, Vol. (1979), pp. 1–43.

[37] Kant, *Kritik der Urteilskraft*, p. 419. Bernard, p. 268.

[38] *Ibid.*, p. 420.

[39] Christoph Girtanner, *Über das Kantische Prinzip für die Naturgeschichte* (Göttingen, 1796), p. 2.

[40] Blumenbach, *Beiträge zur Naturgeschichte* (Göttingen, 1790), pp. 20–23.

[41] Blumenbach, 'Specimen archaeologiae telluris terrarumque imprimis Hannoveranarum', *Comm. Gott. Soc.*, p. 1801.

[42] Reil, 'Von der Lebenskraft', *Archiv für die Physiologie*, Vol. 1 (1795), p. 44.

[43] *Ibid.*, pp. 46–47.

[44] *Ibid.*, p. 46.

[45] *Ibid.*, pp. 25–26.

[46] *Ibid.*, p. 26.

[47] *Ibid.*, p. 70. "The entire process of growth and the formation of animal matter is therefore a chemical process resting upon the laws of affinity of organic matter. If only our

researchers would investigate the relations of animal matter in the same way they have studied the chemical nature of fossils!" (p. 73.)

48 *Ibid.*, p. 85.
49 *Ibid.*, p. 63.
50 *Ibid.*, p. 81.
51 *Ibid.*, p. 79. Reil also departed from Blumenbach on the question of the transmutability of species, which he denied. (p. 81)
52 In spite of the importance of Kielmeyer for early nineteenth century German biology, to which his contemporaries as well as modern historians have called attention, no major biography exists of the man. For biographical details on Kielmeyer the following works should be consulted: William Coleman, 'Kielmeyer, Carl Friedrich', in *Dictionary of Scientific Biography* (Charles Scribner's Sons, New York, 1973), ed., Charles C. Gillispie, Vol. VII, pp. 366–69. Dorthea Kuhn, 'Die naturwissenschaftliche Unterricht an die Hohen-Karlsschule', *Medizin historisches Journal*, 11 (1876), pp. 319–334. Heinrich Balss, 'Kielmeyer als Biologe', *Sudhoffs Archiv*, 23 (1930), pp. 268–288. Felix Buttersack, 'Karl Friedrich Kielmeyer', *Sudhoffs Archiv*, 23 (1930), pp. 236–247. Dorthea Kuhn, 'Karl Friedrich Kielmeyers System der organischen Kräfte. Uhrwerk oder Organismus?' *Nova Acta Leopoldina*, New Series, 36 (1970), pp. 157–168. Ingrid Schumacher, 'Karl Friedrich Kielmeyer. Ein Wegbereiter neuer Ideen. Der Einfluss seiner Methode des Vergleichens auf die Biologie der Zeit', *Medizin historisches Journal*, 14 (1979), pp. 81–89.
53 Christian Heinrich Pfaff, *George Cuviers Briefe an C.H. Pfaff aus den Jahren 1788–1792.* Nebst eine biographischen Notiz von Pfaff (Kiel, 1845). See especially February 1790, p. 141, February 19, 1791, p. 213 and August 1792, pp. 283–84.
54 Cf.Reinhard Löw, *Die Pflanzenchemie Zwischen Lavoisier und Liebig.* (Straubing and Munich, 1977), particularly Chapter 3.
55 Blumenbach, *Handbuch der Naturgeschichte* (1797), p. 24n.
56 *Ibid.*, p. 24.
57 Kielmeyer, 'Entwurf zu einer vergleichenden Zoologie', in *Gesammelte Schriften*, ed. by F.H. Holler (Berlin, 1938), pp. 17–19. The importance of this organ for the classification of animals was emphasized by Blumenbach as well as later by Meckel, Agassiz, and Johannes Müller.
58 *Ibid.*, pp. 25–26.
59 *Ibid.*, p. 26.
60 *Ibid.*, p. 27.
61 *Ibid.*, pp. 28–29.
62 Kielmeyer, 'ideen zu einer allgemeineren Geschichte und Theorie der Entwickelungserscheinungen der Organizationen', in *Schriften*, p. 107. In a Kantian vein he goes on to indicate the importance of temporal relations for indicating lawlike patterns of phenomena as the only possible sign of the internal organization of things.
63 *Ibid.*, pp. 122–123.
64 *Ibid.*, p. 123.
65 'Ideen zu einer Entwickelungsgeschichte der Erde und ihrer Organizationen' in loc. cit.: pp. 205–206.
66 *Ibid.*, pp. 207–208.
67 Cf. Blumenbach, *Handbuch der Naturgeschichte* (1802), pp. 8–9.
68 Kielmeyer, loc. cit., note 62 above, p. 209.

[69] *Ibid.*, p. 210.
[70] *Ibid.*, p. 209.
[71] For further discussion of this point Cf. William Coleman, 'Limits of the Recapitulation Theory: Carl Friedrich Kielmeyer's Critique of the Presumed Parallelism of Earth History, Ontogeny and the Present Order of Organisms', *Isis*, 64 (1973), pp. 341–350.
[72] Carl Friedrich Kielmeyer, 'Über die Verhältnisse der organischen Kräfte untereinander in der Reihe der verschiedenen Organizationen: Die Gesetze und Folgen dieser Verhältnisse', in *Gesammelte Schriften*, p. 63.
[73] *Ibid.*, pp. 66–67.
[74] *Ibid.*, p. 67.
[75] *Ibid.*, p. 71.
[76] *Ibid.*, p. 75.
[77] *Ibid.*, p. 77.
[78] *Ibid.*, p. 78.
[79] *Ibid.*, p. 80.
[80] *Ibid.*, p. 85.
[81] *Ibid.*, p. 86.
[82] *Ibid.*, p. 89.
[83] *Ibid.*, p. 89.
[84] *Ibid.*, pp. 91–92.
[85] Kielmeyer, 'Allgemeine Zoologie, oder Physik der organischen Körper', unpublished manuscript in the Kielmeyer Nachlass at the Stuttgart Würtembergische Landesbibliothek, Cod. Med. et Phys. 4o 69d, p. 3.
[86] *Ibid.*, p. 9.
[87] *Ibid.*, p. 10.
[88] *Ibid.*, pp. 10–11.
[89] *Ibid.*, p. 11.
[90] *Ibid.*, p. 12.
[91] *Ibid.*, p. 13.
[92] *Ibid.*, p. 17.
[93] *Ibid.*, p. 18.
[94] *Ibid.*, p. 18.
[95] On vital force and the problems of the use of Kantian methodology, see James Larson, 'Vital Forces: Regulative Principles or Constitutive Agents?' *Isis*, 70 (1979), pp. 235–249.

CHAPTER 2

[1] Friedrich Tiedemann, *Physiologie des Menschen* (Darmstadt, 1830), p. 41.
[2] Friedrich Tiedemann and Leopold Gmelin, *Die Verdauung nach Versuchen*, 2 volumes (Heidelberg and Leipzig, 1826–27).
[3] Johann Friedrich Meckel, *Abhandlungen aus der menschlichen und vergleichenden Anatomie* (Halle, 1806), pp. 1–3.
For a discussion of Meckel's work on teratology see: Owen E. Clark, 'Contributions of J.F. Meckel the Younger to the Science of Teratology', *Journal of History of Medicine and Allied Sciences*, 24 (1969), pp. 310–322. The argument that vivisection disrupts the interconnection with the whole was made by Cuvier. See *Leçons d'anatomie comparée* (Paris, 1800), Vol. I, V.

CHAPTER 2

[4] *Ibid.*, pp. 7–8.
[5] *Ibid.*, p. 8.
[6] *Ibid.*, pp. 174–175.
[7] *Ibid.*, pp. 176–177.
[8] *Ibid.*, p. 260.
[9] Meckel, *Deutsches Archiv für die Physiologie* (1815), pp. iii–viii.
[10] Meckel, *Beiträge zur vergleichenden Anatomie* (Halle, 1808–1812), Vol. 2, p. 123.
[11] Kielmeyer, 'Ideen zu einer allgemeineren Geschichte und Theorie der Entwickelungserscheinungen der Organizationen', in *Kielmeyers gesammelte Schriften*, ed. by F.H. Höller (Berlin, Keiper, 1938), pp. 107ff.
[12] Meckel, *Abhandlungen*, 2, pp. 110–111.
[13] *Ibid.*, Vol. 2, Part II, pp. 11–12. This section was written by Meckel's younger brother, but included by Meckel because he found it exactly in agreement with his own views.
[14] See Stephen J. Gould, *Ontogeny and Phylogeny* (Cambridge, Mass., Harvard University Press, 1977), pp. 45–46.
[15] See the discussion of Reichert's examination of the principles of developmental morphology treated below.
[16] See Carl Asmund Rudolphi, *Beiträge zur Anthropologie und Naturgeschichte* (Göttingen, 1806), especially 83, pp. 95–106.
I have discussed Rudolphi's theory of types based on organization of the nervous system in my article, 'The Göttingen School and the Origins of Transcendental Naturphilosophie in the Romantic Era', *Studies in History of Biology*, 5 (1981), pp. 111–205.
[17] Cuvier's teleology and its relationship to developments in Germany, particularly the work of Kielmeyer and Kant, is treated by William Coleman, *Georges Cuvier. Zoologist* (Cambridge, Mass., Harvard University Press, 1964), pp. 26–43. Also see Julius Schuster, 'Kielmeyer und Cuvier', in *Mitteilungen zur Geschichte der Medizin und Naturwissenschaft und der Technik*, 37 (1938), pp. 309–310.
[18] Georges Cuvier, *Leçons sur l'anatomie comparée* (Paris, 1800), translated by John Allen, London, 1801, pp. 6–8.
[19] Cuvier, *Le règne animal* (Paris, 1817), translated by H. McMurtrie, New York, 1831, pp. 3–5.
[20] Cuvier, Leçons sur l'anatomie comparée, p. 65.
[21] *Ibid.*, p. 19.
[22] *Ibid.*, p. 51.
[23] *Ibid.*, p. 52.
[24] *Ibid.*, p. 66.
[25] Georges Cuvier, 'Sur un nouveau rapprochement á etablir entre les classes qui composent le règne animal', *Annales du museum*, Vol. 19 (1812), pp. 73–84.
[26] Although Cuvier was forced to admit the recent appearance of man. See Martin S. Rudwick, *The Meaning of Fossils. Episodes in the History of Paleontology* (Neale Watson, New York, 1976, Second Edition), pp. 145–96.
[27] Cuvier's static view of nature was consistent with – and probably resulted from – his conservative theological and political views. See William Coleman, *Georges Cuvier: Zoologist*, p. 17.
[28] Friedrich Wilhelm Joseph Schelling, *Von der Weltseele* (1798) in *Schellings Werke*, (Beck, Munich, 1927), ed. by Manfred Schröter, Vol. I, pp. 635–637.
This problem cannot be the subject of a full treatment here, but it is central to understanding

the biological theories of the other major tradition competing with the Göttingen School, namely *romantische Naturphilosophie*. I intend to devote a separate study to this tradition. The core of the philosophical issues are treated in Robert Spaemann and Reinhard Löw, *Die Frage Wozu? Geschichte und Wiederentdeckung des teleologischen Denkens*. (Münich, Piper Verlag, 1981), pp. 147–186.

[29] Ignaz Döllinger, *Über die Metamorphose der Erd-und-Steinarten aus der Kieselreihe* (Erlangen, 1803), Vorrede.

[30] *Ibid.*, p. 46.

[31] *Ibid.*, p. 85.

[32] Ignaz Döllinger, *Grundriss der Naturlehre des menschlichen Organizmus* (Bamberg and Würzburg, 1805), Vorrede.

[33] Döllinger, *Was ist Absonderung und wie geschieht sie?* (Würzburg, 1819), pp. 15–16.

[34] See, for instance, *ibid.*, p. 12.

[35] See for instance Georg Goldfuss, *Über die Entwickelungstufe der Thiere* (Marburg, Basilisken Presse, 1979, reprint of the original 1817 edition) especially the excellent introduction by Hans Querner.

[36] See Frederic L. Holmes, *Claude Bernard and Experimental Chemistry* (Cambridge, Mass., Harvard University Press, 1974), pp. 1–47.

[37] Ignaz Döllinger, 'Von den Fortschritten Welche die Physiologie seit Haller gemacht hat' (1824), in *Festreden der baierischen Akademie der Wissenschaften* (1822–1825) no. 7–11: p. 16.

[38] On speculative or *romantische Naturphilosophie*, the following works should be consulted: Heinrich Knittermeyer, *Schelling und die romantische Schule* (Munich, 1919), particularly chapter 4, 'Die romantische Naturphilosophe'. Werner Leibrand, *Die spekulative Medizin der Romantik* (Hamburg, 1956). Brigitte Hoppe, 'Polarität, Stufung, und Metamorphose in der spekulativen Biologie der Romantik', *Naturwissenschaftliche Rundschau*, 20 (1967), pp. 380–383. E. Mende, 'Der Einfluss von Schellings 'Prinzip' auf die Biologie und Physik der Romantik', *Philosophia Naturalis*, 15 (1975), pp. 461–485. Hans Querner, 'Gotthilf Heinrich von Schubert und die Biologie der Romantik', *Jahrbuch für fränkische Landesforschung*, 30 (1970), pp. 273–286. Nelly Tsouyopolos, 'Die Auffassung der klinischen Medizin als Wissenschaft unter dem Einfluss der Philosophie im fruhen 19. Jahrhundert', *Berichte zur Wissenschaftsgeschichte*, 1 (1978), pp. 87–100. Dietrich von Engelardt, *Hegel und die Chemie* (Wiesbaden, Guido Pressler Verlag, 1976). Gunther B. Risse, 'Kant, Schelling and the early Search for a Philosophical 'Science' of Medicine in Germany', *Journal for the History of Medicine and Allied Sciences*, 27 (1972), pp. 145–158. K.E. Rothschuh, 'J. Görres und die romantische Physiologie', *Medizinische Monatsschrift* 5 (1951), pp. 128–131.

[39] Döllinger, *Grundriss*, Vorrede.

[40] Döllinger, 'Von den Fortschritten welche die Physiologie seit Haller gemacht hat', p. 8, p. 10.

[41] *Ibid.*, p. 12.

[42] *Ibid.*, pp. 14–15.

[43] *Ibid.*, p. 21.

[44] Boris E. Raikov, *Karl Ernst von Baer 1792–1876. Sein Leben und Sein Werk* (Leipzig, Barth, 1968), translated from the original Russian by Heinrich von Knorre in *Acta Historica Leopoldina*, 5 (1968), p. 23. Parrot was also a friend of Döllinger. Döllinger noted that it was Parrot, "whose exemplary precision in observation is widely celebrated", who inspired

his work on secretions. Parrot was a frequent visitor at Döllinger's house in Würzburg. Cf. Döllinger, *Was ist Absonderung und wie geschieht sie?*, p. 9.

[45] Karl Ernst von Baer, *Nachrichten über Leben und Schriften des Herrn Geheimrathes Dr. Karl Ernst von Baer mitgetheilt von ihm Selbst* (St. Petersburg, 1865), pp. 118–119. Also see Raikov, p. 24.

[46] See Karl Friedrich Burdach, *Die Physiologie als Efahrungswissenschaft* (Leipzig, 1826), Vol. 1, p. 5.

[47] Von Baer, *Leben*, p. 184. Raikov, p. 39.

[48] Von Baer, *Leben*, pp. 167–169. Raikov, p. 36.

[49] Raikov, p. 133.

[50] Karl Friedrich Burdach, *Über die Aufgabe der Morphologie* (Leipzig, 1817), p. 7.

[51] *Ibid.*, pp. 19–21.

[52] *Ibid.*, p. 21.

[53] Cf. *Ibid.*, pp. 60–64.

[54] See the classic study by Georg Uschmann, *Der morphobiologische Vervollkommnungsbegriff bei Goethe und seine problemgeschichtlichen Zusammenhänge* (Fischer, Jena, 1939). See also E.S. Russell, *From and Function*, pp. 45–51 and the related discussion on Geoffroy Saint-Hilare, pp. 74–78.

[55] Lorenz Oken, *Über die Bedeutung der Schädelknochen* (Jena, 1807); Carl Gustav Carus, *Von den Ur-Theilen des Knochen und Schalen-Gerustes* (Leipzig, 1828).

[56] Burdach, *Morphologie*, p. 27.

[57] *Ibid.*, p. 28.

[58] Burdach, *Morphologie*, p. 62.

[59] Karl Ernst von Baer, *Vorlesungen über Anthropologie* (Königsberg, 1824), p. 10.

[60] *Ibid.*, p. 12.

[61] *Ibid.*, p. 12.

[62] *Ibid.*, p. 14.

[63] *Ibid.*, p. 14.

[64] *Ibid.*, p. 15.

[65] *Ibid.*, p. 15.

[66] *Ibid.*, pp. 15–16.

[67] *Ibid.*, p. 18.

[68] Cf. Döllinger, *Grundriss*, pp. 32–33. It is interesting to note that in this work Döllinger describes the constitution of animal matter as resulting from the activity of the *Bildungstrieb*. Non-animal substances taken up into the body are transformed into appropriate animal substances by the *Bildungstrieb*, through what Döllinger describes as an 'Einverleiben'. Von Baer does not refer to a 'vitalization' through the activity of a hyperphysical *Lebenskraft*. the advances in organic chemistry, particularly the conception of animal chemistry advanced by Berzelius had permitted the reformulation of this concept in terms of *order* within a functional arrangement, thereby eliminating the strong vitalistic element of Döllinger's early thought. See below, Chapter 4.

[69] Karl Ernst von Baer, *Untersuchung über die Entwickelungsgeschichte der Fische* (Leipzig, 1835), p. 3, note.

[70] *Ibid.*, p. 20.

[71] *Ibid.*, p. xiii.

[72] It is equally clear, however, that this entire period was one of constant questioning

concerning the foundations of biology, and that von Baer was under the influence of Schelling and Oken at this time. (Raikov, pp. 389–94). By 1825, he appears to have awoken from his 'Naturphilosophic slumbers', (Raikov, pp. 58ff). In my view, what remained in von Baer's approach to nature after he began to excise its Naturphilosophic dross were the elements common to vital materialism. My interpretation of von Baer's approach disagrees with that of Raikov. Raikov interprets von Baer as a lifelong follower of Schelling: Von Baer along with Döllinger, Kielmeyer, and Burdach belonged to the 'moderate' Naturphilosophen (Raikov, pp. 394, note 1056). While rejecting Schelling's flights of fancy, the 'moderates' acquired from Schelling a sense of how to view nature in terms of developmental criteria (Raikov, p. 395). All its other truly magnificent qualities as a piece of impeccable scholarship notwithstanding, this aspect of Raikov's interpretation of von Baer is, in my view, inaccurate.

[73] These manuscripts, preserved in the von Baer Nachlass at the University of Dorpat (now Tartu), have never been published. The only access to them in print is provided by the invaluable, extensive excerpts supplied by Raikov.

[74] Raikov, pp. 60–61.

[75] Quoted from Raikov, p. 63.

[76] Ibid., pp. 64–65.

[77] Ibid., p. 66.

[78] Ibid., p. 83.

[79] Ibid., pp. 83–84.

[80] Ibid., p. 84.

[81] Ibid., p. 84–85.

[82] Von Baer, *Über die Entwickelungsgeschichte der Thiere* (Königsberg, 1828), p. vii. The best discussion of von Baer's embryology has been provided by Jane Oppenheimer, 'K.E. von Baer's beginning Insights into Causal-Analytical Relationships during Development', *Essays in the History and Biology* (M.I.T. Press, Cambridge, Mass. 1967), pp. 295–307. Stephen J. Gould, *Ontogeny and Phylogeny*, pp. 59–63.

[83] Ibid., p. xxii.

[84] Ibid., pp. 139–140.

[85] Ibid., p. 157.

[86] Ibid., pp. 147–148.

[87] For a penetrating analysis of teleology, see Reinhard Löw, *Zum Begriff des Organischen bei Kant* (Suhrkamp Verlag, Frankfurt am Main, 1980).

[88] For some outstanding exceptions see the papers in *Folia Baeriana*, I (1975). Several of these papers provide a more balanced interpretation of von Baer's teleology. See in particular, M. Valt and L. Valt, 'K.E. v. Baeri Filosoofiliste Vaadete Hindamises Alustest', pp. 7–27. K. Poldvere, 'K.E. v. Baer und die Grundprobleme der Embryologie', pp. 30–39. M. Valt, 'Monigaid Probleme Seoses K.E. v. Baeri Töödes Esitatud Tüüpideteooria Marfo-Taksonoomilise Süsteemi Evolutsionistliku Tolgendamisega', pp. 47–71.

[89] Von Baer, *Entwickelungsgeschichte*, Scholium V.

[90] Ibid., p. xxii. He noted, that "the tree from which his cradle will be made has not yet bloomed'.

[91] Von Baer, 'Über die Entwickelungsgeschichte der niedern Thiere', *Nova Acta Physicomedica*, Vol. 13 (1826), p. 762.

[92] *Entwickelungsgeschichte*, Vol. 1, p. 8ff.

[93] Ibid., p. 12. Perhaps it is in this context that we are to interpret von Baer's extensive

readings in chemistry, particularly electrochemistry and physics while a student at Dorpat. Cf. *Folia Baeriana* I, pp. 72–85. Von Baer seems to have studied Scherer's works carefully, and his whole conception of chemical affinities in the organic context fares well with the 'dynamic' system attempted by Scherer. Scherer was a Kant student who attempted to put chemistry on a mathematical foundation in his *Stochiometrie*.

[94] *Ibid.*, pp. 12–13. In Vol. II, the first 36 pp. of which had been sent to the printer in 1828 but which were not published until 1837, von Baer noted that physiological phenomena are to be understood in terms of physical conditions just as "electrical phenomena always give rise to magnetism and vise versa". Cf. von Baer, *Entwickelungsgeschichte*, Vol. II, pp. 3–4.

[95] *Ibid.*, p. 88.

[96] *Ibid.*, pp. 88–89.

[97] *Ibid.*, p. 208.

[98] *Ibid.*, p. 233. The relationship of von Baer's embryology to problems of classification and to evolutionary theory have been discussed by Jane Oppenheimer, 'An Embryological Enigma in the Origin of Species', in *Essays in the History of Embryology and Biology* (M.I.T. Press, Cambridge, Mass, 1967), pp. 221–225. Stephen J. Gould, *Ontogeny and Phylogeny*, pp. 59–63 and elsewhere. Also see Dov Ospovat, 'Embryos, Archtypes, and Fossils: von Baer's Embryology and British Paleontology in the Mid-Nineteenth Century', Ph. D. Dissertation, Harvard University, 1974. Gavin R. de Beer, *Embryos and Ancestors* (Clarendon Press, Oxford, 1958).

[99] *Entwickelungsgeschichte*, Vol. II, p. 46.

[100] *Ibid.*, p. 67.

[101] Carl Reichert, though agreeing with von Baer in all other respects, criticized his explanation of the generation of the 'fundamental organ' for the central nervous system. Since this is the most important of the fundamental organs in the vertebrate type, it should have its own independent existence. Von Baer, however, claimed that the fundamental organ for the nervous system is 'precipitated' from the cutis and muscle layers. This hardly accords with its 'fundamental' status. See C.B. Reichert, *Beiträge zur Kenntniss des Zustandes der heutigen Entwickelungsgeschichte* (Berlin, 1843), pp. 81–117, especially p. 111.

[102] *Ibid.*, p. 219.

[103] Von Baer, 'Beiträge zur Kenntniss der niedern Thiere', *Nova Acta Leopoldina*, 13 (1827), p. 740.

[104] *Ibid.*, p. 740.

[105] *Ibid.*, pp. 742–743.

[106] Von Baer never explicitly introduces a specific term corresponding to 'homology', but that he had such a concept clearly in mind will be evident from the following.

[107] *Entwickelungsgeschichte*, p. 236.

[108] *Ibid.*, p. 237.

[109] Wilhelm Haberling, *Johannes Müller. Das Leben des rheinischen Naturforschers* (Leipzig, 1924), p. 103.

[110] Heinrich Rathke, *Untersuchungen über den Kiemenapparat und das Zungenbein* (Dorpat and Riga, 1832), pp. iv–v.

[111] *Ibid.*, p. 127.

[112] As we have already observed, in the teleomechanist scheme addition of new characters can never occur; hence the type must be identified from its most complex manifestation. On the other hand implicit in this approach is the possibility that the most complex

representative of the type may not yet have appeared in nature, that it may appear at a later time. This results from the assumption that the set of structural and systemic elements forming the basis of organization, although permanently fixed, nonetheless is accorded a sphere of potential adaptive responses.

[113] Rathke, *Untersuchungen*, p. 102.
[114] *Ibid.*, p. 102.
[115] *Ibid.*, p. 103.
[116] *Ibid.*, pp. 10–11.
[117] *Ibid.*, p. 12.
[118] *Ibid.*, pp. 105–106.
[119] *Ibid.*, p. 106.
[120] *Ibid.*, pp. 106–107.
[121] *Ibid.*, pp. 73–75.
[122] *Ibid.*, pp. 31–34.
[123] *Ibid.*, p. 47.
[124] *Ibid.*, p. 128.
[125] *Ibid.*, pp. 111–112.
[126] Johannes Müller, 'Gedächtnissrede auf Carl Asmund Rudolphi', *Abhandlungen der Preussischen Akademie der Wissenschaften* (1837), pp. 1–24. A. Waldeyer, 'Carl Asmund Rudolphi und Johannes Müller', *Forschen und Wirken. Festschrift zur 150 Jahr-Feier der Humboldt-Universität zu Berlin*, I (Berlin, 1960), pp. 97–115.
[127] Johannes Müller, 'Einleitung' to *Zur vergleichenden Physiologie des Gesichtssinnes des Menschen und der Thiere* (Leipzig, 1826), p. 12.
[128] *Ibid.*, p. 19.
[129] *Ibid.*, p. 22.
[130] See Martin Müller, 'Johannes Müllers Grundriss der Vorlesungen über Physiologie', *Historische Studien und Skizzen zu Natur und Heilwissenschaft* (1930), p. 132.
[131] *Ibid.*, p. 132.
[132] See Jakob von Uexküll, *Der Sinn des Lebens* (Scheideweg, Stuttgart, 1977). Also Theure von Uexküll, 'Die Physiologie Johannes Müller und die Moderne Medizin', *Ärtztliche Wochenschrift*, 13 (1958), pp. 613–18.
[133] *Ibid.*, p. 28.
[134] Johannes Müller, *Bildungsgeschichte der Genitalien aus anatomischen Untersuchungen an Embryonen des Menschen und der Thiere* (Düsseldorf, 1830), pp. ix–xiii.
[135] *Ibid.*, p. 9.
[136] *Ibid.*, p. 12.
[137] *Ibid.*, p. 111.
[138] An important lacuna in Müller's extensive embryological work was investigation of the Wolffian body in fish. Müller acknowledged this and encouraged others to complete the study he had begun.
[139] Müller, *Bildungsgeschichte*, pp. 11–14.
[140] *Ibid.*, pp. 51ff.
[141] *Ibid.*, pp. 27–28.
[142] *Ibid.*, p. 26, p. 28.
[143] *Ibid.*, p. 35.
[144] *Ibid.*, pp. 56–57.
[145] *Ibid.*, pp. 33–35, p. 95.

[146] *Ibid.*, pp. 36–38.
[147] *Ibid.*, p. 31, pp. 95–96.
[148] *Ibid.*, pp. 59–61, pp. 117–118.
[149] *Ibid.*, pp. 62–68, especially p. 62, p. 96.

CHAPTER 3

[1] Karl Ernst von Baer, *Über die Bildung des Eies der Saugethiere und des Menschen* (Leipzig, 1827), p. 39.
[2] *Ibid.*, p. 20.
[3] *Ibid.*, p. 22.
[4] *Ibid.*, p. 23.
[5] *Ibid.*, p. 24.
[6] *Ibid.*, p. 28.
[7] *Ibid.*, p. 28.
[8] *Ibid.*, p. 28.
[9] The paper was delivered at the celebration of the 50th anniversary of Blumenbach's doctorate held in 1825 in Göttingen.
[10] *Ibid.*, pp. 36–37.
[11] A classic piece of embryology on a par with the work of Müller and Rathke demonstrating the intermediate steps between the primitive stages of the splanchnocranium and the middle ear auditory ossicles of mammals was Weber's work, *De aure et auditu hominis et animalium* (Leipzig, 1820).
[12] This was particularly obvious in the case of his experiments on threshold sensibility which led to the formulation of the Weber-Fechner psycho-physico law. See Weber, 'Tastsinn und Gemeingefühl', in Rudolph Wagner, *Handwörterbuch der Physiologie*, Vol. 3 (Braunschweig, 1846).
[13] See especially E.H. Weber and Wilhelm Weber, *Wellenlehre auf die Lehre vom Kreislauf des Blutes und ins besondere auf die Pulslehre* (Leipzig, 1850).
[14] Cf. Friedrich Hildebrandt, *Lehrbuch der Physiologie* (Erlangen, 1796) five editions, 5th edition 1828, and *Anfangsgründe der dynamischen Naturlehre* (Erlangen, 1807–1808), 2 Vols., 2nd edition 1828.
[15] Johannes Müller, *Handbuch der Physiologie des Menschen* (Berlin, 1838), 3rd. edition, Vol. 1, pp. 20–21 and elsewhere. The first edition of Müller's *Handbuch* appeared in 1833.
[16] E.H. Weber (ed.). *Friedrich Hildebrandts Lehrbuch der Anatomie des Menschen* (Braunschweig, 1830) 2 Vols., Vol I, p. 64.
[17] *Ibid.*, p. 65.
[18] *Ibid.*, p. 67.
[19] *Ibid.*, p. 107.
[20] *Ibid.*, p. 108.
[21] Theodor Schwann, *Mikroskopische Untersuchungen über die Übereinstimmung in der Struktur und dem Wachstume der Thiere und Pflanzen* (Berlin, 1839), pp. 221–222. Schwann's cell theory has been the subject of numerous scholarly treatments. For my purposes here the most important are the following: Marcel Florkin, *Naissance et derivation de la theorie cellulaire dans l'oeuvre de Theodore Schwann* (Paris, Hermann, 1960). Rembert Watermann, *Theodor Schwann: Leben und Werk* (Düsseldorf, L. Schwann, 1960). Everett Mendelsohn, 'Schwann's Mistake', *Actes 10th Congress International*

d'Histoire des Sciences (Paris, 1964), pp. 967–970. Everett Mendelsohn, 'Physical Models and Physiological Concepts: Explanation in Nineteenth Century Biology', *British Journal for the History of Science*, Vol. 2 (1965), pp. 201–219. Georges Canguilhem, 'La theorie cellulaire', in *La connaisance de la vie* (Vrin, Paris, 1965), pp. 43–80. Erwin H. Ackerknecht, *Rudolph Virchow, Doctor, Statesman, Anthropologist* (University of Wisconsin Press, Madison, Wisconsin, 1953). The best evaluation of the relationship between Müller and Schwann concerning the development of the cell theory is to be found in: Brigitte Lohff, 'Johannes Müllers Rezeption der Zellenlehre in seinem "Handbuch der Physiologie des Menschen"', *Medizin historisches Journal* Vol. 13 (1978), pp. 247–258. My interpretation is essentially in agreement with Lohff's analysis.

[22] *Ibid.*, p. 222.
[23] *Ibid.*, p. 222.
[24] *Ibid.*, p. 224.
[25] I have not treated the work of Schleiden in this context, but it too follows the same pattern. See Gerd Buchdahl, 'The Leading Principles of Induction: The Methodology of Matthias Schleiden', in *Foundations of Scientific Method: The Nineteenth Century*, ed. by Ronald M. Giere and Richard S. Westfall (Indiana University Press, Bloomington, Indiana, 1973), pp. 23–52.
[26] *Ibid.*, p. 229.
[27] *Ibid.*, pp. 205ff.
[28] *Ibid.*, p. 212. Compare here also pp. 46ff especially pp. 51–52.
[29] Pander, *Beiträge zur Entwickelungsgeschichte des Hühnchens im Ei* (Wurzburg, 1817), pp. 6–7.
[30] Schwann, op. cit., p. 63.
[31] *Ibid.*, p. 226.
[32] *Ibid.*, p. 231.
[33] *Ibid.*, pp. 233–234.
[34] *Ibid.*, pp. 234–235n.
[35] *Ibid.*, pp. 255–256.
[36] *Ibid.*, p. 237.
[37] Frederick Gregory, *Scientific Materialism in Nineteenth Century Germany* (D. Reidel Publishing Co., Dordrecht and Boston, 1977), pp. 54–56. Chapter III of Gregory's book provides an excellent biography of Vogt.
[38] Liebig's distinctive approach to vital phenomena will be treated in the next chapter.
[39] Cf. Gregory, p. 57. Agassiz had similar dealings with Schimper, Forbes, and Charpentier concerning the theory of ice ages. Cf. Edward Lurie, *Louis Agassiz: A Life in Science* (University of Chicago Press, Chicago, 1960), pp. 102–111.
[40] Virchow, *Die Cellularpathologie in ihrer Begrundung auf physiologische und pathologische Gewebelehre* (Berlin, 1858), 4th ed., 1871, pp. 22–24.
[41] Martin Barry, 'Research in Embryology. First, Second, and Third Series', *Philosophical Transactions of the Royal Society*, Vol. 128 (1838), pp. 301–341; Vol. 129 (1839), pp. 307–380; Vol. 130 (1840), pp. 527–593. It is relevant to the thesis being advanced throughout my study to note that Barry studied anatomy and physiology in Heidelberg with Tiedemann during 1833. In 1837 he published a series of articles on the perspective of developmental morphology advocated by von Baer and the German embryological school. These articles appeared in the *Edinburgh New Philosophical Journal*. Barry's work on the cell theory and its relation to the problem of inheritance are treated by Arleen Tuchman, 'Epigenetic

Development and the Problem of Heredity: An Analysis of Martin Barry's Solution to the Problem', unpublished Master's Thesis, University of Wisconsin, 1980. I am deeply indebted to Ms. Tuchman for calling my attention to this important aspect of Barry's work, and for many stimulating discussions on other problems related to themes I have explored in the pressent study.

[42] Vogt, *Histoire naturelle des poissons de l'eau douce* (Neuchâtel, 1842), pp. 6–7.
[43] *Ibid.*, pp. 26–28.
[44] *Ibid.*, p. 37.
[45] *Ibid.*, pp. 35–36.
[46] *Ibid.*, p. 16.
[47] *Ibid.*, p. iii.
[48] Vogt's view of the problem of organic form, later detailed by him in his *Physiologische Briefe* (1845–47), encorporated all the characteristic elements of the revision of the teleomechanist program attributed below to the 'functional morphologists'. He defended, for instance, the reality of the morphological type, which he viewed as grounded in the structure and combination of the component parts of the egg (p. 460). He also argued that the morphotype [Organizations-Typus] was recapitulated in the developing embryo (p. 451). Like Leibig and Lotze he argued that while form is to be explained only in terms of material forces, the combination and order among inorganic and organic constituents, how the morphotypes first came to be constituted, must lie beyond the limits of physical explanation. In explaining the transformation of species within the limits set by the morphotype, Vogt adopted the mechanism of generic preformationism first developed by Kant. This same model was developed by Ludwig Büchner in his *Natur und Geist* (1857), p. 250, and in his *Der Fortschritt in Natur und Geschichte* (1868), pp. 19–20. Although both Vogt and Büchner supported Darwin's theory, they rejected the view that chance variation and natural selection could explain the generation of form. This problem will be discussed in detail in the following chapters. See the excellent discussion in Gregory, *Scientific Materialism*, pp. 168–172, pp. 176–177, pp. 182–184. Also of interest is John Farley, 'The Spontaneous Generation Controversy (1859–1880): British and German Reaction to the Problem of Abiogenesis', *Journal of the History of Biology*, Vol. 5 (1972), pp. 285–319.
[49] Döllinger, *Grundzüge der Physiologie der Entwickelung des Zell-und-Knochen und Blutsystem* (Joseph Mainze, Regensberg, 1842), p. 43.
[50] *Ibid.*, p. 82.
[51] *Ibid.*, pp. 51–52.
[52] *Ibid.*, p. 82.
[53] *Ibid.*, p. 84.
[54] *Ibid.*, p. 62.
[55] See Marcel Florkin, 'Schwann, Theodor Ambrose Hubert' in *Dictionary of Scientific Biography,* ed. by Charles C. Gillispie, (Charles Scribner's Sons, New York, 1975), Vol. XII, pp. 240–245, especially p. 243. In his treatment of Schwann's cell theory Florkin argues that Schwann constructed his theory in direct opposition to Müller's vitalism. See Florkin, *Naissance et dérivation de la theorie cellulaire dans l'oeuvre de Theodore Schwann*, p. 74. For more balanced view, see Brigitte Lohff, loc. cit., note 21 above.
[56] Cf. Müller, *Über die krankhaften Geschwülste*, p. 3.
[57] *Ibid.*, pp. 3–4. Müller also emphasized the importance of Schwann's discovery in his 'Bericht über die Fortschritte der mikroskopischen Anatomie im Jahre 1838' in his *Archiv* (1839), especially p. cxcix, where Müller describes the cell theory as the "Grundlage für die Theorie des organischen Wesens".

[58] *Ibid.*, p. 7, p. 23.
[59] *Ibid.*, p. 10.
[60] *Ibid.*, p. 10.
[61] *Ibid.*, p. 11.
[62] *Ibid.*, p. 9.
[63] *Ibid.*, p. 27.
[64] *Ibid.*, p. 41.
[65] Müller, *Handbuch der Physiologie des Menschen* (Berlin, 1837), p. 8.
[66] *Ibid.*, p. 9.
[67] For a fuller discussion of the doctrines of 'organic spherelets', 'organic molecules' and the relation of these theories to *Romantische Naturphilosophie* see the treatment of the work of Stark, Schönlein, and Röschlaub by Johanna Bleker, *Die naturhistorische Schule 1825-1845. Ein Beitrag zur Klinischen Medizin in Deutschland* (Gustav Fischer Verlag, Stuttgart, 1981).
[68] Müller, *Handbuch der Physiologie des Menschen* (Berlin, 1844), p. 7.
[69] *Ibid.*, p. 19.
[70] *Ibid.*, p. 18.
[71] *Ibid.*, pp. 28–29.
[72] *Ibid.*, p. 24.
[73] *Ibid.*, p. 3.
[74] *Handbuch*, Vol. II (1840), p. 597.
[75] *Ibid.*, p. 597.
[76] *Ibid.*, pp. 616–617.
[77] *Ibid.*, pp. 630–631.
[78] *Ibid.*, p. 688.
[79] *Ibid.*, p. 692.
[80] *Ibid.*, pp. 612–613.
[81] *Ibid.*, pp. 612–613.
[81] *Ibid.*, p. 655.
[82] *Ibid.*, p. 647.
[83] *Ibid.*, p. 655.
[84] *Ibid.*, p. 655.
[85] *Ibid.*, p. 769.
[86] *Ibid.*, p. 790.
[87] *Ibid.*, p. 791.
[88] *Ibid.*, pp. 761–769.

CHAPTER 4

[1] For a discussion of the role of surgery in the rise of experimental physiology, see Oswei Temkin, 'The Role of Surgery in the Rise of Modern Medical Thought', *Bulletin of the History of Medicine*, Vol. 25 (1951), pp. 248–259. Reprinted in Oswei Temkin, *The Double Face of Janus* (Johns Hopkins University Press, Baltimore, 1977), pp. 487–496. Toby Gelfand, 'The Training of Surgeons in Eighteenth Century Paris and its Influence on Medical Education', Ph.D. dissertation (Johns Hopkins, 1973).

[2] Magendie's work on absorption has been discussed by Frederic Lawrence Holmes, *Claude Bernard and Animal Chemistry. The Emergence of a Scientist* (Harvard University

Press, Cambridge, Mass., 1974), pp. 126–128. The role of surgery in these developments is also treated in detail by John Lesch, 'The Origins of Experimental Physiology and Pharmacology and its Influence on Medical Education, 1790–1820: Bichat and Magendie', Ph.D. dissertation, Princeton. 1977.

[3] Friedrich Tidemann and Leopold Gmelin, *Die Verdauung nach Versuchen* (Heidelberg, 1826–27), The work of Tiedemann and Gmelin is discussed in detail by F.L. Holmes, *Claude Bernard*, pp. 149–159.

[4] These inconsistencies in Reil's treatment of the Lebenskraft were discussed in detail by L.F. Koch in his article 'Über Teile und Lebenskraft', in Meckel's *Archiv für Anatomie und Physiologie*, Vol. 13 (1828), pp. 225–337, especially pp. 270–290.

[5] Berzelius's contributions to organic chemistry as well as the philosophical background of his work are discussed in Reinhard Löw, *Die Pflanzenchemie zwischen Lavoisier und Liebig* (Donau Verlag, Straubing/Munich, 1977), pp. 273–282, pp. 284–291.

[6] Jakob Berzelius, *Bericht über die Fortschritte der physischen Wissenschaften* (Leipzig, 1836), p. 238.

[7] *Ibid.*, p. 243.

[8] *Ibid.*, p. 245.

[9] Liebig's views on fermentation are discussed by Timothy O. Lipman, 'Vitalism and Reductionism in Liebig's Physiological Thought', *Isis*, Vol 58 (1967), pp. 170–173. Also see Joseph Fruton, *Molecules and Life: Historical Essays on the Interplay of Chemistry and Biology*, (John Wiley & Sons, New York, 1972), pp. 49–50.

[10] The dispute between Liebig and Berzelius over the 'catalytic force' is discussed by Joseph Fruton, *Molecules and Life*, pp. 47–49.

[11] Justus Liebig, *Die organische Chemie in ihrer Anwendung auf Physiologie und Pathologie* (Braunschweig, 1842), p. 1. In order to gain support for his enterprise Liebig dedicated the book to Berzelius. On the stormy relationship between Liebig and Berzelius and the latter's response to Liebig's *Organische Chemie*, see Frederic L. Holmes, 'Introduction', to Liebig, *Animal Chemistry* (Johnson Reprint Corporation, New York, 1964), pp. lviii–lxv, and Frederic L. Holmes, 'Liebig, Justus von', in *Dictionary of Scientific Biography* (Charles Scribner's Sons, New York, 1973), ed. by Charles Gillispie, Vol. VIII, pp. 329–350.

[12] *Ibid.*, p. 29.

[13] *Ibid.*, p. 23.

[14] *Ibid.*, pp. 11–12.

[15] *Ibid.*, p. 237.

[16] *Ibid.*, pp. 207–210.

[17] *Ibid.*, p. 204.

[18] *Ibid.*, p. 206.

[19] *Ibid.*, p. 207.

[20] *Ibid.*, p. 207.

[21] *Ibid.*, p. 208.

[22] *Ibid.*, p. 209.

[23] *Ibid.*, p. 213.

[24] *Ibid.*, p. 227.

[25] Doubts might arise concerning the originality of Liebig's ideas considering the fact that after being turned down by *Poggendorfs Annalen* Mayer submitted his paper on energy conservation to Liebig's *Annalen der Chemie*. Mayer's paper, 'Bemerkungen über die Kräfte der unbelebten Natur,' was published in 1842, the same year that Liebig's own

formulation of the problem appeared. But the paper Mayer sent Liebig was a much improved formulation of the problem in that in the revision he used MV² for quantity of motion instead of MV. If Liebig knew Mayer's work before his own book went to press, he did not profit from it. For as we see he used MV for quantity of motion. See R. Steven Turner, 'Mayer, Julius Robert', *Dictionary of Scientific Biography* (Charles Scribner's sons, New York, 1974), ed. by Charles C. Gillispie, Vol. XI, 235–44, especially p. 236 and p. 239.

[26] *Ibid.*, T.L.W. Bischoff, 'Theorie der Befruchtung und über die Rolle, welche die Spermatozoiden dabei spielen', *Archiv für Anatomie, Physiologie und wissenschaftlichen Medizin* (Berlin, 1847), p. 426.

[27] *Ibid.*, p. 427.
[28] *Ibid.*, especially p. 431.
[29] *Ibid.*, p. 435.
[30] *Ibid.*, pp. 435–437.
[31] Herman Lotze, 'Lebenskraft,' *Handwörterbuch der Physiologie* (Göttingen, 1842), ed. by Rudolph Wagner, Vol. 1, pp. xv–xvi.
[32] *Ibid.*, p. xv.
[33] *Ibid.*, p. xv.
[34] *Ibid.*, p. xxxvi.
[35] *Ibid.*, p. xxxvii.
[36] *Ibid.*, p. xxvii.
[37] *Ibid.*, p. xxiv.
[38] *Ibid.*, p. xxii.
[39] *Ibid.*, p. xlvii.
[40] *Ibid.*, p. xlix.
[41] Bergmann (1814–1865) was born and educated in Göttingen, where he studied with the aged Blumenbach, Langenbeck, Berthold, Wöhler, and Wilhelm Weber. Bergmann moved to Rostock in 1852. The only biography of Bergmann is E. Backes, 'Carl Georg Lukas Christian Bergmann. Eine biographisch-wissenschaftliche Studie zur anatomisch-physiologischer Forschung', *Ergebnisse der Anatomie und Entwickelungsgeschichte*, 24 (1923), pp. 686–744. Leuckart (1822–1898), who was by several years the junior member in the collaboration, attended the University of Göttingen from 1842–1845, where he studied with Rudolph Wagner. After serving as Wagner's lecture assistant from 1845–47 he became a lecturer in zoology in 1847. In 1850, at the age of twenty-eight, he was appointed as professor of zoology at the University of Giessen. In 1869 he moved to Leipzig, where in 1877 he became rector of the University. For biographical details see: H. Schadewaldt, 'Leuckart, Karl Georg Friedrich Rudolph', *Ditionary of Scientific Biography* (Charles Scribner's Sons, New York, 1973), ed. by Charles C. Gillispie, Vol. VIII, pp. 269–271. Klaus Wunderlich, *Rudolph Leuckart. Weg und Werk* (Fischer, Jena, 1978).
[42] Carl Bergmann and Rudolph Leuckart, *Anatomisch-physisch Übersicht des Thiereiches. Vergleichende Anatomie und Physiologie. Ein Lehrbuch für den Unterricht und zum Selbststudium* (Stuttgart, 1852), pp. 3–4.
[43] *Ibid.*, p. 6.
[44] *Ibid.*, pp. 7–8.
[45] *Ibid.*, pp. 22–23.
[46] *Ibid.*, p. 23.
[47] *Ibid.*, p. 23.

⁴⁸ *Ibid.*, p. 18.
⁴⁹ *Ibid.*, p. 390.
⁵⁰ *Ibid.*, p. 391.
⁵¹ *Ibid.*, p. 391
⁵² This point will be discussed more fully below in reference to Bergmann's paper 'Über die Verhältnisse der Wärmeökonomie der Thiere zu ihrer Grösse'.
⁵³ *Ibid.*, p. 393.
⁵⁴ *Ibid.*, p. 393.
⁵⁵ *Ibid.*, p. 393.
⁵⁶ *Ibid.*, pp. 396–397.
⁵⁷ *Ibid.*, pp. 304–305.
⁵⁸ See Berzelius, *Bericht über die Fortschritte der physischen Wissenschaften* (Leipzig, 1836), pp. 243–244. Liebig, *Die organische Chemie in ihrer Anwendung auf Physiologie und Pathologie* (Braunschweig, 1842), pp. 248–249.
⁵⁹ *Ibid.*, p. 342. Emphasis added.
⁶⁰ *Ibid.*, pp. 616–617.
⁶¹ *Ibid.*, pp. 623–624.
⁶² Carl Bergmann, *Über die Verhältnisse der Wärmeökonomie der Thiere zu ihrer Grösse* (Göttingen, 1848). This treatise first appeared in the *Göttingen Studien* in 1847, pp. 495–708, and was published as a separate monograph the following year. I am citing the monograph here. This work is discussed in detail in an excellent article by William Coleman, 'Bergmann's Rule: Animal Heat as a Biological Phenomenon', *Studies in History of Biology*, Vol. 3 (1979), pp. 67–88. The present discussion is indebted to Professor Coleman's paper for several points.
⁶³ *Ibid.*, p. 21.
⁶⁴ *Ibid.*, pp. 10–11.
⁶⁵ *Ibid.*, p. 9.
⁶⁶ *Ibid.*, pp. 30–31.
⁶⁷ *Ibid.*, p. 12.
⁶⁸ *Ibid.*, p. 28.
⁶⁹ *Ibid.*, p. 28.
⁷⁰ *Ibid.*, p. 34–37.
⁷¹ For a detailed discussion of this part of Bergmann's work see the paper by Coleman, loc. cit., note 62 above.
⁷² *Ibid.*, pp. 114–115.
⁷³ *Ibid.*, pp. 104–107.
⁷⁴ Leuckart to Karl Ernst von Baer, December 1, 1865, p. 2. This letter is contained in the von Baer Nachlass in the Handschriften Abteilung of the Universität Giessen, which has kindly granted permission to quote passages from the letter.
⁷⁵ *Ibid.*, p. 2. Von Baer may have been planning to visit Leuckart during 1866. Although no details exist of von Baer's trip to Germany during that year, it is clear that the problems of evolution were on his mind. His paper on the role of teleological thinking in biology, to be discussed in the following chapter, was written in Berlin during 1866.
⁷⁶ Rudolph Leuckart, 'Zeugung', in *Handwörterbuch der Physiologie*, edited by Rudolph Wagner, Vol. IV (1853), p. 950.
⁷⁷ *Ibid.*, p. 51.
⁷⁸ *Ibid.*, p. 951.

[79] *Ibid.*, pp. 961–963.
[80] 'Pregeanstalten' is a term invented for this context by Leuckart. I know of only one other occurrence of the term; it is the term used by Ludwig in his treatment of cell formation. Leuckart was here turning the reductionists' language on its head. See below Chapter 5. One 'pregt' a coin, for example: "eine Munze pregen". 'Anstalten', of course, most commonly signifies 'institutions', but it carries the sense of organizational arrangement in this context.
[81] Von Baer was a member of a discussion group in St. Petersburg which included Dimitrij Mendeleev, K.F. Kessler, P.H. Ovsjannikov and others. The group met periodically to discuss scientific topics. Von Baer had set forth his views for that group during 1860–61 in a series of sessions devoted to Darwin's theory. Von Baer also discussed Darwin's theory publicly in the sitting of the Academy on June 21, 1861. Cf. Boris Raikov, *Karl Ernst von Baer 1792–1876, Sein Leben und Sein Werk* (*Acta Historica Leopoldina*, Barth; Leipzig, 1968), p. 365. In addition to von Baer's writings Leuckart may have learned about von Baer's thoughts on evolution from one of the Russian students studying with him in Giessen. He reported the progress of one of these students to von Baer in another part of the letter from which the passage quoted here is taken.
[82] Leuckart to Karl Ernst von Baer, April 26, 1866, pp. 3–4.
[83] *Ibid.*, p. 4. Leuckart may have been referring here to a pilgrimage to Berlin to see von Baer. See note 75 above.

CHAPTER 5

[1] T.L.W. Bischoff, *Über Johannes Müller und sein Verhältniss zum jetzigen Standpunkt der Physiologie* (Bayrische Königliche Akademie der Wissenschaften, Munich, 1858), pp. 25–26.
[2] See F.L. Holmes, 'Introduction', to Liebig's *Animal Chemistry* (Johnson Reprint Corporation, 1964), pp. lxvii–lxx.
[3] See Joseph Fruton, *Molecules and Life: Essays on the Interplay of Biology and Chemistry* (John Wiley & Sons, New York, 1972), p. 51. Fruton interprets this paper as an attempt to confirm and refine Schwann's work. When taken in the total context of the vitalist-mechanist debate then getting under way, much more was at stake in Helmholtz's paper.
[4] Helmholtz, 'Über das Wesen der Fäulniss und Gährung', *Archiv für Anatomie und Physiologie* (1843), 459.
[5] *Ibid.*, p. 459.
[6] *Ibid.*, p. 460.
[7] *Ibid.*, p. 462.
[8] *Ibid.*, p. 462.
[9] In preparing his scientific papers for publication in his *Wissenschaftliche Abhandlungen* Helmholtz was forced to admit that the argument in this paper had been called into question by the germ theory of disease. In a Zusatz of 1882 Helmholtz wrote: "Recent investigation of this subject makes it probable that the process which I here treated as pure putrefaction really differs from fermentation only in the types and size of the developing organisms. The penetrating organisms must either be able to get through a wet membrane, which withstands constant water pressure, and thus they must be capable of exerting a considerable action while being very small; or they must be able to grow through the membrane. Accordingly the types of penetrating organisms is considerably limited." *Wissenschaftliche Abhandlungen* (Leipzig, 1882), Vol. 2, p. 734.

[10] Helmholtz, 'Über den Stoffverbrauch bei der Muskelaction', *Archiv für Anatomie und Physiologie* (1845), p. 72.
[11] *Ibid.*, p. 72.
[12] Carl G. Lehmann, 'Untersuchungen über den menschlichen Harn', *Journal für praktische Chemie*, Vol. xxvii (1842), pp. 256–274. Johann, F. Simon, 'Über die Harnsedimente' *Hufeland's Journal der praktischen Arzneykunde*, Vol. xcii (1841), pp. 73–88.
[13] *Ibid.*, p. 78.
[14] *Ibid.*, p. 82.
[15] *Ibid.*, p. 83.
[16] Helmholtz, 'Wärme, physiologisch', *Handwörterbuch der medicinischen Wissenschaften* (Berlin, 1845), p. 542.
[17] *Ibid.*, p. 542.
[18] *Ibid.*, p. 542.
[19] *Ibid.*, pp. 542–543.
[20] *Ibid.*, p. 545.
[21] *Ibid.*, p. 547.
[22] *Ibid.*, p. 553.
[23] *Ibid.*, p. 560.
[24] *Ibid.*, p. 561.
[25] See L. Koenigsberger, *Hermann von Helmholtz* (Braunschweig, 1902/03), translated by Francis A. Welby (Dover, New York, 1965), pp. 34–38 for the complex chronology of events surrounding the writing of these two intimately related papers.
[26] Helmholtz, 'Über die Wärmeentwickelung bei der Muskelaction', *Archiv für Anatomie und Physiologie* (1848), pp. 158–164.
[27] L. Koenigsberger, *Hermann von Helmholtz*, translated Welby, p. 35.
[28] *Ibid.*, p. 38.
[29] Unable to get the work published in Poggendorff's *Annalen*, Helmholtz published the paper as a separate monograph bearing the full title: *Über die Erhaltung der Kraft. Eine physikalische Abhandlung* (Georg Reimer, Berlin, 1847).
[30] Helmholtz, *Über die Erhaltung der Kraft* (Berlin, 1847), pp. 4–5.
[31] *Ibid.*, p. 8.
[32] *Ibid.*, p. 12.
[33] *Ibid.*, p. 14.
[34] *Ibid.*, p. 17.
[35] See Peter M. Heimann, 'Conversion of Forces and the Conservation of Energy', *Centaurus*, Vol. 18 (1974), pp. 147–161. Also of interest is Heimann's paper 'Helmholtz and Kant: The Metaphysical Foundations of "Über die Erhaltung der Kraft"', *Studies in the History and Philosophy of Science*, Vol. 5 (1974), pp. 205–238. Thomas Kuhn, 'Energy Conservation as an Example of Simultaneous Discovery', in *Critical Problems in the History of Science*, ed. by M. Clagett (Madison, 1959), pp. 321–356. Yehuda Elkana, *The Discovery of the Conservation of Energy* (Harvard University Press, Cambridge, Mass., 1974), pp. 97–145, is also particularly relevant to the present discussion. Elkana interprets a transitional period in the formation of Helmholtz's ideas during which he accepted Liebig's approach to vital phenomena. While he endorsed Liebig's methods and those aspects of his approach to physiology which fit within a reductivist framework it is my view that Helmholtz never endorsed the teleomechanical program of Liebig, Lotze, Bergmann, and Leuckart.

[36] DuBois-Reymond, *Untersuchungen über thierische Elektrizität* (Berlin, 1848), p. xxxvi.
[37] *Ibid.*, p. xxxii.
[38] *Ibid.*, p. xxxix.
[39] *Ibid.*, pp. xliii–xliv.
[40] *Ibid.*, p. xliv. DuBois-Reymond set forth this position in detail in *Über die Grenzen des Naturerkennens und die Sieben Welträtsel* (Liepzig, 1871), p. 30 and pp. 84–85.
[41] See Frederick Gregory, *Scientific Materialism in Nineteenth Century Germany* (D. Reidel Publishing Co. Dordrecht/Boston, 1977), especially pp. 145–188.
[42] Ludwig's review appeared in *Schmidts Jahrbücher für in-und-ausländische Medizin* (1849), part 6.
[43] Rudolph Leuckart, 'Ist die Morphologie denn wirklich so ganz unberechtigt?' *Zeitschrift für wissenschaftliche Zoologie*, Vol. 2 (1850), pp. 271–275.
[44] *Ibid.*, p. 273.
[45] *Ibid.*, p. 275.
[46] Carl Ludwig, *Lehrbuch der Physiologie des Menschen*, Heidelberg, 1852), p. 2.
[47] Ludwig, *Lehrbuch der Physiologie des Menschen*, Vol. II, pp. 144ff.
[48] Carl Reichert, 'Bericht über die Fortschritte der mikroskopischen Anatomie im Jahre 1855', *Archiv für Anatomie, Physiologie und wissenchaftliche Medicin* (1856), p. 9.
[49] Reichert, 'Bericht über die Fortschritte der mikroskopischen Anatomie im Jahre 1854', *Archiv für Anatomie, Physiologie und Wissenschaftliche Medicin* (1855), p. 9.
[50] *Ibid.*, p. 10.
[51] *Ibid.*, p. 5.
[52] *Ibid.*, p. 6.
[53] *Ibid.*, p. 6.
[54] Reichert, 'Bericht', (1856), p. 5.
[55] *Ibid.*, p. 5.
[56] Reichert, 'Bericht', (1855), p. 11.
[57] *Ibid.*, p. 12.
[58] *Ibid.*, pp. 11–12.
[59] *Ibid.*, p. 5.
[60] Reichert, 'Bericht', (1856), p. 3. For interesting discussions of the political theme in the debate over the cell theory, see: Oswei Temkin, 'Metaphors of Human Biology', in *The Double Face of Janus* (Johns Hopkins University Press, Baltimore, 1977), pp. 271–283, especially pp. 272–274. Everett Mendelsohn, 'Revolution and Reduction: The Sociology of Methodological and Philosophical Concerns in Nineteenth Century Biology', in *The Interaction between Science and Philosophy*, ed. by Yehuda Elkana (Humanities Press, Atlantic Highlands, New Jersey, 1974), pp. 407–426. Frederick Gregory, *Scientific Materialism in Nineteenth Century Germany*, pp. 189–214.
[61] Rudolph Virchow, *Die Cellularpathologie in ihrer Begrundung auf physiologische und pathologische Gewebelehre* (Berlin, 1858), p. 18.
[62] *Ibid.*, p. 24.
[63] *Ibid.*, pp. 382–287.
[64] *Ibid.*, p. 392.
[65] *Ibid.*, p. 493.
[66] *Ibid.*, p. 486.
[67] *Ibid.*, p. 4.
[68] *Ibid.*, p. 12

[69] T.L.W. Bischoff, *Über Johannes Müller und sein Verhältniss zum jetzigen Standpunkt der Physiologie*, (Munich, 1858), pp. 22–23.

[70] This objection was raised by Clausius in an attack on Helmholtz's derivation of the conservation of energy in a paper published in *Poggendorff's Annalen* in 1853. For further discussion of this problem in Helmholtz's memoire, see M. Norton Wise, 'German Concepts of Force, Energy, and the Electromagnetic Ether: 1845–1880', in *The History of Ether Theories in Modern Physics*, ed. by Geoffry Cantor and Jonathan Hodge (Cambridge University Press, Cambridge, 1980), pp. 269–307.

[71] Helmholtz, 'Über das Ziel und die Fortschritte der Naturwissenschaft' delivered at the Naturforscherversammlung in Innsbrück, 1869, in *Vortäge und Reden* (5th ed., Braunschweig, 1903), Vol. 2, p. 374. Quoted from *Selected Writings of Hermann von Helmholtz*, ed. by Russel Kahl, (Wesleyan University Press, Middleton, Connecticut, 1971), p. 228.

[72] See M. Norton Wise, 'German Concepts of Force, Energy and the Electromagnetic Ether', in *History of Ether Theories in Modern Physics*, ed. by Geoffry Cantor and Jonathan Hodge (Cambridge University Press, Cambridge, 1980), pp. 269–307.

[73] Helmholtz, 'Über die Wechselwirkung der Naturkräfte und die darauf bezüglichen neuesten Ermittlungen der Physik' (1854), *Vorträge und Reden* Vol. I, quoted from *Popular Scientific Lectures*, ed. by Morris Klein, (Dover, New York, 1962), p. 73.

[74] *Ibid.*, pp. 73–74.

[75] Gregory, *Scientific Materialism in Nineteenth Century Germany*, pp. 180–187, has demonstrated this concern for progressive evolution among 'scientific materialists' such as Czolbe and Büchner, both of whom accepted the conservation of energy but thought it must be united with a purposively organized universe in order to account for the transmutation of species. Of interest in the question of order within biophysical systems are two papers published by Helmholtz in 1882: 'Die Thermodynamik chemischer Vorgänge', in *Wissenschaftliche Abhandlungen*, Vol. II, pp. 958–978, and 'Zur Thermodynamik chemischer Vorgänge', in *Wissenschaftliche Abhandlungen*, Vol. II, pp. 979–992. In a discussion of Clausius' law of entropy, Helmholtz noted the intrinsically more difficult problem of converting heat into mechanical work in contrast to electrical and mechanical forces. In order to distinguish the portion of the total energy which is convertible to work from the portion which is manifested as heat only, Helmholtz introduced the notions of 'organized' and 'disorganized' motions. The former corresponded to the *vis viva* or kinetic energy of mechanics, while entropy is, he noted, a measure of disorganization. The notion of 'order' turned out to be related to spatial arrangement, so that, for example, a system of particles capable of 'organized motion' (i.e. capable of generating work efficiently) is one in which the velocity components of each particle have definite relations to the velocity components of the other particles which are characterizable in terms of continuously differentiable functions of spatial co-ordinates. By contrast, disorganized motion, such a thermal motion, is characterized by the lack of specifiability and randomness in the spatial interrelations of the particles in the system. In a pregnant passage Helmholtz noted that: "For our machines (which are coarse in comparison with molecular structure) it is organized motion alone that is freely convertible into other forms of work". And in a footnote to this statement he adds, "Whether such transformation is actually impossible in view of the fine structure of living organic tissues appears to me still to be an open question, the importance of which in the economy of nature is plainly obvious". *Wissenschaftliche Abhandlungen*, Vol. II, p. 972.

[76] Helmholtz, 'Über das Ziel und die Fortschritte der Naturwissenschaft', *Vorträge und Reden*, Vol. II, pp. 385–386.
[77] *Ibid.*, p. 385. Quoted from *Selected Writings of Hermann von Helmholtz*, ed. by Russell Kahl, p. 237.
[78] *Ibid.*, p. 385. Quoted from Kahl, op. cit., p. 237.
[79] *Ibid.*, p. 385. Quoted from Kahl, op. cit., p. 237.
[80] *Ibid.*, p. 386. Quoted from Kahl, op. cit., p. 238.
[81] *Ibid.*, p. 386. Quoted from Kahl, op. cit., p. 238.
[82] *Ibid.*, p. 386. Quoted from Kahl, op. cit., p. 238.
[83] *Ibid.*, p. 389. Quoted from Kahl, op. cit., p. 239.
[84] See Joe D. Burchfield, *Lord Kelvin and the Age of the Earth* (Neale Watson, New York, 1975), pp. 72–74.
[85] On this point see Martin J. Rudwick, *The Meaning of Fossils: Episodes in the History of Paleontology* (Neale Watson; New York, 1976), especially Chapter 5. Michael Ruse, *The Darwinian Revolution. Science Red in Tooth and Claw* (University of Chicago Press, Chicago, 1979), Chapters 4 and 5.
[86] See Peter Heimann, 'Molecular Forces, Statistical Representation, and Maxwell's Demon', *Studies in History and Philosophy of Science*, Vol. I (1970), pp. 189–211. Heimann shows that Maxwell ruled out the possibility of reducing statistical laws to the effects of Newtonian mechanics. Like Boltzmann, Max Planck also hoped to replace probability theory in physics with deeply hidden causal factors. The problem of reconciling statistical explanation and classical mechanics is also discussed in several important papers by Martin J. Klein. See Martin J. Klein, 'Boltzmann, Monocycles and Mechanical Explanations', *Boston Studies in the Philosophy of Science*, Vol. II (1974), pp. 155–175. Martin J. Klein, 'Mechanical Explanation at the End of the Nineteenth Century', *Centaurus*, Vol. 17 (1972), pp. 58–82.
[87] Rudolph Virchow, *Vier Reden über Leben und Kranksein* (Berlin, 1862), p. 12.
[88] *Ibid.*, p. 12.
[89] *Ibid.*, p. 14.
[90] Matthias Schleiden, *Über den Materialismus der neueren deutschen Naturwissenschaft* (Engelmann, Leipzig, 1863), p. 41.
[91] *Ibid.*, p. 43.
[92] See for instance Helmholtz, 'The Origin and Meaning of Geometrical Axioms (II)', *Mind*, Vol 3 (1878), p. 213.
[93] Du-Bois-Reymond, *Über Neo-Vitalismus*, ed. by Erick Metz, (Brachwede i. W., 1913), from the frontespiece.
[94] Hans Driesch, *Die Biologie als selbständige Grundwissenschaft* (Leipzig, 1893).
[95] DuBois-Reymond, *Über Neo-Vitalismus*, p. 24.
[96] Driesch, *Die Biologie als selbständige Grundwissenschaft*, p. 58.
[97] DuBois-Reymond, *Über Neo-Vitalismus*, p. 24.
[98] Karl Ernst von Baer, 'Welche Auffassung der lebenden Natur ist die Richtige?' (reprint of the original 1860 lecture in St. Petersburg by the Peter Presse, Darmstadt, 1970), pp. 39–40.

CHAPTER 6

[1] Von Baer, *Leben*, pp. 321–322. Raikov, pp. 140–141.
[2] This is made explicit by von Baer in 'Über Darwins Lehre', *Reden*, Vol. 2, p. 252.
[3] *Ibid.*, p. 277.

⁴ *Ibid.*, p. 287. Bronn's views have been discussed by Peter J. Bowler, *Fossils and Progress* (Neale Watson, New York, 1976), pp. 107–110.
⁵ *Ibid.*, p. 288.
⁶ *Ibid.*, p. 289. Huber's argument appears in his *Die Lehre Darwins, kritisch betrachtet* (1871).
⁷ On Wagner's theory of geographical isolation see: Peter Vorzimmer, *Charles Darwin: The Years of Controversy. The Origin of Species and its Critics 1859*–1882 (Temple University Press, Philadelphia, 1970). Frank J. Sulloway, 'Geographical Isolation in Darwin's Thinking: The Vicissitudes of a Crucial Idea', *Studies in History of Biology*, Vol. 3 (1979), pp. 23–65.
⁸ Von Baer, 'Über Darwins Lehre', p. 289.
⁹ *Ibid.*, p. 358.
¹⁰ *Ibid.*, p. 360.
¹¹ *Ibid.*, p. 363.
¹² *Ibid.*, p. 366. A similar situation obtained on the island of St. Helena, which is inhabited by several genera of land snails found nowhere else. In von Baer's view, "They all appear to have independently emerged or come into being on the island'. (p. 366).
¹³ *Ibid.*, p. 383.
¹⁴ *Ibid.*, p. 374.
¹⁵ *Ibid.*, p. 375.
¹⁶ See, for instance, p. 239. Von Baer makes numerous parallels throughout his work to the attempt by the Naturphilosophen to construct an absolutistic philosophy of nature and the modern reductionistic or monistic philosophies of nature, among which Darwin's hypothesis and Haeckel's elevation of it to a dogma were prime examples. In light of the persistent criticism von Baer makes of Schelling's philosophy of nature throughout his career, it is surprising to see Raikov and others supporting the view that von Baer always remained a (moderate) follower of Schelling. There were, to be sure, numerous parallels in expressions used by von Baer in his philosophical writings and formulations of the young Schelling. And I have not denied a genuine influence of Schelling on the early von Baer. But the evidence of his biological thought and his insistent criticism of the professed followers of Schelling in the same terms used by others I have considered in this study points overwhelmingly in the direction of a Kantian, teleological approach to biological organization. Von Baer himself explicitly confirms this view, as we shall see shortly.
¹⁷ *Ibid.*, p. 378.
¹⁸ *Ibid.*, p. 378.
¹⁹ *Ibid.*, p. 379.
²⁰ *Ibid.*, pp. 395–396.
²¹ *Ibid.*, p. 396.
²² *Ibid.*, p. 383.
²³ Karl Ernst von Baer, 'Entwickelt sich die Larve der einfachen Ascidien in der ersten Zeit nach dem Typus der Wirbeltiere?' *Memoires de l'academie Imperiale des sciences de St. Petersbourg*, VII Series, Vol. XIX (1873), No. 8, p. 1.
²⁴ *Ibid.*, p. 7.
²⁵ *Ibid.*, p. 8.
²⁶ *Ibid.*, p. 8. For an excellent discussion of this problem in its connection with Kowalewsky's work from the point of modern neural morphology, cf. Hartwig Kuhlenbeck, *The Central Nervous System of the Vertebrates* (J. Karger, New York, 1967), Vol. 2, pp. 296–297.

[27] *Ibid.*, p. 10.

[28] *Ibid.*, p. 19.

[29] They would not have overlooked it had they kept in mind the dynamic 'vital axes' which von Baer schematized in his *Entwicklungsgeschichte*. These functional 'axes' provide the coordinate framework within which homologies are to be traced among interrelated systems of organs throughout the animal kingdom. See above. Chapter 3, pp. 87–88.

[30] *Ibid.*, p. 20, 21, and 31.

[31] *Ibid.*, pp. 20–21.

[32] *Ibid.*, p. 33. My italics. Von Baer's German is as follows: Die Lehre von der Transmutation der Tierformen principiell nicht abgeneigt, sondern eher zugeneigt, verlange ich doch vollständigen Beweis, bevor ich an eine Umwandlung des Wirbeltier – Typus in den der Mollusken glauben kann.

[33] *Ibid.*, p. 35. Von Baer has been joined in his point of view concerning the ascidians by more than one noted morphologist. In spite of his belief in the Darwinian theory of evolution, Hartwig Kuhlenbeck, after many years of careful embryological research, concluded that: "Yes, as can be seen, my present phylogenetic interpretation does not thereby significantly differ from that of more than 36 years ago. Despite the large amount of additional data and speculations that have accumulated since that time, I can here repeat, without any modifications my statement of 1927 (p. 20); Wenn auch zum Mindesten die Möglichkeit, um nicht zu sagen die Wahrscheinlichkeit einer solchen allgemeinen Ableitung der Wirbeltiere von Wirbellosen zugegeben werden muss, so bleibt doch zunächst jeder speziellere Ableitungsversuch eine auf sehr unsicheren Grundlagen ruhende theoretische Konstruktion". cited from *The Central Nervous System of Vertebrates* (S. Karger, New York, 1967), Vol. 2, p. 346.

[34] Von Baer, 'Über Darwins Lehre', *Reden*, Vol. 2, pp. 388–387.

[35] *Ibid.*, pp. 389–392.

[36] *Ibid.*, pp. 400–402.

[37] *Ibid.*, pp. 413–414.

[38] *Ibid.*, p. 418.

[39] *Ibid.*, p. 422. Most scholars have doubted von Baer's sincerity on this point. See S.J. Homles, 'K.E. von Baer's Perplexities over Evolution', *Isis*, (1947), pp. 7–14, especially p. 11. Also see Jane Oppenheimer, 'An Embryological Enigma in the Origin of Species', *Essays in the History of Embryology and Biology*, (M.I.T. Press, Cambridge, Mass., 1967), p. 231. Raikov demonstrates conclusively that von Baer had nothing but contempt for organized Christianity. He regarded von Baer's religious views, if indeed he can be said to have had any, as closer to a form of pantheism. See Raikov, *Karl Enrst von Baer*, pp. 418–419.

[40] *Ibid.*, p. 425.

[41] *Ibid.*, p. 530. The 1834 paper was not the only place in which von Baer had discussed his evolutionary views. He had also presented a limited theory of evolution and applied it directly to man in several papers, but particularly in a paper published shortly before the appearance of Darwin's *Origin*. See von Baer, 'Über Papuas und Alfuren', *Memoires de l'academie Imperiale des sciences de St. Petersbourg*, Vol. VI, Series 10 (1859), pp. 269–346. Von Baer applied the limited theory of evolution to man, concluding that while man has evolved, he has not evolved from the apes, but had rather a separate, independent origin. Von Baer's anthropological work is treated extensively by Raikov, pp. 260–290, and pp. 409–418.

42 *Ibid.*, p. 439.
43 *Ibid.*, p. 439.
44 *Ibid.*, p. 440.
45 *Ibid.*, pp. 420–421.
46 *Ibid.*, pp. 440–441.
47 *Ibid.*, p. 451. On various uses of pedogenesis, heterochrony, neoteny and progenesis as evolutionary mechanisms, *see* Stephen J. Gould, *Ontogeny and Phylogeny*, pp. 267–351.
48 *Ibid.*, pp. 448–449.
49 *Ibid.*, p. 453.
50 *Ibid.*, p. 423.
51 *Ibid.*, p. 442.
52 *Ibid.*, p. 442.
53 *Ibid.*, p. 273.
54 Ernst Haeckel, *Generelle Morphologie der Organismen* (Berlin, 1866), Vol. 1, p. 101.
55 Von Baer, 'Über Zielstrebigkeit in den organischen Körper insbesondere', *Reden*, Vol. 2, p. 231.
56 *Ibid.*, p. 180.
57 *Ibid.*, p. 180.
58 *Ibid.*, p. 180.
59 *Ibid.*, p. 232.
60 Von Baer, 'Über den Zweck in den Vorgängen in der Natur', *Reden*, Vol. 2, p. 65.
61 Von Baer, 'Über Zielstrebigkeit...', pp. 186–187.
62 *Ibid.*, p. 188.
63 *Ibid.*, p. 189.
64 Von Baer, 'Über den Zweck...', p. 71.
65 *Ibid.*, p. 88.
66 *Ibid.*, p. 58.
67 Von Baer, 'Über Zielstrebigkeit...', pp. 228–229.

EPILOGUE

1 Joseph Ben-David, *The Scientist's Role in Society: A Comparative Study* (Prentice Hall, Englewood Cliffs, New Jersey, 1971). Fritz Ringer, *The Decline of the German Mandarins: the German Academic Community 1890–1933* (Harvard University Press, Cambridge, Mass., 1969). R. Steven Turner, 'The Growth of Professorial Research in Prussia, 1818 to 1848 – Causes and Context', *Historical Studies in the Physical Sciences*, Vol. 3 (1971), pp. 137–182. Charles E. McClelland, *State, Society, and University in Germany, 1700–1914* (Cambridge University, Cambridge, 1980).
2 Charles McClelland, *State, Society, and University in Germany, 1700–1914*, pp. 151–189.
3 *Ibid.*, pp. 164–169.
4 *Ibid.*, p. 167.

NAME INDEX

Aristotle 6, 10, 25, 129, 195, 242
Autenrieth, Johann H. 59

Baer, Karl Ernst von 3, 4, 6, 14, 16, 17, 38, 50, 54, 59, 65, 67, 72–95, 103, 105, 106, 114, 115–121, 124, 126, 129, 146, 184, 189–194, 215, 219, 220, 223, 246–275, 277, 278
Barrande, Joachim 261
Barry, Martin 4, 137
Beaumont, Elie de 156
Bergmann, Carl 3, 10, 15, 158, 172, 175, 186, 189, 229, 278
Bernard, Claude 10, 156
Berzelius, Jöns Jacob 79, 122, 157, 160, 161, 180
Bichat, Xavier 9, 70
Bischoff, Theodor Ludwig Wilhelm 167, 168, 195, 196, 229
Blondlot, René-Prosper 156
Blumenbach, Johann Friedrich 3, 9, 14, 17, 18, 29, 38, 39, 43, 54, 56, 58, 61, 65, 70, 83, 114, 120, 121, 126, 129, 131, 152
Boltzmann, Ludwig 240
Bonnet, Charles 45
Bronn, Heinrich G. 260
Brücke, Ernst 195
Buffon, Georges Louis Leclerc, Comte de 18, 21
Burdach, Karl Friedrich 72, 95, 116

Carpenter, William 4
Carus, Gustav 59, 70, 76, 146
Clapeyron, Benoit-Pierre-Émile 215
Clausius, Rudolph 240
Cuvier, Georges 38, 54, 56, 57, 61–65, 86, 181

Darwin, Charles 1, 3, 4, 246–275

Daubenton, Louis, Jean-Marie 58
Davy, Edmund 160
Davy, Humphrey 160
Democritus 6
Döbereine, Johann W. 160
Döllinger, Ignaz 38, 55, 61, 65–71, 73, 77, 103, 115, 116, 140–143, 146
Driesch, Hans 243
DuBois-Reymond, Emil 5, 15, 195, 207, 211, 216, 217, 242, 243
Dulong, Pierre Louis 160

Esenbeck, Nees von 70

Fichte, Johann G. 66
Foucroy, Antoine François de 122
Frank, Johann Peter 56

Gall, Franz Joseph 55
Gatlin, Lila 8
Gay-Lussac, Louis Joseph 79
Girtanner, Christoph 34, 121
Gmelin, Leopold 38, 56, 157
Goerres, Joseph von 70
Goethe, Johann Wolfgang von 17, 75, 76, 103

Haeckel, Ernst 140, 268, 271
Haller, Albrecht von 9, 18, 19, 68, 70, 115
Hegel, Georg Wilhelm Friedrich 10, 11, 12, 75, 242
Helmholtz, Hermann 5, 15, 115, 195, 197–215, 232, 234, 237, 240, 242
Hertwig, Oscar 243
Hildebrandt, Friedrich 121, 159
Holtzmann, O. V. 215
Humboldt, Alexander von 17, 55, 56, 70
Hume, David 240
Huxley, T. H. 4, 246

Illiger, Johannes 17

Jenkin, Fleeming 238
Joule, James 215

Kant, Immanuel 2, 4, 13, 14, 22, 25, 31, 50, 52, 61, 72, 81, 126, 128, 129, 148, 152, 221, 240, 242, 271
Kielmeyer, Carl Friedrich 3, 5, 6, 14, 17, 37, 40–53, 54, 56, 58, 59, 61, 62, 65, 70, 71, 77, 81, 83, 114, 120, 126, 159, 215, 278
Koestler, Arthur 8
Kölliker, Albert 268
Köreuter, Joseph G. 19
Kowalewsky, Alexander 247, 253, 257
Kuhlenbeck, Hartwig 3
Kuhn, Thomas 5

Lamarck, Jean Baptiste 1, 43, 55
Leuckart, Rudolph 3, 10, 15, 158, 167, 172, 175, 189–194, 217, 268, 278
Lichtenberg, Georg Christoph 3, 38
Liebig, Justus von 15, 115, 135, 156, 158–168, 172, 180, 196, 204, 209, 229
Link, Heinrich Friedrich 17, 70, 114
Linneaus, Carl 18
Lotze, Hermann 14, 147, 158, 168, 172, 229
Ludwig, Carl 15, 217
Lyonet, Pierre 45

Magendie, François 103, 104, 156
Marsh, O. C. 253
Maxwell, James Clerk 240
Mayer, Julius Robert 209
Meckel, Johann Friedrich 3, 6, 14, 17, 54, 56, 59, 65, 68, 70, 77, 88, 89, 106
Mitscherlich, Eilhardt von 160
Monod, Jacques 8, 11
Müller, Fritz 268
Müller, Johannes 3, 4, 6, 14, 54, 95, 96, 103–111, 114, 121, 127, 143–155, 159, 195, 196, 215, 219, 220, 229, 278

Newton, Isaac 11, 21, 62

Oken, Lorenz 59, 70, 76, 106, 124, 146, 147
Owen, Richard 4, 246, 255

Paley, William 4
Pallas, Pyotr Simon 58
Pander, Christian 65, 86, 95, 131
Parrot, Georg Friedrich 72
Pfaff, Christian Heinrich 38
Plato 66, 195, 242
Polanyi, Michael 7, 8
Prochaska, Georg 73
Purkinje, Evangelista 96, 131, 135

Rathke, Heinrich 14, 54, 95–102, 103, 106
Reichert, Carl 147, 219, 220
Reil, Johann Christian 9, 14, 35–37, 54, 56, 58, 61, 70, 81, 114, 126, 149, 159
Rudolphi, Carl Asmund 61, 103, 149
Russell, E. S. 3, 8, 147

Saint-Hilaire, Geoffroy 55, 76, 147
Schelling, Friedrich Wilhelm Joseph von 18, 59, 65, 68, 124, 234
Schleiden, Matthias 114, 195, 242
Schwann, Theodor 14, 114, 124–134, 149, 150, 195
Sömmering, Samuel Thomas 55, 70
Spinoza, Benedict 66
Spix, Johann B. 146
Sprengel, Kurt 58
Stahl, Georg Ernst 9, 234
Steffens, Heinrik 66

Temkin, Oswei 14
Thenard, Louis Jacques 79, 160
Thompson, D'Arcy 8, 10
Thomson, William (Lord Kelvin) 234, 238
Tiedemann, Friedrich 17, 55, 77, 157
Treviranus, Gotthelf Reinhold 1, 3, 5, 6, 17, 56, 70, 71

Valentin, Gabriel 135
Virchow, Rudolph 14, 114, 195, 219, 225, 229, 241

Vogt, Carl 14, 114, 134–140

Wagner, Moritz 251
Wagner, Rudolph 17, 95, 114, 131, 137, 251
Weber, Eduard 186
Weber, Ernst Heinrich 96, 126, 159, 186, 215

Weiss, Paul 7
Werner, Abraham G. 66
Wigner, Eugene P. 11
Wilberforce, Samuel 4
Windischmann, Karl J. 38, 70
Wöhler, Friedrich 150, 156
Wolff, Caspar Friedrich 21, 106
Wrisberg, Heinrich A. 70

SUBJECT INDEX

Animal Heat 180–186, 203–209

Bergmann's Rule 231
Biogenetic Law 41, 49, 56, 59–60
Biology 1, 28, 37, 277

Catalysis 160
Catalytic Force 160–161, 168, 192
Causality 2, 25–28, 170, 172–176, 239––240
Cell Theory 112, 130, 143, 145, 222–228
    Cell differentiation 141–142, 150-151
    Relationship to generation 150-154
Chain of Being 43, 59, 69
Comparitive anatomy 33, 38–39, 61–64, 76
Conservation of Energy 15, 163–165, 211–215, 231–232
Crystallization 122, 132, 145

Darwinian Evolution 233, 246–275, 269–270
Developmental Morphology 14, 54, 60, 70–71, 84–85, 86–88, 98–102, 106, 257–261

Embryology 40–42, 57, 65, 70
    and developmental mechanics 59–60, 86–88
    and developmental grades 67, 90–95, 96, 180
    and functional explanation 14
    of the Ascidians 257–261
    of the endocrine system 57–58
    of the gill arches 96–102
    of the urogenital system 106–111
Entropy 234–236

Epigenesis 18–21, 37

Fermentation 197–199
Functional Morphology 14–15, 156, 172–176
Functionalism 9–10, 26, 63–64, 176–188

Germ Layers 89, 95, 119, 223

Homologies 95, 98–102

Kantian Research Tradition 6, 22–35

Mammalian Ovum 115–120
Material Conversions 164–167, 171–172, 180–181, 199–215
Mechanism 168–172, 215–216, 241 (also see Teleology)
Methodology of Scientific Research Programs 12–15
Morphotypes 14–15, 33–34, 60, 61–64, 84–85, 98
Müller Ducts 109
Muscle Action 200–209

Natural Selection 233, 237, 249, 251
Newtonianism 20, 228, 238–239
Non-Darwinian Evolution 82–95, 254–255, 261, 262–270

Organic Chemistry 22, 38, 57, 78–80, 121
Organization 21–22, 30–31, 34–37, 42, 61, 103, 112–114, 145, 169–171

Physiology 33, 40, 57, 67–69, 75–81, 104–106
Physiological Chemistry 38–39, 68–69, 78–79, 103–104, 157–158, 160–162, 197–209

Phylogeny 253
Plans of Organization 31–33, 64, 180–188, 218
Preformation Theory 19, 21
    Generic preformationism 31, 36–37, 81, 134–140, 152, 168
Putrefaction 197–199

Races 19, 31, 82–83
Reductionism 6–7, 11, 78, 126, 146, 150, 163, 174, 197–209, 215–216, 217–219
Romantische Naturphilosophie 3, 5, 6, 10, 26, 55, 59, 67, 69, 73, 103, 120, 124, 146–147, 215, 254

Species 19, 30–32, 40, 43, 176, 261–262
Specific Sense Energies 103
Systematics 39, 89–95, 257

Teleology
    and Adaptation 4, 22, 33–35
    and Bionomic Laws 7, 10, 62–64, 117, 180–183, 231
    and Design 4
    and functional morphology 14, 156, 172–176
    and Functionalism 9–10, 63–64
    and Holism 7, 10, 25, 219–228
    and mind 10
    and Natural Purposes 25, 62–63, 81, 168–169, 176, 241, 270–275
    and Reductionism 6, 219–228
    and regulative principles 22, 27–28, 52–53, 159
    and Bergmann and Leuckart 174–175
    and Cuvier 61–64
    and Darwin 4, 32–33
    and Kant 6, 24–35
    and Lotze 168–172
    and Müller 105, 11, 145–150
    and Reichert 220–223
    and Schwann 127–128
    and Virchow 225–228, 241–242
    and von Baer 80–81, 85, 117, 270–275
    as Theory of Causal Relations in Biology 2, 24–27, 168–172, 270–275

Teleomechanism 12, 24–30, 112, 126, 158, 168, 172–176, 229–230
Transformation of Species 83, 251, 253, 261–262

Variation 19, 31, 40, 82, 238, 250–251, 269–270
Vitalism
    and teleology 9
    vital materialism 9, 126
Vital Forces 9, 12, 20, 35–37, 45, 51, 118, 125, 162–167, 191, 215–216
    Bildungsbetrieb 12, 20, 24, 29, 37
    Gestaltungkraft 14, 18, 24, 29, 37, 118–121, 191–192, 215
    Lebenskraft 50–51, 148–149, 163, 200, 209, 216
    Moule interieur 21
    vis essentialis 21

Wolffian Body 106–111